Ioan Dzitac · Simona Dzitac ·
Florin Gheorghe Filip · Janusz Kacprzyk ·
Misu-Jan Manolescu · Horea Oros
Editors

Intelligent Methods in Computing, Communications and Control

Proceedings of the 8th International Conference on Computers Communications and Control (ICCCC) 2020

Springer

Editors
Ioan Dzitac
Agora University of Oradea
Oradea, Romania

Simona Dzitac
University of Oradea
Oradea, Romania

Florin Gheorghe Filip
Academia Romana
Bucharest, Romania

Janusz Kacprzyk
Polish Academy of Sciences
Systems Research Institute
Warszawa, Poland

Misu-Jan Manolescu
Agora University of Oradea
Oradea, Romania

Horea Oros
University of Oradea
Oradea, Romania

ISSN 2194-5357 ISSN 2194-5365 (electronic)
Advances in Intelligent Systems and Computing
ISBN 978-3-030-53650-3 ISBN 978-3-030-53651-0 (eBook)
https://doi.org/10.1007/978-3-030-53651-0

This Springer imprint is published by the registered company Springer Nature Switzerland AG
The registered company address is: Gewerbestrasse 11, 6330 Cham, Switzerland

Contents

Advances in Intelligent Systems and Computing

Volume 1243

The series "Advances in Intelligent Systems and Computing" contains publications on theory, applications, and design methods of Intelligent Systems and Intelligent Computing. Virtually all disciplines such as engineering, natural sciences, computer and information science, ICT, economics, business, e-commerce, environment, healthcare, life science are covered. The list of topics spans all the areas of modern intelligent systems and computing such as: computational intelligence, soft computing including neural networks, fuzzy systems, evolutionary computing and the fusion of these paradigms, social intelligence, ambient intelligence, computational neuroscience, artificial life, virtual worlds and society, cognitive science and systems, Perception and Vision, DNA and immune based systems, self-organizing and adaptive systems, e-Learning and teaching, human-centered and human-centric computing, recommender systems, intelligent control, robotics and mechatronics including human-machine teaming, knowledge-based paradigms, learning paradigms, machine ethics, intelligent data analysis, knowledge management, intelligent agents, intelligent decision making and support, intelligent network security, trust management, interactive entertainment, Web intelligence and multimedia.

The publications within "Advances in Intelligent Systems and Computing" are primarily proceedings of important conferences, symposia and congresses. They cover significant recent developments in the field, both of a foundational and applicable character. An important characteristic feature of the series is the short publication time and world-wide distribution. This permits a rapid and broad dissemination of research results.

**** Indexing: The books of this series are submitted to ISI Proceedings, EI-Compendex, DBLP, SCOPUS, Google Scholar and Springerlink ****

More information about this series at http://www.springer.com/series/11156

Instead of Preface

Redesign of a Conference from In-Person to Online. Case Study: ICCCC

Ioan Dzitac[1(✉)], Simona Dzitac[2], Florin Gheorghe Filip[3], and Misu-Jan Manolescu[1]

[1] Agora University of Oradea, Piata Tineretului 8, 410526 Oradea, Romania
professor.ioan.dzitac@ieee.org
[2] University of Oradea, Universitatii 1, Oradea, Romania
[3] Romanian Academy, 125 Calea Victoriei, Bucharest, Romania

Abstract. "Intelligent Methods for Computing, Communications and Control" volume, published by Springer in "Advances in Intelligent Systems and Computing" Series, is in fact the Proceedings of the 8th International Conference on Computers Communications and Control (ICCCC) 2020. The ICCCC has been founded in 2006 by Ioan Dzitac, Florin Gheorghe Filip and Misu-Jan Manolescu, and was organized every even year by Agora University of Oradea, under the aegis of the Information Science and Technology Section of Romanian Academy. The first seven editions were organized face to face (in-person, traditional). Due to the COVID-19 pandemic, the 8th edition, ICCCC2020, which was designed initial to be in-person, we had redesign it as an online event (remotely). In this article we will present our study and conclusions regarding a parallel between the two types of conferences, traditional vs. online, each with advantages and disadvantages.

Keywords: Conference organization · ICCCC · In-person · COVID-19 · Online · Remote · Accessibility · Costs · Comfort · Socialization · Zoom Pro

1 Introduction

The 8th International Conference on Computers Communications and Control (ICCCC) 2020, planned to be held in-person at Hotel President from Băile Felix, Oradea, in the period May 11-15, 2020, was redesigned on March 19, due to the COVID-19 pandemic, for Online Host (remotely): Agora University of Oradea, Romania, in the same period.

Scope and Topics of ICCCC2020. The goal of this conference is to bring together international researchers, scientists from academia and industry to present and discuss in a friendly environment their latest research findings on a broad array of topics in computer networking and control. The Program Committee is soliciting paper describing original, previously unpublished, completed research, not currently under review by another conference or journal, addressing state-of-the-art research and development in all areas related to computer networking and control. Topics of ICCCC2020 were:

I. Dzitac et al. (Eds.): ICCCC 2020, AISC 1243, pp. 3–17, 2021.
https://doi.org/10.1007/978-3-030-53651-0_1

- Theory for Computing and Communications,
- Integrated Solutions in Computer-based Control,
- Computational Intelligence and Soft Computing,
- Decision Making and Support Systems.

A good organization of a conference is not a simple thing, but it involves a quality managerial and organizational activity, which involves planning, economic evaluation and a good knowledge of social psychology (Figs. 1 and 2).

(a) (b)

Fig. 1. Registration: (a) ICCCC2018 (in-person) vs. (b) ICCCC2020 (online)

(a) (b)

Fig. 2. Official opening: (a) ICCCC2018 (in-person) vs. (b) ICCCC2020 (online)

When organizing a traditional international conference we must pay an interactive attention to the following:

(1) Choosing an attractive, accessible and safe location,
(2) Planning a suitable period for the conference,

(3) Evaluation of the costs and identification of funding sources,
(4) Scientific attractiveness of the conference (keynote speakers, publications).

2 Choosing an Attractive, Accessible, and Safe Location

For attractiveness, many traditional conferences choose their location in exotic areas. Many of these areas, however, present risks to personal health or safety. Others are not accessible by plane, not having an airport nearby. All previous editions of ICCCC took place at Băile Felix, an attractive tourist resort, located 10 km from Oradea, a city where you can reach by plane, train or car transport. From Baile Felix we organized trips in the mountains, visiting the Bear Cave, the Glacier Cave, Turda Salt Mine, etc.

In online organization the location is not important. Important is the good quality of the technical resources [2].

For ICCCC2020 online we are used the resources from home of the General Chair (Ioan Dzitac, Oradea, Romania): a good connection to the Internet (via optical fiber link from RDS-RCS company), good quality of the host devices (Lenovo PC, Microsoft video - audio camera), video conference room online (Zoom Pro). All worked perfectly during the five days of the online conference (Fig. 3).

Fig. 3. Location: (a) ICCCC2018 (Baile Felix) vs. (b) ICCCC2020 (Zoom Pro at Home)

To make up for the social fun of the in-person model, we introduced, in each of the five days, a two-hour program of regular paper presentations, a one-hour break, and a two-hour keynote lecture. During the break, we broadcast entertainment: videos about Romania, Oradea, collages with memories from previous editions of ICCCC, excursions filmed through magnificent areas of Romania. We also transmitted local folk customs from Bihor County (costumes, songs, dances), as well as songs from the countries of the participants of that day.

The advantage of this online conference was that people who were not planned in the classic version could also participate: all co-authors, former participants or some people with physical disabilities.

For example, in the end of official opening of ICCCC2020 we had receives several messages, online or video registered, from *former invited speakers at ICCCC*, such as:

1. Fuad ALESKEROV, National Research University Higher School of Economics, Moscow, Russia (ICCCC2016)
2. Barnabas BEDE, DigiPen University, USA (ICCCC2018)
3. Pierre BORNE, Ecole Centrale de Lille, France (ICCCC2008)
4. Amlan CHAKRABATI, University of Calcutta, India (ICCCC2018)
5. Kaoru HIROTA, Beijing Institute of Technology, China/Tokyo Institute of Technology, Japan (ICCCC2006)
6. Arturas KAKLAUSKAS, Gediminas Vilnius Technical University, Lithuania (ICCCC2018)
7. Gang KOU, Southwestern University of Finance and Economics, Chengdu, China (ICCCC2012, ICCCC2016)
8. George METAKIDES, University of Patras, Greece & Digital Enlightenment, Bruxelles (ICCCC2008)
9. Gheorghe PĂUN, Romanian Academy, Romania (ICCCC2008, ICCCC2018)
10. Yong SHI, University of Chinese Academy of Sciences, China/University of Nebraska at Omaha, USA (ICCCC2012, ICCCC2014)
11. Athanasios STYLIADIS, International Hellenic University, Greece (ICCCC2006)
12. Mincong TANG, Beijing Jiatong University, China (ICCCC2014)
13. Horia-Nicolai TEODORESCU, Romanian Academy, Romania (ICCCC2008)
14. Dan TUFIȘ, Romanian Academy, Romania (ICCCC2006, ICCCC2008)
15. Zenonas TURSKIS, Gediminas Vilnius Technical University, Lithuania (ICCCC2016)

Another example of online facility is the case of Marius Șucan, a brilliant digital artist and computer programmer, with severe physical disabilities (without fingers and with Epidermolysis Bullosa), presented a series of his works (paintings and software products). Marius had the chance to attend a conference and presents a lecture for the first time in his life, which he could not have done in the case of an in-person conference (Fig. 4).

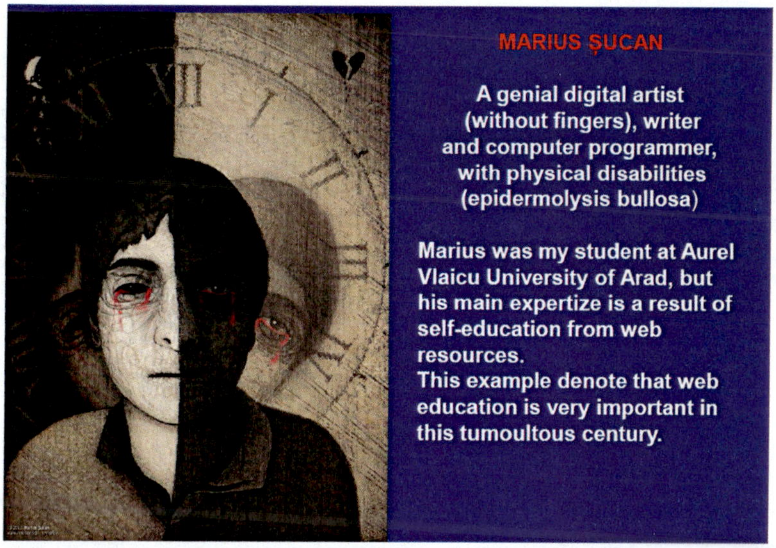

Fig. 4. Presentation by Ioan Dzitac of a digital artist, Marius Şucan (self-portrait) http://robodesign.ro/

3 Planning a Suitable Period for the Conference

In period planning, similar as in traditional in-person variant, in online variant we need to consider to be out of religious holidays, taking into account all religions, to be out of exam sessions, not to be in a period when other similarly attractive conferences take place, not to be in the middle of the summer season, to be in a season with a pleasant climate etc. (Figs. 5, 6 and 7).

(a) (b)

Fig. 5. Period: (a) ICCCC2018 (May 08-12) vs. (b) ICCCC2020 (May 11-15)

(a) (b)

Fig. 6. Socialization: (a) ICCCC2018 (In-Person) vs. (b) ICCCC2020 (Online)

Fig. 7. Socialization in excursion at ICCCC2018

4 Evaluation of the Costs and Identification of Funding Sources

ICCCC in-person not was very expensive but it is clear that for the participants in traditional version is much expensive than the online one. If a participant do not have a funding source then her/his participation in-person at conference is very difficult (Table 1).

For organizers is also much expensive in traditional vs. online version, because must cover the travel, accomodation, and daily meals costs for invited speakers. In online version no any costs in this matter (Fig. 8).

5 Scientific Attractiveness of the Conference

5.1 Scientific Reputation of the Organizers, Program Committee Members, and of Invited Speakers

Staff of ICCCC2020:

1. General Chair: Ioan DZITAC, Aurel Vlaicu University of Arad & Agora University of Oradea, Romania
2. IPC Chair: Florin Gheorghe FILIP, Romanian Academy, Romania
3. Conference Chair: Simona DZITAC, University of Oradea, Romania

Table 1. Conference Cost: (a) ICCCC2018 (in-person) vs. (b) ICCCC2020 (online)

Services	(a) ICCCC2018	(b) ICCCC2020
Including registration fee, accommodation at 4 stars hotel for 4 nights and daily meals	800 EUR 700 EUR –for students	–
Including registration fee and daily meals	600 EUR 500 EUR – for students	–
Registration fee	–	400 EUR 300 EUR for students
Transport cost	Depend of distance to conference location (100–1,500 EUR)	–
Total cost	500-2,300 EUR	300-400 EUR

Fig. 8. The invited speakers at ICCCC2016

4. Co-Chair: Misu-Jan MANOLESCU, Agora University of Oradea, Romania
5. Proceedings Chair: Horea OROS, University of Oradea, Romania
6. Online Chair: Domnica DZITAC, New York University Adu Dhabi, UAE

IPC: International Program Committee (ICCCC2020):

1. Răzvan ANDONIE, Central Washington University, USA

2. Valentina BALAS, Aurel Vlaicu University of Arad, Romania
3. Barnabas BEDE, DigiPen University, USA
4. Valeriu BEIU, Aurel Vlaicu University of Arad, Romania
5. Pierre BORNE, Ecole Centrale de Lille, France
6. Dominic BUCERZAN, Aurel Vlaicu University of Arad, Romania
7. Felisa CORDOVA; Universiy of Finis Terrae, Chile
8. Gabriela CRISTESCU, Aurel Vlaicu University of Arad, Romania
9. Antonio DI NOLA, University of Salerno, Italy
10. Yezid DONOSO, Universidad de los Andes, Colombia
11. Gintautas DZEMYDA, University of Vilnius, Lithuania
12. Ömer EĞECİOĞLU, University of Santa Barbara, USA
13. Florin Gheorghe FILIP, Romanian Academy, Romania
14. Enrique HERERRA-VIEDMA, University of Granada, Spain
15. Kaoru HIROTA, Tokyo Institute of Technology, Japan
16. Arturas KAKLAUSKAS, Vilnius Gediminas Technical University, Lithuania
17. Gang KOU, Southwestern University of Finance and Economics, Chengdu, China
18. Ioana MOISIL; Lucian Blaga University of Sibiu, Romania
19. Radu NICOLESCU, The University of Auckland, New Zealand
20. Sorin NADABAN, Aurel Vlaicu University of Arad, Romania
21. Mariana NAGY, Aurel Vlaicu University of Arad, Romania
22. Shimon Y. NOF, Purdue University, USA
23. Stephan OLARIU, Old Dominion University, USA
24. Gheorghe PĂUN, Romanian Academy, IMAR, Romania
25. Yi PENG, University of Electronic Science and Technology of China, China
26. Mario de J. PEREZ-JIMENEZ, University of Seville, Spain
27. Lorena POPA, Aurel Vlaicu University of Arad, Romania
28. Radu-Emil PRECUP, Politehnica University of Timisoara, Romania
29. Imre J. RUDAS, Óbuda University, Hungary (Member of IEEE)
30. Yong SHI, University of Chinese Academy of Sciences
31. Bogdana STANOJEVIC, Mathematical Institute of the Serbian ASA, Serbia
32. Milan STANOJEVIC, University of Belgrad, Serbia
33. Codruța STOICA, Aurel Vlaicu University of Arad, Romania
34. Athanasios D. STYLIADIS, Kavala Institute of Technology, Greece
35. Ioan Alexandru ŞUCAN, Google [x], USA
36. Gheorghe TECUCI, George Mason University, USA
37. Horia-Nicolai TEODORESCU, Technical Univ. Gh. Asachi Iasi, Romania
38. Dan TUFIŞ, Romanian Academy – Institute of Artificial Intelligence "Mihai Draganescu", Romania
39. Zenonas TURSKIS,Vilnius Gediminas Technical University, Lithuania

OC: Organizing Committee ICCCC2020:

1. Dan BENTA, Agora University of Oradea, Romania
2. Gabriela BOLOGA, Agora University of Oradea, Romania
3. Casian BUTACI, Agora University of Oradea, Romania
4. Domnica DZITAC, New York University Abu Dhabi, UAE

5. Viorina JUDEU, Agora University of Oradea, Romania
6. Adriana MANOLESCU, Agora University of Oradea, Romania
7. Ioana MOISIL, "Lucian Blaga" University of Sibiu, Romania
8. Astrid ODDERSHEDE, University of Santiago de Chile, Chile
9. Horea OROS, University of Oradea, Romania
10. Marius SINCA, Agora University of Oradea, Romania
11. Ramona URZICEANU, Agora University of Oradea, Romania

Reviewers of the Manuscripts (ICCCC2020):

1. Adriana ALEXANDRU, National Inst. for R & D in Informatics, Romania
2. Răzvan ANDONIE, Central Washington University, USA
3. Valeriu BEIU, Aurel Vlaicu University Arad, Romania
4. Dominic BUCERZAN, Aurel Vlaicu University Arad, Romania
5. Cristian CIUREA, Bucharest University of Economic Studies, Romania
6. Charles J. COLBOURN, University of Toronto, Canada
7. Gabriela CRISTESCU, Aurel Vlaicu University Arad, Romania
8. Leonard DAUS, Technical University of Civil Engineering Bucharest, Romania
9. Yezid DONOSO, Universidad de los Andes, Colombia
10. Florin DRAGOMIR, Valahia University of Târgoviște, Romania
11. Horațiu DRAGOMIRESCU, Bucharest University of Economic Studies, Romania
12. Vlad DRĂGOI, Aurel Vlaicu University Arad, Romania
13. Florin Gheorghe FILIP, Romanian Academy, Romania
14. Arpad GELLERT, Lucian Blaga University of Sibiu, Romania
15. Angela IONIȚĂ, RACAI, Romania
16. Gang KOU, Southwestern University of Finance and Economics, China
17. Gabriel NEAGU, National Inst. for R & D in Informatics, Romania
18. Marilena IANCULESCU, National Inst. for R & D in Informatics, Romania
19. Ioana MOISIL, Lucian Blaga University of Sibiu, Romania
20. Nageswara Rao MOPARTHI, KL University, India
21. Mariana NAGY, Aurel Vlaicu University Arad, Romania
22. Sorin NĂDĂBAN, Aurel Vlaicu University Arad, Romania
23. Radu-Emil PRECUP, Politehnica University of Timisoara, Romania
24. Constanta Zoie RADULESCU, National Inst. for R & D in Informatics, Romania
25. Ioan Ștefan SACALA, Politehnica University of Bucharest, Romania
26. Bogdana STANOJEVIC, Math. Inst. of the Serbian Academy of Sci. and Arts
27. Csaba SZANTO, Babeş-Bolyai University of Cluj-Napoca, Romania
28. Ioan ȘUȘNEA, University of Galați, Romania
29. Bogdan ȚIGĂNOAIA, Politehnica University of Bucharest, Romania
30. Doina ZMARANDA, University of Oradea, Romania

In 2018 were the same organizers like in 2020. At ICCCC 2018 was seven keynote speakers from six countries (Colombia, India, Israel, Lithuania, Romania, and USA), and at ICCCC2020 was nine keynote speakers from eight countries (Chile, Colombia, Israel, Lithuania, Poland, Romania, Spain, and USA) (Figs. 9 and 10).

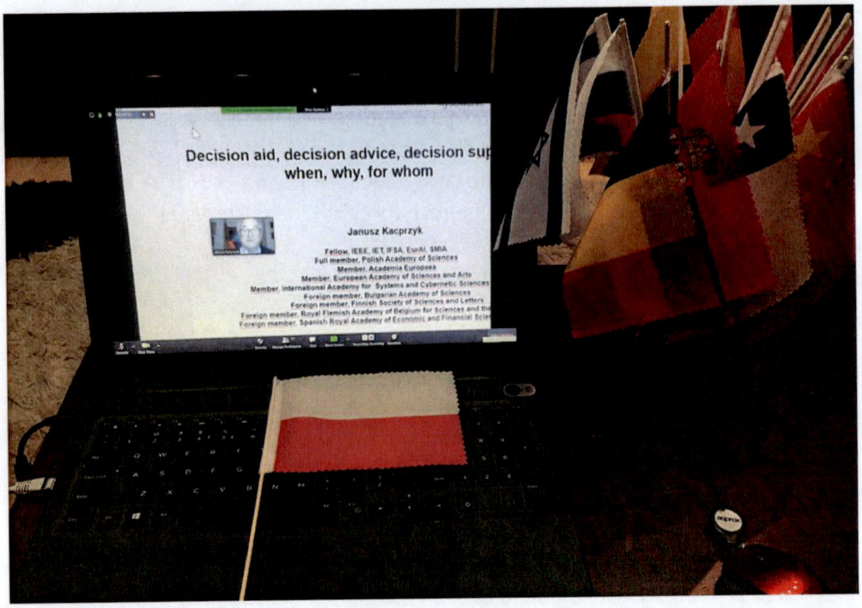

Fig. 9. ICCCC2020 Online Capture: Keynote Lecture of Acad. Janusz Kacprzyk

5.2 Papers Publication

Proceedings:

- IEEE Proceedings for ICCCC2018 papers [1],
- Springer Proceedings for ICCCC2020 papers [3].

International Journal of Computers Communications & Control (IJCCC) for several selected papers publication (extended variants) (Fig. 11):

Best Paper Award (ICCCC2020 Online). Sponsor for Best Paper Award was the International Journal of Computers Communications & Control (IJCCC), http://univagora.ro/jour/index.php/ijccc/.

The Editorial Team of IJCCC and chairpersons of conference sessions grant the Best Paper Award for following papers included in this volume:

Experimenting with Beta Distributions for Approximating Hammocks' Reliability, *by Simon Cowell, Sorin Hoara and Valeriu Beiu*
Automatons Immersed in Ocean Currents for Transformation of Biomass Into Fuel, *by Lucio Cañete Arratia, Felisa Córdova and Andrés Pérez de Arce*
Empirical versus Analytical Solutions to Full Fuzzy Linear Programming, *by Bogdana Stanojević and Milan Stanojević*

Fig. 10. ICCCC2020 Poster: Chairs and Keynote Speakers http://univagora.ro/en/icccc2020/program/

Fig. 11. IJCCC was an atractive indexed journal for ICCCC2018 and ICCCC2020

6 Conclusions

International participation at ICCCC2020 online, via Zoom Pro, was very good. Our attendees were from 20 countries and from over 50 universities and other academic institutions or industrial companies. All were very impressed, excited and enthusiastic. Please see bellow a summary.

Countries (20): Chile, China, Colombia, Greece, Iran, Israel, Japan, Jordan, France, Lithuania, Mexico, Poland, Portugal, Romania, Russia, Serbia, Spain, Switzerland, UAE, USA.

Academies (4): Lithuanian Academy of Sciences, Polish Academy of Sciences, Romanian Academy, Serbian Academy of Sciences and Arts.

Universities of Romania (10): Agora University of Oradea, Aurel Vlaicu University of Arad, Babes-Bolyai University of Cluj-Napoca, Bucharest University of Economic Studies, Lucian Blaga University of Sibiu, Politehnica University Timisoara, Transilvania University of Brasov, University Politehnica of Bucharest, University of Oradea, Technical University of Civil Engineering Bucharest.

Universities of Abroad (19): Central Washington University, Gediminas Technical Vilnius University, National Research University Higher School of Economics, Pontificia Universidad Catolica de Valparaiso, Princess Sumaya University for Technology, Universities Technion of Haifa, University of Belgrade, University of the Basque Country, Universidad del País Vasco, University of Cádiz, University of Zacatecas, University of Granada, Universidad Antofagasta Chile, Universidad Antofagasta, Université

Jean Monnet, Universidad Finis Terrae, Universidad de Santiago de Chile, Universidad Mayor, Vilnius University.

R & D Institutes (3): Institute of Mathematical Statistics and Applied Mathematics Gheorghe Mihoc - Caius Iacob, Mathematical Institute of the Serbian Academy of Sciences and Arts, National Institute for Research and Development in Informatics.

Companies & NGO (4): Association Cluster of Scientific Research, Innovation and European Studies from Oradea, Cercetare Dezvoltare Agora (Romania), LITIS (France), Trade Plus S.A. (Romania).

Appreciations from Attendees of the Virtual ICCCC2020. During the five days of the conference join online all presenters of regular papers, all keynote speakers, several co-authors, many former ICCCC special guests, and as well lot of other occasional guests or collaborators. Some feedback messages are presented below.

"I would like to thank you and all your colleagues for organizing and running this beautiful conference." (Ahmat Hiasat, participant with an research article at ICCCC2020, first time at ICCCC, Jordan).

"I wish to congratulate you for such a great conference. I participated in recent weeks at a couple of online conferences and your had the only one which was really running, in the sense of full session, and numerous speakers, and also quality. At other conferences, there were all kind of problems, lack of connections, absence of confirmed speakers, etc. Thanks so much for your hospitality which was remarkable in spite of a lack of physical contact. I strongly believe that we will be able to meet in the future when the situation will change." (Janusz Kacprzyk, invited speaker at ICCCC2020, first time at ICCCC, Poland).

"The conference was particularly successful in all respects. You have induced the spirit of previous editions in this online edition as well. Thank you for giving us the opportunity to participate. Thank you very much for the diploma and for the nomination. All participants were present with the soul not only with the slides in all sections. The presentations were interesting and instructive. And the emotional surprises. Congratulations to you, Simona and Domnica for the conception and organization. I have not participated in any online conference, but this conference did not lack anything and we do not see what could have been better!" (Bogdana & Milan Stanojevic, participants at several previous editions of ICCCC and at ICCCCC2020, Serbia).

"Congratulations dear Ioan, Simona, Domnica, Agora University team, congratulations and thanks for these great moments. The ICCCC2020 online conference is truly successful and I am very happy to be able to take part in it. I am certain that the days to come will offer us great presentations. You are great!! Congratulations from the bottom of my heart. You are fantastic! And many thanks to all the organizers: The Golden Dzitac Trio: Ioan Dzitac, Simona Dzitac and Domnica Dzitac, Rector Misu-Jan Manolescu and Adriana Manolescu and the Agora University, Florin Gheorghe Filip and the Romanian Academy, the CCSISEO Association and their talented friends, and the Aurel Vlaicu University of Arad. And all my thanks to the participants for the interesting presentations. It was an enriching experience and it gave us hope in these hard times. I must confess that there were moments when the discussions were so interesting and dynamic that I almost forget we were online." (Ioana Moisil, former invited speaker at ICCCC2016, Romania).

"Excellent Organization, truly wonderful experience!" (Amlan Chakrabarti, former invited speaker at ICCCC2018, India).

"I enjoyed the virtual conference very much; interesting talks, nice participants and wonderful social program during breaks. I wanted to thank you very much for your excellent management of what has turned out to be a successful and highly stimulating conference. I am sure that everyone must have told you this. Congratulations!" (Valentina Balas, former invited speaker at ICCCC2008, Romania).

"Dear friend, you made the best decision and did everything possible in such a situation." (Razvan Andonie, former invited speaker at ICCCC2010, keynote speaker at ICCCC2020, USA) (Table 2).

Table 2. Online vs. In-Person in ICCCC Organization

Type/Indicator	Online	In-Person	Remarks
Accessibility	High	Low-Medium	In online version can participate all co-authors and other people, people with physically disability. In-Person participation depends of geographical distance
Cost	Low	Medium-High	For remote participation is not necessary transport, accommodation, meals etc.
Comfort	Medium	Low	In online we can attend the conference from home but the time zone can be a problem. In traditional variant the travel can be a problem
Sessions	Serial for ICCCC 2020	Plenary & Parallel	In online we can attend all presentation but duration is bigger than in classic variant. In online are possible also parallel sessions with multiple hosts
Socialization	Low-Medium	High	Face-to-face meeting is better vs. Online

Several final conclusions about online conference versus in-person conference are listed below:

1) In online organization the location is not important. Important is a good quality of the technical resources and we had demonstrate that all our digital resources worked perfectly during the five days of the online conference.
2) It is clear that the online version is much cheaper than the traditional one (for online is not necessary transport, accommodation, meals etc.).
3) For social contact traditional face-to-face meeting is better that a virtual one.
4) In online variant we can attend the conference from home but the time zone can be a problem if the timetable is not correct planed.
5) In traditional variant the travel can be a problem and accessibility is low (in pandemic time is impossible).
6) Persons with physically disabilities can participate easy in online variant.

Our final conclusion is that the mixed organization, in-person & online (remotely), will be a good solution in the future.

References

1. Dzitac, I., Filip, F.G., Manolescu, M.J., Dzitac, S., Oros, H., Dzitac, D.: 2018 7th International Conference on Computers Communications and Control (ICCCC): Preface, IEEE, pp. 1–11 (2018). https://doi.org/10.1109/icccc.2018.8390429
2. Reshef, O., et al.: How to organize an online conference. Nat. Rev. Mater. **18**, 1–4 (2020). https://doi.org/10.1038/s41578-020-0194-0
3. 2020 8th International Conference on Computers Communications and Control (ICCCC) (Online). http://univagora.ro/en/icccc2020/

Theory for Computing and Communications

On Recursively Defined Combinatorial Classes and Labelled Trees

Ali Chouria[1,2], Vlad-Florin Drăgoi[2,3](✉), and Jean-Gabriel Luque[2]

[1] ISLAIB, University of Jendouba, Jendouba, Tunisia
[2] LITIS, University of Rouen Normandie Avenue de l'université, 76801 Saint-Étienne-du-Rouvray, France
{ali.chouria1,vlad-florin.dragoi}@univ-rouen.fr
[3] Department of Mathematics and Computer Sciences, "Aurel Vlaicu" University of Arad, Arad, Romania

Abstract. We define and prove isomorphisms between three combinatorial classes involving labeled trees. We also give an alternative proof by means of generating functions.

Keywords: Combinatorial classes · Catalan numbers · Labeled trees · Rooted trees · Generating functions

1 Introduction

Based on the theory of combinatorial species (see *e.g.* [1]), Flajolet and Sedjewick [4] wrote a reference book on combinatorial analysis. In particular in the first part, they provided a list of basic constructions for exponential generating functions. Mainly, complex combinatorial structures are obtained by combining the following three combinatorial classes: $\mathrm{SET}, \mathrm{SEQ}$, and CYC. For instance, they described *surjections* ($\mathrm{SEQ}(\mathrm{SET})$), *set partitions* ($\mathrm{SET}(\mathrm{SET})$), *alignments* ($\mathrm{SEQ}(\mathrm{CYC})$) and *permutations* ($\mathrm{SET}(\mathrm{CYC})$). In this article, we focus on the later one because it has the remarkable property of having the same generating function as the combinatorial class of sequences. More precisely, our starting point consists in giving an explicit bijection between the class of set-of-cycles and the class of sequences (see Sect. 3.4). Our goal is to study an example of a class defined inductively by a combinatorial class equation. We chose the equation $\mathrm{SET}(\mathrm{CYC}(\mathcal{C})) = \mathcal{C}$ because the underlying combinatorics reveal a world rich in interpretation and provide fruitful perspectives. In particular, this equation reveals an isomorphism between sets of necklaces of planar labelled trees, forests of labelled trees, and rooted labelled trees (see Sect. 4). At the heart of the equation we study here are the Catalan numbers. They are involved in the enumeration of numerous classes of combinatorial objects of prime importance in computer science, *e.g.* Dyck paths, binary trees, non-crossing partitions *etc.* [4,5,16]. Notice that more than 60 possible enumerations are listed in [16] and the enumerated objects lead to several applications in computer science,

I. Dzitac et al. (Eds.): ICCCC 2020, AISC 1243, pp. 21–34, 2021.
https://doi.org/10.1007/978-3-030-53651-0_2

such as sorting techniques based on binary trees [6]. In our case, Catalan numbers are representing ordered trees. The article ends with Sect. 4.4, where we propose other recursive tree-like combinatorial classes.

2 Background and Notations

We recall here well known definitions and results concerning combinatorial classes and generating functions. The material contained in this section mainly refers to [4].

2.1 Combinatorial Classes

In the most general context, a *combinatorial class* is a triplet $(\mathcal{O}, \mathcal{P}, \omega)$ where $\mathcal{O}$ is the discrete set of the combinatorial objects we want to enumerate, $\mathcal{P}$ is the discrete set of the properties in regard to which you want to enumerate our objects, and $\omega : \mathcal{O} \to \mathcal{P}$ is a map such that for every $p \in \mathcal{P}$ the preimage $\omega^{-1}(p)$ is finite, which is a minimal requirement in order to be able to enumerate the objects of $\mathcal{O}$ with respect to the properties of $\mathcal{P}$.

We consider the restricted context where $\mathcal{P} = \mathbb{N}$ and the preimage of 0 contains only one element. More formally, a combinatorial class is a pair $\mathcal{C} = (\mathcal{O}, \omega)$ where $\omega : \mathcal{O} \to \mathbb{N}$ is such that $\mathrm{card}(\omega^{-1}(n)) < \infty$ for any integer n. For the sake of simplicity, and when there is no ambiguity, we use the same name for a class and the set of its objects. Then, we denote by $|\mu|$ the *degree* (or *weight*) $\omega(\mu)$ of $\mu \in \mathcal{C}$ and we set $\mathcal{C}_n = \{\mu \in \mathcal{C} \mid \omega(\mu) = n\}$. If $\#(\omega^{-1}(0)) = 1$ then we denote by ϵ the unique element of $\mathcal{C}$ of weight 0. We also set $\mathcal{C}^+ := \mathcal{C} \setminus \{\epsilon\}$.

2.2 Labelled Combinatorial Classes

Recall that the symmetric group $\mathcal{S}_n$ is the group of bijections of $\{1, \ldots, n\}$. It is a group of order $n!$ whose each element is denoted by the word of its images. For instance, the cycle sending 1 to 2, 2 to 3, and 3 to 1 is denoted by 231. Obviously, the set of permutations is closed by composition and all its elements are invertible.

Formally, a *labelled combinatorial class* is a combinatorial class endowed with a sequence $(\rho_n)_{n\in\mathbb{N}}$ of representations ρ_n of the symmetric group $\mathcal{S}_n$ (*i.e.* an application associating a map $\rho_n(\sigma) : \mathcal{C}_n \to \mathcal{C}_n$ to each permutation $\sigma \in \mathcal{S}_n$ in such a way that $\rho_n(\sigma \circ \sigma') = \rho_n(\sigma) \circ \rho_n(\sigma')$). An equivalent way (see *e.g.* [4]) to define labelled combinatorial class consists in considering that each element of $\mathcal{C}_n$ is a graph whose vertices are labelled by numbers from 1 to n; the image of a permutation by the underlying representation is just the permutation of the labels.

Let $(\mathcal{C}, \omega)$ and $(\mathcal{C}', \omega')$ be two labeled combinatorial class. If the sets $\mathcal{C}$ and $\mathcal{C}'$ are disjoint, then we define the class $\mathcal{C} \boxplus \mathcal{C}' = (\mathcal{C} \cup \mathcal{C}', \omega'')$ with $\omega''(e) = \omega(e)$ if $e \in \mathcal{C}$ and $\omega(e) = \omega'(e)$ if $e \in \mathcal{C}'$. One extends to the case where $\mathcal{C} \cap \mathcal{C}' \neq \emptyset$ by replacing $\mathcal{C}'$ by a copy which is disjoint of $\mathcal{C}$ in the definition of $\mathcal{C} \boxplus \mathcal{C}'$.

We also define $\mathcal{C} \boxtimes \mathcal{C}'$, i.e., the combinatorial class such that the elements of $(\mathcal{C} \boxtimes \mathcal{C}')_n$ are the pairs (e, e') where e is obtained by relabeling an element of $\mathcal{C}_i$ and e' is obtained by relabeling an element of $\mathcal{C}'_j$, with $i + j = n$ such that the set of the labels in (e, e') is $\{1, \dots, n\}$ and each relabeling preserves the initial order on the vertices. The degree of (e, e') in $\mathcal{C} \boxtimes \mathcal{C}'$ is the sum of the degree of the respective preimage of e and e' in $\mathcal{C}$ and $\mathcal{C}'$. As a special case, for each labeled class $\mathcal{C}$, we denote $\mathcal{C}^\bullet = \bullet \boxtimes \mathcal{C}$, where $\bullet$ is the class of the unique element of which, denoted also by $\bullet$, has degree 1.

2.3 The Exponential Generating Function of a Combinatorial Class

The exponential generating function (EGF) of a combinatorial class $\mathcal{C}$ is the exponential generating function of the numbers $C_n = \#(\mathcal{C}_n)$, in other words

$$S_{\mathcal{C}}(x) = \sum_{n \geq 0} C_n \frac{x^n}{n!} = \sum_{\mu \in \mathcal{C}} \frac{x^{|\mu|}}{|\mu|!}.$$

We say that two classes are *isomorphic* if their EGF are equal

$$\mathcal{C} \equiv \mathcal{C}' \Leftrightarrow S_{\mathcal{C}} = S_{\mathcal{C}'} \Leftrightarrow \forall n \in \mathbb{N}, C_n = C'_n.$$

Classically, we have [4]

$$S_{\mathcal{C} \boxplus \mathcal{C}'} = S_{\mathcal{C}} + S_{\mathcal{C}'} \text{ and } S_{\mathcal{C} \boxtimes \mathcal{C}'} = S_{\mathcal{C}} S_{\mathcal{C}'}.$$

2.4 Labelled Sequences

If $\mathcal{C}^+$ is a labelled combinatorial class such that $\mathcal{C}_0^+ = \emptyset$ then we define, up to an equivalence, the class $\text{SEQ}(\mathcal{C}^+)$ of labelled sequences by the equation

$$\text{SEQ}(\mathcal{C}^+) \equiv [\,] \boxplus (\mathcal{C}^+ \boxtimes \text{SEQ}(\mathcal{C}^+)),$$

where $[\,]$ denotes the class having a single element ϵ which is degree 0.

It is easy to show that such a class exists and that its associated exponential generating function is

$$S_{\text{SEQ}(\mathcal{C}^+)}(x) = \frac{1}{1 - S_{\mathcal{C}^+}(x)}.$$

From a combinatorial point of view, the elements of $\text{SEQ}(\mathcal{C}^+)_n$ are k-tuple $[\mu_1, \dots, \mu_k]$ where each μ_i is obtained by an order preserving relabelling of an element of $\mathcal{C}^+_{j_i}$,in such a way that the whole set of labels in $[\mu_1, \dots, \mu_k]$ is $\{1, \dots, n\}$ (as a consequence one has $\sum_{i=1}^k j_i = n$). So to any element $s = [\mu_1, \dots, \mu_k] \in \text{SEQ}(\mathcal{C}^+)_n$ we associate an ordered partition $\Pi = [\Pi_1, \dots, \Pi_k]$ of size n such that each Π_i is the set of the labels of μ_i.

Let us be a little more precise. A *labelled* list of elements of $\mathcal{C}^+$ is a list $L = [\mu_1, \mu_2, \dots, \mu_k] \in \text{SEQ}(\mathcal{C}^+)$ with each μ_i associated to a set Ω_i of non-negative integers with $\text{card}(\Omega_i) = |\mu_i|$ and $\Omega_i \cap \Omega_j = \emptyset$ for all i, j. A *standard* labelled list of elements of $\mathcal{C}^+$ is a labelled list L of elements of $\mathcal{C}^+$ such that the set of all the labels of L is $\{1, \dots, |L|\}$.

2.5 Labelled Sets and Labelled Cycles

We consider the labelled combinatorial classe $\text{SET}(\mathcal{C}^+)$ such that $\text{SET}(\mathcal{C}^+)_n$ is the quotient of the set $\text{SEQ}(\mathcal{C}^+)_n$ by the relation $[\mu_1, \ldots, \mu_k] \equiv_S [\mu_{\sigma(1)}, \ldots, \mu_{\sigma(k)}]$ for any permutation $\sigma \in \mathcal{S}_k$. Straightforwardly, each element of $\text{SET}(\mathcal{C})$ can be represented by a set of (order preserved) relabelled elements of $\mathcal{C}^+$. The exponential generating function

$$S_{\text{SET}(\mathcal{C}^+)}(x) = \exp\{S_{\mathcal{C}^+}(x)\}.$$

is easily deduced from the construction.

If one consider the equivalence relation generated by $[\mu_1, \ldots, \mu_k] \equiv_C [\mu_2, \ldots, \mu_k, \mu_1]$, the one obtains an other labelled combinatorial class $\text{CYC}(\mathcal{C}^+)$ whose elements can be represented by necklace of (order preserved) relabelled elements of $\mathcal{C}^+$. We denote a necklace by $(\mu_1, \ldots, \mu_k) = (\mu_2, \ldots, \mu_k, \mu_1)$. Again, the generating series

$$S_{\text{CYC}(\mathcal{C}^+)}(x) = \log\left\{\frac{1}{1 - S_{\mathcal{C}^+}(x)}\right\}.$$

is deduced from the construction.

3 Set Partitions and Related Constructions

3.1 Three Constructions Based on Set Partitions

A *set partition* of size n is a set $\pi = \{\pi_1, \ldots, \pi_k\}$ such that $\pi_1 \cup \cdots \cup \pi_k = \{1, \ldots, n\}$ and $\pi_i \cap \pi_j = \emptyset$ for any two indices $1 \leq i \neq j \leq k$. The set of set partitions $\mathcal{S}_p$ endowed with the size is a combinatorial classes satisfying $\mathcal{S}_p \equiv \text{SET}(\text{SET}(\bullet)^+)$, and so, $S_{\mathcal{S}_p}(x) = \exp(\exp(x) - 1)$. The numbers $\mathtt{Sp}_n = 1, 1, 2, 5, 15, 52, 203, 877, 4140 \ldots$ are the well known Bell numbers (see sequence A000110 in [15]).

If $\mathcal{C}^+$ is a labeled combinatorial class such that $\mathcal{C}_0^+ = 0$ and $\mathcal{P} \equiv \text{SET}(\mathcal{C}^+)$ then the definitions above allows to associate to each element $p = \{p_1, \ldots, p_k\} \in \mathcal{P}_n$ a set partitions $\pi(p) = \{\mathtt{labels}(p_1), \ldots, \mathtt{labels}(p_k)\}$ of size n where for each $1 \leq i \leq k$, $\mathtt{labels}(p_i)$ denotes the set of the labels of p_i.

An *ordered partition* of size n is a sequence $\Pi = [\Pi_1, \ldots, \Pi_k]$ of non empty sets such that $\{\Pi_1, \ldots, \Pi_k\}$ is a set partition of $\{1, \ldots, n\}$. The set of ordered partitions $\mathcal{O}p$ endowed with the size is a combinatorial classes satisfying $\mathcal{O}p \equiv \text{SEQ}(\text{SET}(\bullet)^+)$ and $S_{\mathcal{O}p}(x) = \frac{1}{2-\exp(x)}$.

The numbers $\mathtt{Op}_n = 1, 1, 3, 13, 75, 541, 4683, 47293, 545835, \ldots$ are the Fubini numbers (see sequence A000670 in [15]). If $\mathcal{L} \equiv \text{SEQ}(\mathcal{C}^+)$ then the definitions above allows to associate to each element $\ell = [\ell_1, \ldots, \ell_k] \in \mathcal{L}_n$ a set partition $\Pi(\ell) = [\mathtt{labels}(\ell_1), \ldots, \mathtt{labels}(\ell_k)]$ of size n.

A *cyclic partition* of size n is a necklace $\mathfrak{p} = (\mathfrak{p}_1, \ldots, \mathfrak{p}_k)$ such that $\{\mathfrak{p}_1, \ldots, \mathfrak{p}_k\}$ is a set partition of size n. The set of ordered partitions $\mathcal{C}p$ endowed with the size is a combinatorial classes satisfying $\mathcal{C}p \equiv \text{CYC}(\text{SET}(\bullet)^+)$ and $S_{\mathcal{C}p}(x) = \log(\frac{1}{2-\exp(x)})$.

The numbers $\mathtt{C}p_n = 1, 1, 3, 13, 75, 541, 4683, 47293, 545835, \ldots$ are listed in sequence A000670 [15]. If $\mathcal{N}e \equiv \text{CYC}(\mathcal{C}^+)$ then the above definitions allow us to associate to each element $c = (c_1, \ldots, c_k) \in (\mathcal{N}e)_n$ a cyclic partition $\mathfrak{p}(c) = (\texttt{labels}(c_1), \ldots, \texttt{labels}(c_k))$ of size n.

3.2 An Explicit Isomorphism

From the generating series we have $\text{SET}(\mathcal{C}p) \equiv \mathcal{O}p$. Indeed, this equality translates in terms of generating function as $\exp\left(\log\left(\frac{1}{2-e^x}\right)\right) = \frac{1}{2-e^x}$. In order to understand a more general identity introduced later in the paper, we make explicit this bijection. Assume that $c = \{c^{(1)}, \ldots, c^{(k)}\} \in \text{SET}(\mathcal{C}p)$. If $c^{(i)} = (c_1^{(i)}, \ldots, c_{h_i}^{(i)})$ then we consider σ_i the only circular permutation on the indices $\{1, \ldots, h_i\}$ such that $\min \bigcup_j \texttt{labels}(c_{(j)}^{(i)}) = \min \texttt{labels}(c_{\sigma_i^{-1}(1)}^{(i)})$. In other words, if $\ell_i = [\ell_1^{(i)}, \ldots, \ell_{h_i}^{(i)}] = [c_{\sigma_i(1)}^{(i)}, \ldots, c_{\sigma_i(h_i)}^{(i)}]$ then $\min \bigcup_j \texttt{labels}(\ell_{(j)}^{(i)}) = \min \texttt{labels}(\ell_1^{(i)})$. Now, consider the unique permutation $\rho \in \mathcal{S}_k$ such that

$$\min \texttt{labels}(c^{(\rho^{-1}(1))}) > \min \texttt{labels}(c^{(\rho^{-1}(2))}) > \cdots > \min \texttt{labels}(c^{(\rho^{-1}(k))})$$

and set

$$\texttt{stol}(c) = [\ell_1^{(\rho(1))}, \ldots, \ell_{h_{\rho(1)}}^{(\rho(1))}, \ldots, \ell_k^{(\rho(k))}, \ldots, \ell_{h_{\rho(k)}}^{(\rho(k))}] \in \mathcal{O}p. \tag{1}$$

For instance,

$$\texttt{stol}(\{(\{11\}, \{2,5\}, \{10\}), (\{6\}, \{1,3,4\}, \{7,9\}), (\{8,12\})\}) = [\{8,12\}, \{2,5\}, \{10\}, \{11\}, \{1,3,4\}, \{7,9\}, \{6\}].$$

Let $\ell = [\ell_1, \ldots, \ell_k] \in \mathcal{O}p$ and $1 = i_0 \le \cdots \le i_{h-1} < i_h = k+1 \in \{1, \ldots, k+1\}$ be the set of indices satisfying

$$\min \bigcup_{i < i_{j+1}} \texttt{labels}(\ell_i) > \min \bigcup_{i \ge i_j} \texttt{labels}(\ell_i) \tag{2}$$

with h maximal.

For instance, the indices associated to $[\{8,12\}, \{2,5\}, \{10\}, \{11\}, \{1,3,4\}, \{7,9\}, \{6\}]$ are $1 \le 2 \le 5 < 8$. We define

$$\texttt{ltos}(\ell) = \{c_1, \ldots, c_k\} \in \text{SET}(\mathcal{C}p), \tag{3}$$

where c_j denotes the necklace $(\ell_{i_{j-1}}, \ldots, \ell_{i_j - 1})$. For instance,

$$\texttt{ltos}([\{8,12\}, \{2,5\}, \{10\}, \{11\}, \{1,3,4\}, \{7,9\}, \{6\}]) = \{(\{1,3,4\}, \{7,9\}, \{6\}), (\{2,5\}, \{10\}, \{11\}), (\{8,12\})\}.$$

It is easy to check that $\texttt{ltos}(\texttt{stol}(c)) = c$ and $\texttt{stol}(\texttt{ltos}(\ell)) = \ell$. So we have

Proposition 1. *The map* `stol` *is an isomorphism of combinatorial classes and* `ltos` *is its reverse map.*

3.3 About Lyndon Words

Recall that the free monoid (*e.g.* [9]) Σ^* on a set Σ is the monoid whose elements are all the finite sequences endowed with the catenation product $\cdot$ that consists of pasting one sequence to the right of another. The empty sequence plays the role of the identity element. For instance, in the free monoid $\{a, b\}^*$ we have $[a, b, a, a, b] \cdot [b, a, a] = [a, b, a, a, b, b, a, a]$. In literature, brackets and commas are often omitted; the elements of a free monoid are then noted as juxtapositions of letters called words (the empty word, noted by ε, corresponds to the sequence $[\,]$). The name of the free monoid comes from the fact that it fulfills the universal property, that is every monoid having a generating set in bijection with Σ is isomorphic to a quotient of Σ^*.

Any pair of sequences under the form $u \cdot v$ and $v \cdot u$ are said conjugate. In other words, the conjugates of a sequence are all its circular shift. This is obviously an equivalence relation that preserves the periods, *i.e.*, the conjugate sequences of $u^{\cdot k}$ are exactly the sequences $v^{\cdot k}$ where v is conjugate to u. In terms of combinatorial class the free monoid is nothing but $\mathrm{S\small{EQ}}(\Sigma)$ and its quotient by conjugation is $\mathrm{C\small{YC}}(\Sigma)$.

Assume that the alphabet Σ is totally ordered by the order $<$. Then the free monoid is totaly ordered with the lexicographic order $\prec$. The minimal element for the lexicographic order is the empty sequence $[\,]$ and we have $[a] \cdot u \prec [b] \cdot v$ if $a < b$ or $a = b$ and $u \prec v$.

A *Lyndon words* a non periodic sequence which is minimal in its conjugacy class. Their name comes from the mathematician Roger Lyndon who studied them in 1954 [10]. Nevertheless, it should be noted that they had been introduced a year earlier by Anatoly Shirshov [14]. Lyndon words play a very important role for understanding of free groups [2], free associative algebras, and free Lie algebras [13]. Readers may refer to [11] for a rather complete survey.

Among all the properties of Lyndon's words, one of the most interesting is that they play for the free monoid the same role as prime numbers play for integers. This property is that any sequence factorizes as a unique weakly decreasing catenation of Lyndon words [12]. In other words, the free monoid Σ^* is in bijection with the multisets of aperiodic sequences over σ. For instance, if we assume $a < b$ the sequence $u = [a, b, a, b, b, a, b, a, b, a, a, a, b, a, b, a]$ factorizes as $u = [a, b, a, b, b] \cdot [a, b] \cdot [a, b] \cdot [a, a, a, b, a, b] \cdot [a]$. This means that the sequence u is assimilated to the multiset $\{(a, a, a, b, a, b), (a, b, a, b, b), (a, b), (a, b), (a)\}$ (remark the multiplicity of (a, b)). It is interesting to note that this correspondence is precisely the one that is calculated when applying `ltos`. Indeed, let $\ell = [\ell_1, \ldots, \ell_k] \in \mathcal{O}p$, the alphabet $\Sigma = \{\ell_1, \ldots, \ell_k\}$ is totally ordered by $\ell_i < \ell_j$ if and only if $\min \ell_i < \min \ell_j$. In fact, since each numbers of $\{1, \ldots, n\}$ appears only one time in the sequence, only the minimal elements the sets are relevant and all works as if our alphabet be $\{1, \ldots, n\}$. For instance, $[\{8, 12\}, \{2, 5\}, \{10\}, \{11\}, \{1, 3, 4\}, \{7, 9\}, \{6\}]$ is assimilated to $[8, 2, 10, 11, 1, 7, 6]$. The indices of Eq. (2), except the larger which is not relevant, indicate where to catenate in order to apply the complete factorization. In our example, we found the indices $\{1, 2, 5, 8\}$ and, then, we have $[8, 2, 10, 11, 1, 7, 6] = [8] \cdot [2, 10, 11] \cdot [1, 7, 6]$. Notice that, since the components are two by two distinct, the bijection with multisets of cycles class sends the sequences we consider on set of necklaces. For instance, $[8] \cdot [2, 10, 11] \cdot [1, 7, 6] \sim \{(1, 7, 6), (2, 10, 11), (8)\}$. We recover the $\texttt{ltos}(\ell)$ by replacing each integer by the set of which it is the minimum. In our example we have

$$\{(1, 7, 6), (2, 10, 11), (8)\} \rightarrow \{(\{1, 3, 4\}, \{7, 9\}, \{6\}), (\{2, 5\}, \{10\}, \{11\}), (\{8, 12\})\}.$$

Of course, this may seem like a very sophisticated way to revisit the bijection of the previous section. Nevertheless, this remark is valuable because it will allow us to link our constructions to notions of algebras (enveloping algebras, Hopf algebras, Lie algebras of primitive elements etc.) that we will explore in future works.

3.4 Set of Cycles and Sequences

Let $\mathcal{C}^+$ be a labeled combiatorial sequences such that $\mathcal{C}_0^+ = \emptyset$. We define $\mathcal{J} = \text{SET}(\text{CYC}(\mathcal{C}^+))$ and $\mathcal{S} = \text{SEQ}(\mathcal{C}^+)$.

Let us show that the map $\texttt{stol}$ allows us to compute an explicit isomorphism from $\mathcal{J}$ to $\mathcal{S}$. We define $\texttt{jtoset}_{\mathcal{C}^+} : \mathcal{J} \rightarrow \text{SET}(\mathcal{C}p)$ by

$$\texttt{jtoset}_{\mathcal{C}^+}(\{(c_1^{(1)}, \ldots, c_{h_1}^{(1)}), \cdots, (c_1^{(k)}, \ldots, c_{h_k}^{(k)})\}) = \{(\texttt{labels}(c_1^{(1)}), \ldots, \texttt{labels}(c_{h_1}^{(1)})), \cdots, (\texttt{labels}(c_1^{(k)}), \ldots, \texttt{labels}(c_{h_k}^{(k)}))\}.$$

We define also $\texttt{jtoseq}_{\mathcal{C}^+} : \mathcal{J} \rightarrow \mathcal{S}$ such that

$$\texttt{jtoseq}_{\mathcal{C}^+}(\{(c_1^{(1)}, \ldots, c_{h_1}^{(1)}), \cdots, (c_1^{(k)}, \ldots, c_{h_k}^{(k)})\}) = \ell$$

is the unique permutation of the vector $[c_1^{(1)}, \ldots, c_{h_1}^{(1)}, \cdots, c_1^{(k)}, \ldots, c_{h_k}^{(k)}]$ such that $\Pi(\ell) = \texttt{stol}(\texttt{jtoset}_{\mathcal{C}^+}((\{(c_1^{(1)}, \ldots, c_{h_1}^{(1)}), \cdots, (c_1^{(k)}, \ldots, c_{h_k}^{(k)})\})))$.

Let $c = \{(c_1^{(1)}, c_2^{(1)}), (c_1^{(2)}), (c_1^{(3)}, c_2^{(3)})\}$ be a set of labelled graphs as shown in Fig. 1.

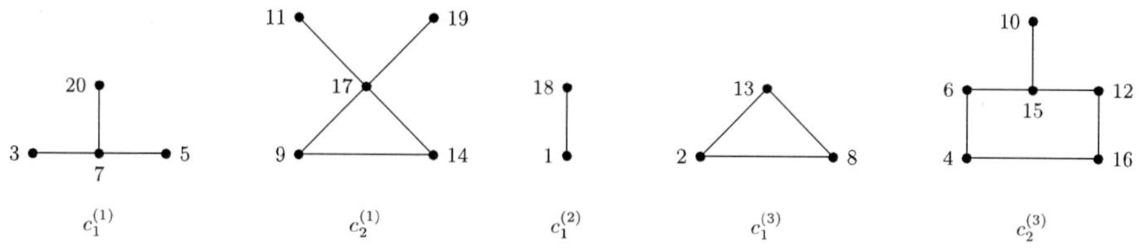

Fig. 1. Five labelled graphs

Then $\texttt{jtoset}_{\mathcal{C}^+}(c)$ equals

$$\{(\{3, 7, 5, 20\}, \{9, 14, 17, 19, 11\}), (\{1, 18\}), (\{2, 8, 13\}, \{4, 16, 12, 15, 10, 6\})\}.$$

Furthermore, $\Pi(l) = \mathtt{stol}\,(\mathtt{jtoset}_{\mathcal{C}^+}(c)) =$

$$[\{3,7,5,20\},\{9,14,17,19,11\},\{2,8,13\},\{4,16,12,15,10,6\},\{1,18\}].$$

Since $\mathtt{stol}$ is one to one, the equality on generating functions allows us to deduce that $\mathtt{jtoseq}_{\mathcal{C}+}$ is an isomorphism of combinatorial classes. The inverse bijection $\mathtt{seqtoj}_{\mathcal{C}+} : \mathrm{SET}(\mathcal{C}p) \to \mathcal{J}$ is defined by

$$\mathtt{seqtoj}_{\mathcal{C}^+}(\ell) = \{(c_1^{(1)},\ldots,c_{h_1}^{(1)}),\cdots,(c_1^{(k)},\ldots,c_{h_k}^{(k)})\}$$

where $[c_1^{(1)},\ldots,c_{h_1}^{(1)},\cdots,c_1^{(k)},\ldots,c_{h_k}^{(k)}]$ is the unique permutation of ℓ such that $\mathtt{ltos}(\Pi(\ell)) = \mathtt{jtoset}_{\mathcal{C}^+}(\{(c_1^{(1)},\ldots,c_{h_1}^{(1)}),\cdots,(c_1^{(k)},\ldots,c_{h_k}^{(k)})\})$.

In the aforementioned example, we deduce the indices of the minimum elements in $\Pi(l)$ being $1 < 3 < 5$. Hence, $\mathtt{ltos}(\Pi(l)) =$

$$\begin{aligned} &= \{(\{1,18\}),(\{2,8,13\},\{4,16,12,15,10,6\}),(\{3,7,5,20\},\{9,14,17,19,11\})\} \\ &= \mathtt{jtoset}_{\mathcal{C}^+}(c). \end{aligned}$$

We summarize the results of this section in the following theorem.

Theorem 1. *The maps which make commuting the following diagram are explicit isomorphisms of combinatorial classes*

$$\mathrm{SET}(\mathrm{CYC}(\mathcal{C}^+)) \underset{\mathtt{seqtoj}_{\mathcal{C}^+}}{\overset{\mathtt{jtoseq}_{\mathcal{C}}^+}{\rightleftarrows}} \mathrm{SEQ}(\mathcal{C}^+). \tag{4}$$

4 Labelled and Unlabelled Trees

We illustrate the previous result by investigating the combinatorial classes $\mathcal{R}$ satisfying

$$\mathrm{SET}(\mathrm{CYC}(\mathcal{R}^\bullet)) \equiv \mathrm{SEQ}(\mathcal{R}^\bullet) \equiv \mathcal{R}. \tag{5}$$

This isomorphism can be translated into the following functional equation:

$$\exp\left\{\log\left\{\frac{1}{1-xS_{\mathcal{R}}(x)}\right\}\right\} = \frac{1}{1-xS_{\mathcal{R}}(x)} = S_{\mathcal{R}}(x). \tag{6}$$

This equation has a unique solution

$$S_{\mathcal{R}}(x) = \frac{1-\sqrt{1-4x}}{2x} \tag{7}$$

which is also the ordinary generating function of the Catalan numbers $\mathsf{C}_n = \frac{1}{n+1}\binom{2n}{n}$ [4, 16]. Hence, $\mathcal{R}$ is unique up to an isomorphism and

$$R_n = \frac{(2n)!}{(n+1)!}. \tag{8}$$

The sequence of R_n is

$$1, 1, 4, 30, 336, 5040, 95040, 2162160 \quad \text{A001761 [15]}.$$

4.1 Labelled Trees from Unlabelled Trees

Definition 1. *A* tree *is a list of trees (possibly empty) connected to a node, called its root, by an edge (also called branch). Notice that this is a valid recursive definition which base case is a root together with an empty list. The* degree $\omega(t)$ *of a tree t is the number of its edges or, equivalently the number of its nodes which are not its root.*

Let $\mathcal{D}$ be the set of trees. There are a finite number of trees having a given degree, so the pair $(\mathcal{D}, \omega)$ is a (unlabelled) combinatorial class. The number D_n is known to be the Catalan number C_n (see *e.g.* [16]). So the ordinary generating function of the class $\mathcal{D}$, *i.e.*,

$$S_{\mathcal{D}}^{ord}(x) = \sum_{n \geq 0} D_n x^n, \tag{9}$$

fulfills the same functional equation (6) as the exponential generating function of $\mathcal{R}$. So each $\mathcal{R}_n$ is in one to one correspondence with $\mathcal{D}_n \times \mathcal{S}_n$. This suggests that one can exhibit an explicit realization of the class $\mathcal{R}$ by labeling the nodes which are not the root of each tree $t \in \mathcal{D}$ by $\{1, \ldots, \omega(t)\}$, without repetition and in any possible way.

4.2 Shifted Structure

The class $\mathcal{R}^{\bullet}$ is isomorphic to the class $\mathcal{R}_r$ of trees with labeled root endowed with the weight ω_r counting the total number of nodes, including the root. More precisely, for a given n, a tree of $(\mathcal{R}_r)_{n+1}$ is obtained by labelling the root of a tree in $\mathcal{R}_n$ with any of the possible value from the set $\{1, \ldots, n+1\}$ and relabel, if necessary, the nodes with respect to the order induced by the initial permutation.

Example 1. Let t be a rooted labelled tree in $\mathcal{R}_5$, and its associated permutation is $\pi = (1, 2, 4, 5, 3)$. Then the set of all rooted labeled trees where the root is labeled obtained from t is given by the set of permutations $\{(6,1,2,4,5,3), (5,1,2,4,6,3), (4,1,2,5,6,3), (3,1,2,5,6,4), (2,1,3,5,6,4), (1,2,3,5,6,4)\}$.

In terms of generating function, this operation leads to

$$S_{\mathcal{R}_r}(x) = S_{\mathcal{R}^{\bullet}}(x) = x S_{\mathcal{R}}(x) = \frac{1 - \sqrt{1-4x}}{2}$$

and so

$$(R_r)_n = n! C_{n-1} = \frac{(2n-2)!}{(n-1)!},$$

for any $n \geq 1$. The sequence of R_{rn} is given by

$$1, 2, 12, 120, 1680, 30240 \quad \mathit{A001813}\ [15].$$

We insists on the fact that at this point $(\mathcal{R}_r)_n$ counts trees having n nodes including the root.

4.3 Hanging Trees in Necklaces

A *labelled necklace of planar trees* is a necklaces on which trees are hung and all the nodes (comprising the roots) are labeled by $\{1, \ldots, n\}$ where n is the total numbers of nodes (comprising roots). We denote by $\mathcal{N}$ the set of such necklaces. The cyclic structure comes from the fact that a necklace is invariant by rotation. The weight $\omega_N(\mathtt{n})$ of a necklace $\mathtt{n}$ is the total number of the nodes, comprising roots, of the trees it contains. The pair $(\mathcal{N}, \omega_N)$ is a labeled combinatorial class that satisfies

$$\mathcal{N} \equiv \textsc{Cyc}(\mathcal{R}^{\bullet}). \tag{10}$$

We depict in Fig. 2 elements of $\mathcal{N}$ that are equivalent under cyclic rotation.

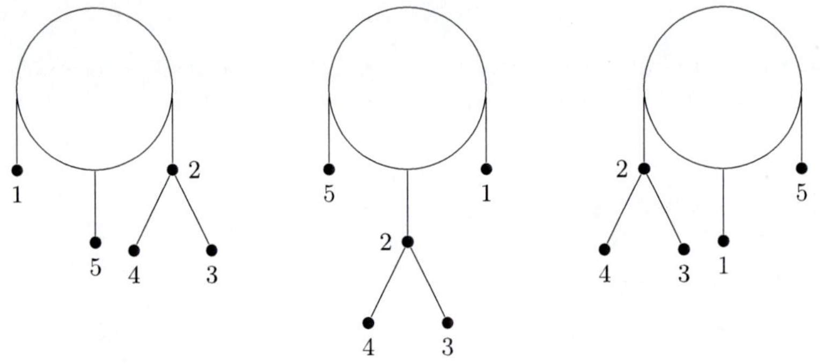

Fig. 2. Three rooted labeled trees equivalent under cyclic rotation

Notices that Labeled necklaces of rooted trees appear under the name of "planar labelled trees" in the work of Miloudi [8]. To be more precise, he studied combinatorial class which is straightforwardly isomorphic to $\mathcal{N}$ and he proved $N_n = (2n-3)!/(n-1)!$ for any $n \geq 2$ and $N_1 = 1$. We recover this result from the interpretation of (10) in terms of generating function. Indeed,

$$S_{\textsc{Cyc}(\mathcal{R}^{\bullet})}(x) = \log \frac{1}{1 - \dfrac{1-\sqrt{1-4x}}{2}} = \log \frac{2}{1+\sqrt{1-4x}} = \log \frac{1-\sqrt{1-4x}}{2x}.$$

Hence,

$$S_{\textsc{Cyc}(\mathcal{R}^{\bullet})}(x) = \log \mathcal{R}(x), \tag{11}$$

and the exact formula for N_n is obtained by expanding the function as a Taylor series.

These numbers are also mentioned by Wolfdieter Lang in [15], see the sequence below

$$1, 1, 3, 20, 210, 3024, 55440, 1235520, 32432400 \quad A006963.$$

We now have all the material to make explicit the isomorphisms suggested by (5). To this aim we consider jewellery boxes which are sets of necklaces and forests which are sequences of trees. More formally, in terms of combinatorial classes we define $\mathcal{J} = \textsc{Set}(\mathcal{N})$ and $\mathcal{F} = \textsc{Seq}(\mathcal{R}^\bullet)$. Let $\texttt{jtof} = \texttt{jtoseq}_{\mathcal{R}^\bullet} : \mathcal{J} \to \mathcal{F}$ and its inverse bijection $\texttt{ftoj} = \texttt{seqtoj}_{\mathcal{R}^\bullet} : \mathcal{F} \to \mathcal{J}$. An explicit isomorphism $\texttt{rtof} : \mathcal{R} \to \mathcal{F}$ is obtained by removing the root to any tree in $\mathcal{R}$. The reciprocal isomorphism $\texttt{ftor} : \mathcal{F} \to \mathcal{R}$ consists in connecting all the trees of a given sequence to an additional node called the root.

All these constructions are summarized in the following result which is a corollary of Theorem 1.

Corollary 1. *The maps which make commuting the following diagram are explicit isomorphisms of combinatorial classes*

$$\mathcal{J} \underset{\texttt{ftoj}}{\overset{\texttt{jtof}}{\rightleftarrows}} \mathcal{F} \underset{\texttt{rtof}}{\overset{\texttt{ftor}}{\rightleftarrows}} \mathcal{R}. \tag{12}$$

In Fig. 3 we illustrate the bijections from (12) using two examples.

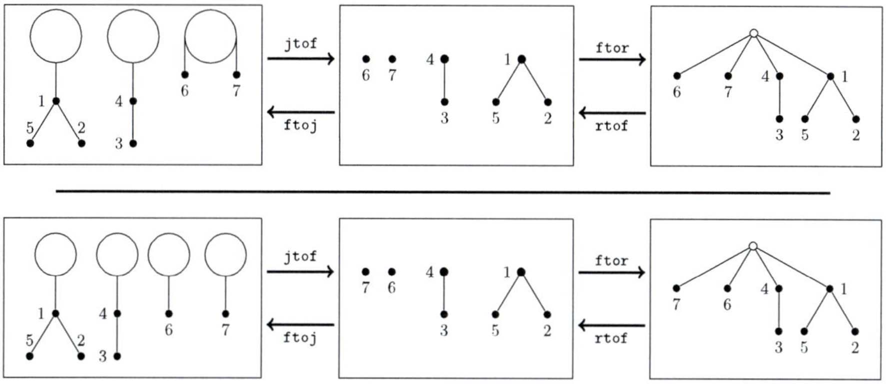

Fig. 3. A set of three labelled necklaces of planar trees (leftmost box) in bijection with a forest of labelled trees (middle box) in bijection with a rooted labelled tree (rightmost box). In the upper part of the figure the cycles are $\{(\{1,5,2\}),(\{3,4\}),(\{6\},\{7\})\}$, whereas in the lower part the cycles are $\{(\{1,5,2\}),(\{3,4\}),(\{6\}),(\{7\})\}$

4.4 Other Recursive Tree-Like Combinatorial Classes

The tree-like structure constructed from sequences is the most rigid one. It involves rooted trees that are embedded in a plan in such a way that the branches are always pointing downwards and so, the order of the sequences of the subtrees is relevant. If we relax the constraint of the orientation of the branches then the trees become invariant by rotation. In that context, a tree is a non-oriented graph without cycle with a privileged vertex called a root. Each root can be seen as a labeled necklace on which the subtrees are hanged, forming a sort of windmill (see Fig. 4 for examples). In other words, a tree is either an isolated root or a root with a cycle of trees. The combinatorial class satisfies the isomorphism

$$\mathcal{W} \equiv \bullet \boxplus (\bullet \boxtimes \mathrm{Cyc}(\mathcal{W})), \tag{13}$$

and its generating function satisfies

$$S_{\mathcal{W}}(x) = x\left(\log\left(\frac{1}{1 - S_{\mathcal{W}}(x)}\right) + 1\right). \tag{14}$$

Expanding both sides of the equation and identifying the coefficients, we get a system, the resolution of which allows us to obtain the first cases of the enumeration:

$$1, 2, 9, 68, 730, 10164, 173838, 3524688, 82627200, \ldots \quad A000169\ [15]$$

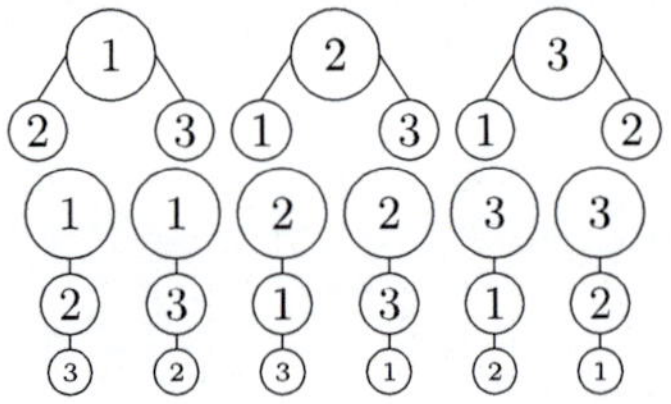

Fig. 4. The nine windmills of degree 3.

This is another example of a tree-like structure studied among others in [1]. Notice that no closed form for the generating function is known but there exists a formula for the coefficients as a combination of Stirling numbers of first kind,

$$W_n = \sum_{i=0}^{n} i! \binom{n}{i} s_{n-1,i}, \tag{15}$$

where $s_{n,i}$ denotes the (unsigned) Stirling number of first kind that counts the number of permutations of n objects with exactly i cycles. Indeed, from (14), $S_{\mathcal{W}}(x)$ is the inverse of $g(x) = \frac{x}{1-\log(1-x)}$ for the composition. The Lagrange inversion theorem [7] is a classical combinatorial tools allowing us to compute the Taylor expansion of inverse function. In our case, the direct application of the Lagrange inversion Theorem implies that

$$W_n = \left(\frac{d}{dx}\right)^{n-1}\left(\frac{x}{g(x)}\right)^n\Bigg|_{x=0} = \left(\frac{d}{dx}\right)^{n-1}\left(1 + \log\left(\frac{1}{1-x}\right)\right)^n\Bigg|_{x=0} \tag{16}$$

is the coefficient of x^{n-1} in $\left(1 - \log\left(\frac{1}{1-x}\right)\right)^n$ multiplied by $(n-1)!$. Knowing that the exponential generating function of Stirling's numbers $s_{i,k}$ (k fixed) is $\frac{1}{k}! \log\left(\frac{1}{1-x}\right)^k = \sum_i s_{i,k} \frac{x^i}{i!}$, an easy computation allows us to deduce (15) from (16).

The last example we consider is the one where no more order constraints are imposed on the sub-trees of the same node. In this context, a tree is a root with a (possibly empty) set of trees. The trees of this kind can be drawn as nesting of disjointed discs with numbered surfaces (see some examples in Fig. 5). Notice that, in these examples nested disk configurations, of degree 3 are as numerous as the windmills of degree 3 (see Fig. 4). Of course, this is not always the case and generally, there are fewer nested discs configurations than windmills. For instance, there are two windmills of degree 4 the roots of which is labeled by 1 with three sub-windmills of degree 1 while there is only one nested discs configuration the root of which is labeled by 1 containing three discs (see Fig. 6).

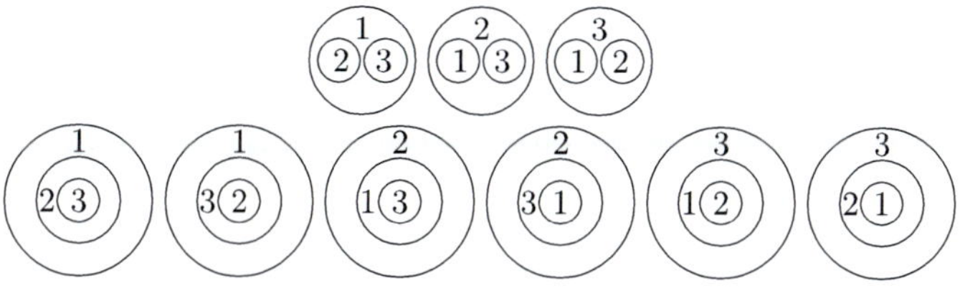

Fig. 5. The nine configurations of nested discs of degree 3.

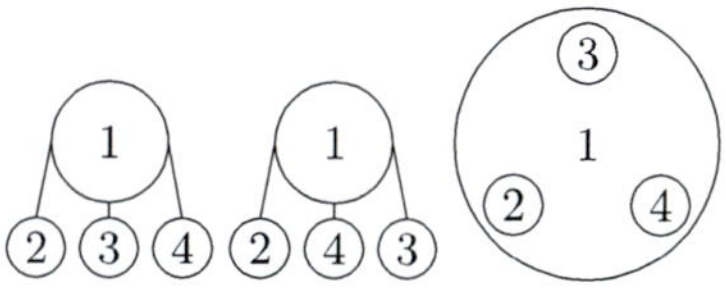

Fig. 6. Two windmills and one nested discs configuration.

The combinatorial class satisfies the isomorphism

$$\mathcal{N}pt = \bullet \boxtimes \textsc{Set}(\mathcal{N}pt),$$

and its generating series satisfies the functional equation

$$S_{\mathcal{N}pt}(x) = xe^{S_{\mathcal{N}pt}(x)}.$$

Solving these equation, one finds

$$S_{\mathcal{N}pt}(x) = -W(-x),$$

where $W(x)$ denotes the Lambert W function that is the principal branch of the functional inverse of $x \to xe^x$ [3]. The Taylor expansion of $W(x)$ is obtained by applying Lagrange inversion Theorem and implies $Npt_n = n^{n-1}$. The sequence of the Npt_n's can also be found in [15]:

$$1, 2, 9, 64, 625, 7776, 117649, 2097152, \ldots \quad A000169.$$

References

1. Bergeron, F., Labelle, G., Leroux, P.: Combinatorial Species and Tree-like Structures. Encyclopedia of Mathematics and its Applications. Cambridge University Press, Cambridge (1997)
2. Chen, K.T., Fox, R.H., Lyndon, R.C.: Free differential calculus, iv. the quotient groups of the lower central series. Ann. Math. **68**(1), 81–95 (1958)
3. Corless, R.M., Gonnet, G.H., Hare, D.E.G., Jeffrey, D.J., Knuth, D.E.: On the lambertW function. Adv. Comput. Math. **5**(1), 329–359 (1996)
4. Flajolet, P., Sedgewick, R.: Analytic Combinatorics, 1st edn. Cambridge University Press, New York (2009)
5. Hivert, F., Novelli, J.-C., Thibon, J.-Y.: The algebra of binary search trees. Theor. Comput. Sci. **339**(1), 129–165 (2005). Combinatorics on Words
6. Knuth, D.E.: The Art of Computer Programming. Sorting and Searching, vol. 3, 2nd edn. Addison Wesley Longman Publishing Co. Inc., Boston (1998)
7. Lagrange, J.-L.: Nouvelle méthode pour résoudre les équations littérales par le moyen des séries. Mémoires de l'Académie Royale des Sciences et Belles-Lettres de Berlin **24**, 251–326 (1770)
8. Leroux, P., Miloudi, B.: Généralisations de la formule d'otter. Annales des sciences mathématiques du Québec **16**(1), 53–80 (1992)
9. Lothaire, M.: Combinatorics on Words. Cambridge Mathematical Library, 2nd edn. Cambridge University Press, Cambridge (1997)
10. Lyndon, R.C.: On burnside's problem. Trans. Am. Math. Soc. **77**(2), 202–215 (1954)
11. Reutenauer, C.: Free Lie Algebras. Handbook of Algebra, vol. 3. North-Holland, Amsterdam (2003)
12. Schützenberger, M.P.: On a factorisation of free monoids. Proc. Am. Math. So. **16**, 02 (1965)
13. Shirshov, A.: On free lie rings. Mat. Sb. **45**, 77–87 (2009)
14. Shirshov, A.: Subalgebras of free lie algebras. Algebra and Logic **11**, 3–13 (2009)
15. Sloane, N.J.A.: Oeis Foundation Inc., The On-line Encyclopedia of Integer Sequences. http://oeis.org/
16. Stanley, R.P.: Catalan Numbers. Cambridge University Press, Cambridge (2015)

Tight Bounds on the Coefficients of Consecutive k-out-of-n:F Systems

Vlad-Florin Drăgoi[1,2(✉)], Simon Cowell[1], and Valeriu Beiu[1]

[1] Department of Mathematics and Computer Sciences, "Aurel Vlaicu" University of Arad, Arad, Romania
{vlad.dragoi,simon.cowell,valeriu.beiu}@uav.ro
[2] LITIS, University of Rouen Normandie Avenue de l'université, 76801 Saint-Étienne-du-Rouvray, France

Abstract. In this paper we compute the coefficients of the reliability polynomial of a consecutive-k-out-of-n:F system, in Bernstein basis, using the generalized Pascal coefficients. Based on well-known combinatorial properties of the generalized Pascal triangle we determine simple closed formulae for the reliability polynomial of a consecutive system for particular ranges of k. Moreover, for the remaining ranges of k (where we were not able to determine simple closed formulae), we establish easy to calculate sharp bounds for the reliability polynomial of a consecutive system.

Keywords: Consecutive systems · Generalized Pascal triangles · Bernstein basis · Reliability polynomial

1 Introduction

A relatively hidden gem of network reliability is represented by the class of consecutive systems. They were introduced in 1980 as r-successive-out-of-n:F systems [13], before being aptly renamed consecutive-k-out-of-n:F systems in 1981 [5]. Clearly, this type of redundancy scheme came reasonably late to the "reliability table," i.e., almost 30 years after the majority-voting and the multiplexing concepts (both gate-level based reliability schemes) were introduced by von Neumann in January 1952.[1] A printed version of those lectures was published in April 1956 [21], followed in September 1956 by the introduction of the hammock networks by Moore and Shannon [15] (the first device-level based reliability scheme). For more information on consecutive systems the interested reader should consult [3,11], while it is worth mentioning that the associated probability problem was proposed and solved as early as 1718 by de Moivre [9]

[1] John von Neumann presented his work in five seminal lectures at the California Institute of Technology (Caltech) in January 1952. They are available, based on the notes taken by R. S. Pierce, at https://sites.google.com/site/michaeldgodfrey/vonneumann/vN_Caltech_Lecture.pdf.

I. Dzitac et al. (Eds.): ICCCC 2020, AISC 1243, pp. 35–44, 2021.
https://doi.org/10.1007/978-3-030-53651-0_3

(see also [8]), with the associated graphs being proven most reliable in the late 90's (see [10]).

Consecutive-k-out-of-n:F systems belong to the class of device-level based reliability schemes (although "devices" might be quite complex entities), and are aimed at communications, as opposed to gate-level based reliability schemes which are targeting computations. Such systems can be abstracted as networks/graphs, network reliability being a field pioneered by [15] and which has significantly evolved ever since (see [4,7,17]). The fundamental problems in network reliability are to determine: two-terminal, k-terminal, and all-terminal reliability of a network, and are all known to be very difficult in general (#P-complete [13,14,19,20]). That is why even the best algorithms are time consuming [6,12,14], and lower and upper bounds were investigated as efficient alternatives to exact but tedious computations. In the particular case of consecutive-k-out-of-n:F systems, bounds have been reported staring from 1981 [5], and improved over time (see [4,7,8,17,18]). A 'midway path' forward is to bound the coefficients of the reliability polynomial [1,2,16], and follow with the exact polynomial computations. All of these different approaches reveal wide trade-offs between accuracy and time-complexity.

In this paper we are investigating a 'midway path' approach for the particular case of consecutive-k-out-of-n:F systems, and we will show that most of the coefficients can be quite easily computed exactly, while only a handful of them are computationally demanding, but can be bounded by reasonably simple formulas.

1.1 Consecutive Systems

A consecutive-k-out-of-n:F system corresponds to a sequence of n independent, identically distributed (i.i.d.) Bernoulli trials, with common probability of success p, in which the system itself is deemed to have failed if the sequence includes a run of at least k consecutive failures, and to have succeeded, otherwise. The *reliability* of the system is the probability $R(k, n; p)$ that it succeeds. We can write this probability as a homogeneous polynomial of degree n in p and q, where $q = 1 - p$, as follows:

$$R(k, n; p) = \sum_{i=0}^{n} N_{n,k,i} p^i q^{n-i}, \tag{1}$$

where $N_{n,k,i}$ is the number of sequences of n trials that include exactly i successes, in which the longest consecutive run of failures has length strictly less than k.

1.2 Standard Multinomial Coefficient

A well known bins-and-balls counting problem that we consider here is the following. What is the number of ways in which n identical balls can be distributed among a sequence of i distinct bins, such that bins may be empty, and no bin may contain more than k balls? The answer to this problem is given by the standard multinomial coefficient, denoted $\binom{i}{n}_k$. The algebraic description of $\binom{i}{n}_k$ is the following

$$(1 + z + z^2 + \cdots + z^k)^i = \sum_{a \geq 0} \binom{i}{a}_k z^a, \tag{2}$$

with $\binom{i}{a}_1$ the usual binomial coefficient and $\binom{i}{a}_k = 0$ for $a > ik$.

More generally, such objects are also known to count the number of A-restricted compositions of an integer n into i parts. That is, the number of ways, $\binom{i}{n}_{(1)_{j \in A}}$, in which n can be written as the sum of a sequence of i integers drawn from a given subset $A \subseteq \{0, 1, \ldots\}$, with replacement (i.e., the order is important). When $A = \{0, \ldots, k\}$, we simply use the $\binom{i}{n}_k$ notation.

2 Results

Theorem 1. *We have*

$$\begin{aligned} N_{n,k,i} &= [z^{n-i}](1 + z + \cdots + z^{k-1})^{i+1} \\ &= [z^i] z^{n-(k-1)(i+1)} \left(\frac{1 - z^k}{1 - x} \right)^{i+1}, \end{aligned} \tag{3}$$

where $[z^t]f(z)$ denotes the coefficient of z^t in the formal power series expansion of $f(z)$ in powers of z.

Proof. Our proof is a combinatorial one, that is, we show that two counting problems are identical. Fixing k, n and i, consider a sequence of n trials that includes exactly i successes and in which all the runs of consecutive failures have length at most $k-1$. We may consider this sequence as a sequence of $i+1$ runs of consecutive failures of lengths between 0 and $k-1$ inclusive, each consecutive pair of such runs separated by a single success, in which the total number of failures is $n-i$. The number of such sequences, which is $N_{n,k,i}$, is therefore also the number of ways in which $n-i$ identical balls can be distributed among a sequence of $i+1$ distinct bins, such that bins may be empty, and no bin may contain more than $k-1$ balls. The first equality in (3) now follows directly from (2), and the second one follows from the identity $[z^{n-i}]f(z) = [z^i](z^n f(1/z))$.

2.1 Properties of the Reliability Polynomials

Theorem 2. *$N_{n,k,i}$ satisfy the following properties:*

$$N_{n,k,i} = 0, \forall i \leq i_{n,k} \triangleq \left\lfloor \frac{n-k+1}{k} \right\rfloor; \tag{4}$$

$$= \binom{n}{n-i}, \forall i \geq n-k+1; \tag{5}$$

$$= \sum_{j=0}^{\lfloor \frac{n-i}{k} \rfloor} (-1)^j \binom{i+1}{j} \binom{n-jk}{i}, \forall i \in \{i_{n,k}+1, \ldots, n-k\}. \tag{6}$$

Corollary 1. *The reliability polynomial of a consecutive-k-out-of-n:F system*

$$R(k,n;p) = \sum_{i=i_{n,k}+1}^{n} \binom{n}{i} p^i q^{n-i} - \sum_{i=i_{n,k}+1}^{n-k} \sum_{j=1}^{\lfloor \frac{n-i}{k} \rfloor} (-1)^{j+1} \binom{i+1}{j} \binom{n-jk}{i} p^i q^{n-i}. \tag{7}$$

Equation (6) gives the full description of the coefficient $N_{n,k,i}$ regardless of the values of k and n. However, by taking a closer look we can deduce simpler expressions for some sub-sets of $\{i_{n,k}+1, \dots, n-k\}$.

Corollary 2.

$$N_{n,k,i} = \binom{n}{i} - (i+1)\binom{n-k}{i}, \forall i \in \{n-2k+1, \dots, n-k\}; \tag{8}$$

$$N_{n,k,i} = \binom{n}{i} - (i+1)\binom{n-k}{i} + \binom{i+1}{2}\binom{n-2k}{i}, \forall i \in \{n-3k+1, \dots, n-2k\}. \tag{9}$$

Relying on these results we will analyze particular cases for a fixed n and k in particular ranges. These analyses will lead to simple formulae for the coefficients, and thus for the reliability of a consecutive system. Let us begin with $k \in \{1, 2, n\}$.

Proposition 1.

- *For* $k = 1$ $N_{n,1,i} = 0, \forall i \neq n$, *and* $N_{n,k,n} = 1$ *and*

$$R(1,n;p) = p^n. \tag{10}$$

- *For* $k = 2$ $N_{n,2,i} = \binom{i+1}{n-i}$ *for* $0 \leq i \leq n$, *and*

$$R(2,n;p) = \sum_{i=0}^{n} \binom{i+1}{n-i} p^i q^{n-i}. \tag{11}$$

- *For* $k = n$ $N_{n,n,i} = \binom{n}{i}, \forall i \geq 1$, *and* $N_{n,n,0} = 0$ *and*

$$R(n,n;p) = \sum_{i=1}^{n} \binom{n}{i} p^i q^{n-i}. \tag{12}$$

Next, we consider the case when $n - 2k < 0$ in (8), and the case when $n - 3k < 0$ in (9).

Proposition 2.

- *For any* $k \geq \lfloor \frac{n}{2} \rfloor$ *we have* $N_{n,k,i} = \binom{n}{i}, \forall i > n-k$, *and* $N_{n,k,i} = \binom{n}{i} - (i+1)\binom{n-k}{i}, \forall i \in \{\frac{n-k+1}{k}, \ldots, n-k+1\}$. *It follows that*

$$R(k,n;p) = \sum_{i=i_{n,k}+1}^{n} \binom{n}{i} p^i q^{n-i} - \sum_{i=i_{n,k}+1}^{n-k} (i+1)\binom{n-k}{i} p^i q^{n-i}. \tag{13}$$

- *For any* $\lfloor \frac{n}{3} \rfloor \leq k < \lfloor \frac{n}{2} \rfloor$ *we have* $N_{n,k,i} = \binom{n}{i}, \forall i > n-k$, $N_{n,k,i} = \binom{n}{i} - (i+1)\binom{n-k}{i}, \forall i \in \{n-2k+1, \ldots, n-k\}$, *and* $N_{n,k,i} = \binom{n}{i} - (i+1)\binom{n-k}{i} + \binom{i+1}{2}\binom{n-2k}{i}, \forall i \in \{\frac{n-k+1}{k}, \ldots, n-2k\}$. *This implies*

$$\begin{aligned} R(k,n;p) = &\sum_{i=i_{n,k}+1}^{n} \binom{n}{i} p^i q^{n-i} - \sum_{i=i_{n,k}+1}^{n-k} (i+1)\binom{n-k}{i} p^i q^{n-i} \\ &+ \sum_{i=i_{n,k}+1}^{n-2k} \binom{i+1}{2}\binom{n-2k}{i} p^i q^{n-i}. \end{aligned} \tag{14}$$

We now use the fact that $N_{n,k,i}$ can be efficiently computed for any n and k when $i \geq \lfloor \frac{n}{3} \rfloor$, to establish new bounds on the remaining coefficients.

Proposition 3. *For any* $k < \lfloor \frac{n}{3} \rfloor$ *and* $\forall i \in \{i_{n,k}+1, \ldots, n-3k\}$

$$\begin{aligned} N_{n,k,i} &\leq \binom{n}{i} - (i+1)\binom{n-k}{i} + \binom{i+1}{2}\binom{n-2k}{i} \\ N_{n,k,i} &\geq \binom{n}{i} - (i+1)\binom{n-k}{i}. \end{aligned} \tag{15}$$

Straightforward, we now define for any $k < \lfloor \frac{n}{3} \rfloor$ and $\forall i \in \{i_{n,k}+1, \ldots, n-3k\}$ the upper and lower bounds as

$$U_{n,k,i} \triangleq \min\left\{\binom{n}{i}, \binom{n}{i} - (i+1)\binom{n-k}{i} + \binom{i+1}{2}\binom{n-2k}{i}\right\} \tag{16}$$

$$L_{n,k,i} \triangleq \max\left\{0, \binom{n}{i} - (i+1)\binom{n-k}{i}\right\}. \tag{17}$$

3 Simulations

We have performed a series of simulations to test our results. We illustrate here only a small part of those, more exactly for $n \in \{16, 32, 64\}$.

In Fig. 1 we plot $R(k,n;p)$ (i.e., $R(k,16;p)$ (1a), $R(k,32;p)$ (1c), and $R(k,64;p)$ (1e), as well as the relative errors of the approximation of $N_{n,k,i}$ using $L_{n,k,i}$ and $U_{n,k,i}$ in Figs. (1b), (1d), and (1f). More precisely, we plot $1 - L_{n,k,i}/N_{n,k,i}$ for $k \geq \lfloor n/2 \rfloor$ (light magenta) and $2 < k < \lfloor n/3 \rfloor$, and $1 - U_{n,k,i}/N_{n,k,i}$ for $\lfloor n/3 \rfloor \leq k < \lfloor n/2 \rfloor$ (dark magenta).

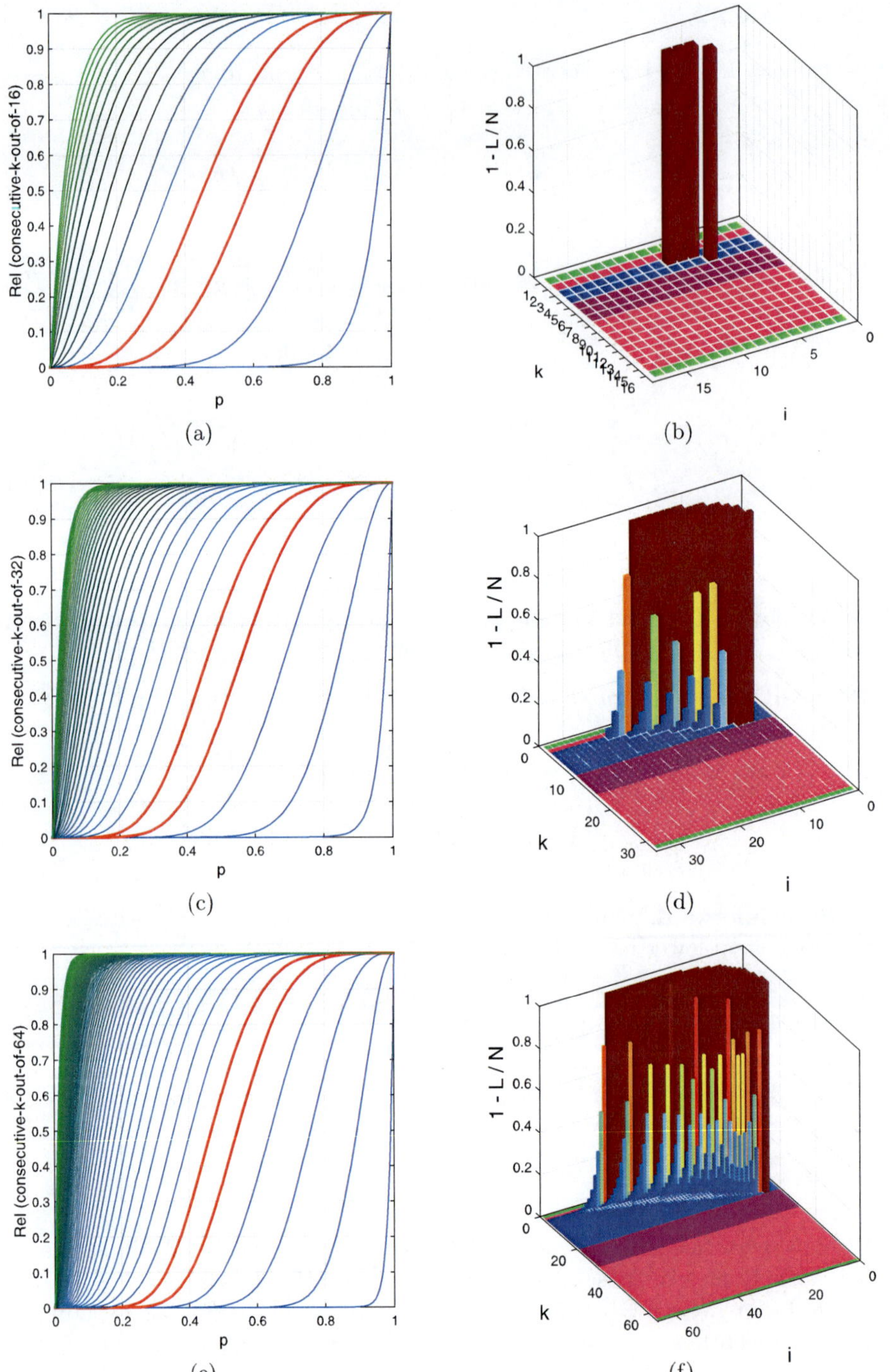

Fig. 1. $R(k, n; p)$ for: (a) $n = 16$, (c) $n = 32$, and (e) $n = 64$, as well as the relative errors for: $N_{n,k,i}$ for (b) $n = 16$, (d) $n = 32$, and (f) $n = 64$.

Remarks

- The flat surfaces in Figs. 1b, 1d, and 1f (green and magenta), show that the coefficients $N_{n,k,i}$ are computed exactly. This is a direct consequence of Proposition 2.
- Focusing our attention on the case $2 < k < \lfloor n/3 \rfloor$, the absolute errors are different than 0 in only a few cases.
- The number of coefficients which are computed exactly (dark blue) is significantly larger than the number of approximated coefficients, e.g., for $n = 32$, almost 81% are computed exactly (187 out of 231).
- The number of approximated coefficients is a decreasing function of k. Hence, as k is approaching $n/3$, the number of exactly computed coefficients increases. For example, for $n = 32$ and $k = 9$ slightly over 90% of the coefficients are computed exactly (30 out of 33).
- The worst approximation with respect to the absolute error $(N_{n,k,i} - L_{n,k,i})$ is achieved for $k = 3$, and any $n \leq 64$.

That is why we have decided to plot the exact reliability polynomial (red) together with the reliability polynomials obtained using the upper (green) and the lower (blue) bounds for $k = 3$ and $n = 16$ (Fig. 2). Notice in Fig. (2a) that from $p \geq 0.5$ the approximations are practically overlapping with the exact reliability, while for smaller values of p the behaviour of the two bounds can be seen in Fig. (2b).

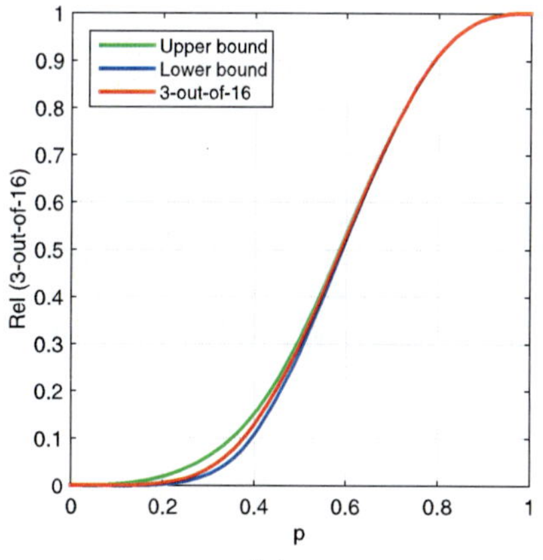

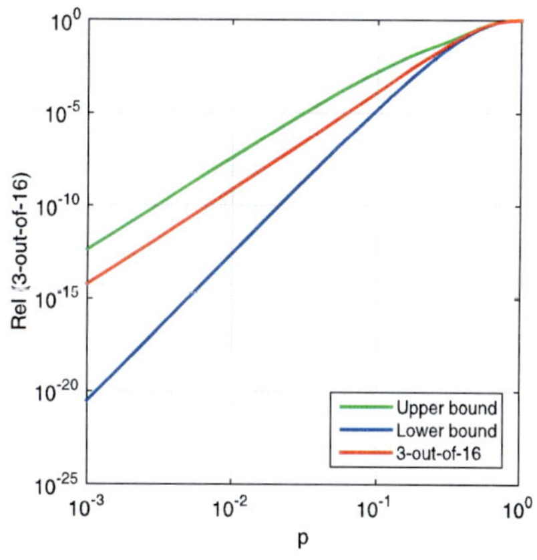

Fig. 2. Reliability of a consecutive-3-out-of-16:F system, and its upper and lower bounds: (a) linear scale, and (b) logarithmic scale.

Finally, Fig. 3 details the exact coefficients (yellow), as well as their lower (blue) and upper (green) bounds, on top of the corresponding binomial coefficients (red), in both linear and logarithmic scales.

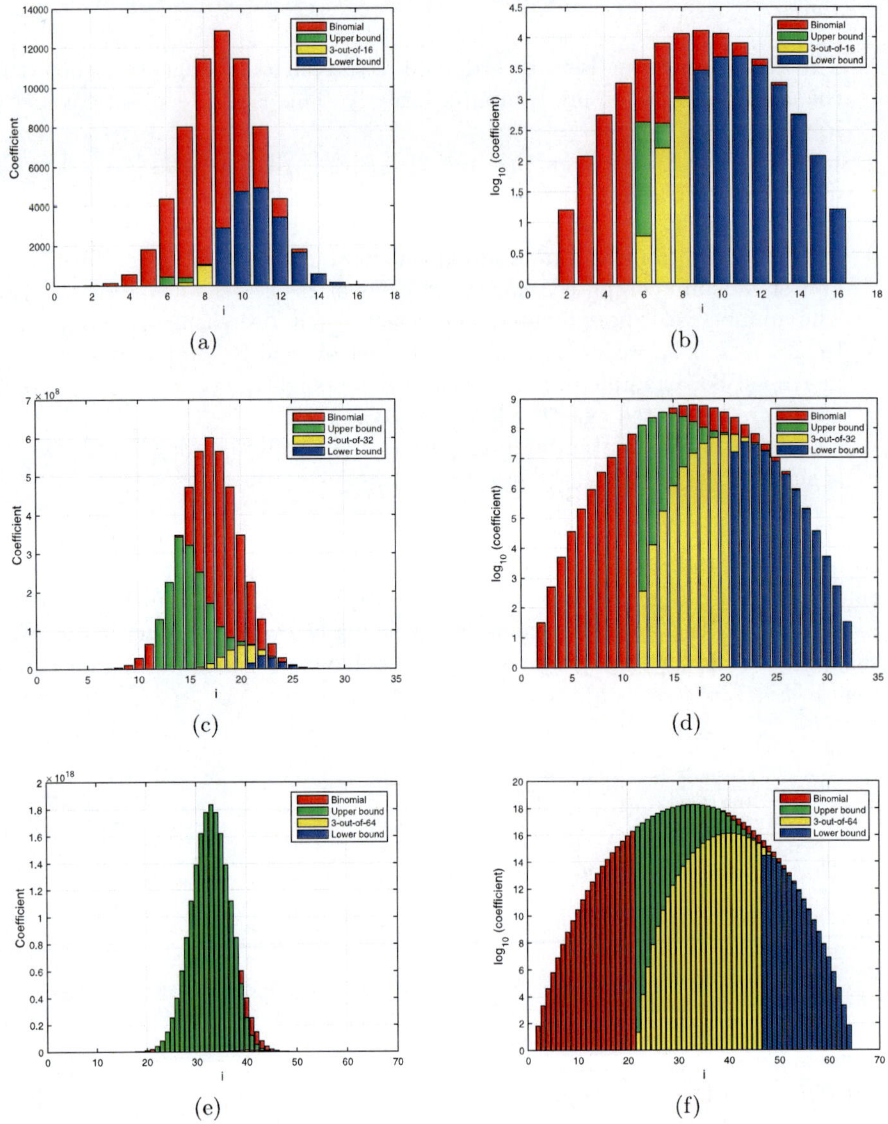

Fig. 3. $N_{n,k,i}, L_{n,k,i}, U_{n,k,i}$, and $\binom{n}{i}$ in linear scale (left) and logarithmic scale (right) for: (a)–(b) $n = 16$, (c)–(d) $n = 32$, and (e)–(f) $n = 64$.

4 Conclusions

In this paper, we have determined closed formulae for the reliability of a consecutive k-out-of-n:F system expressed in the Bernstein basis. Based on the properties of the coefficients, we have proposed simple and easy to compute formulae for all $k \geq \lfloor n/3 \rfloor$. For the remaining range of values, namely for $3 \leq k < \lfloor n/3 \rfloor$, we have proposed lower and upper bounds on the coefficients, and thus bounds on reliability. These bounds have several interesting properties, becoming sharper and sharper as n gets larger, while requiring lower and lower computation work factors.

The approach we have presented here opens the road to a new research direction in the area of consecutive systems. To our knowledge this is the first time bounding/approximating techniques have been used selectively only on a few of the coefficients of a consecutive system, rather than bounding the reliability polynomial. Detailed estimates of the trade-offs between computation complexity and accuracy of approximations have to be evaluated against previously published results for a better understanding of the advantages and disadvantages of the proposed approach (not included due to space limitations).

Acknowledgements. This research was supported by the European Union through the European Regional Development Fund under the Competitiveness Operational Program (*BioCell-NanoART = Novel Bio-inspired Cellular Nano-Architectures*, POC-A1.1.4-E-2015 nr. 30/01.09.2016).

References

1. Beichl, I., Cloteaux, B., Sullivan, F.: An approximation algorithm for the coefficients of the reliability polynomial. Congr. Numer. **197**, 143–151 (2010)
2. Camarda, P.: Bounds evaluation of coefficients in the reliability polynomial. Microelectr. Reliab. **30**(6), 1099–1110 (1990)
3. Chang, G.J., Cui, L., Hwang, F.K.: Reliabilities of Consecutive-k Systems. Kluwer, Dordrecht (2000)
4. Chari, M., Colbourn, C.J.: Reliability polynomials: a survey. J. Combin. Inform. System Sci. **22**(3–4), 177–193 (1997)
5. Chiang, D.T., Niu, S.: Reliability of consecutive-k-out-of-n:F system. IEEE Trans. Reliab. **R–30**(1), 87–89 (1981)
6. Cluzeau, T., Keller, J., Schneeweiss, W.: An efficient algorithm for computing the reliability of consecutive-k-out-of-n:F systems. IEEE Trans. Reliab. **57**(1), 84–87 (2008)
7. Colbourn, C.J.: The Combinatorics of Network Reliability. Oxford University Press, New York (1987)
8. Daus, L., Beiu, V.: Lower and upper reliability bounds for consecutive-k-out-of-n:F systems. IEEE Trans. Reliab. **64**(3), 1128–1135 (2015)
9. de Moivre, A.: The Doctrine of Chances. W. Pearson, London (1971). 3rd edn. (1756). https://archive.org/details/doctrineofchance00moiv/
10. Deng, H., Chen, J., Li, Q., Li, R., Gao, Q.: On the construction of most reliable networks. Discr. Appl. Maths. **140**(1–3), 19–33 (2004)
11. Eryilmaz, S.: Review of recent advances in reliability of consecutive-k-out-of-n and related systems. J. Risk Reliab. **224**(3), 225–237 (2010)
12. Gokdere, G., Gurkan, M., Kilic, M.B.: A new method for computing the reliability of consecutive k-out-of-n:F systems. Open Phys. **14**(1), 166–170 (2016)
13. Kontoleon, J.M.: Reliability determination of a r-successive-out-of-n:F system. IEEE Trans. Reliab. **R–29**(5), 437 (1980)

14. Mohammadi, F., Sáenz-de Cabezón, E., Wynn, H.P.: Efficient multicut enumeration of k-out-of-n:F and consecutive k-out-of-n:F systems. Pattern Recognit. Lett. **102**(1), 82–88 (2018)
15. Moore, E.F., Shannon, C.E.: Reliable circuits using less reliable relays - Part I. J. Frankl. Inst. **262**(3), 191–208 (1956)
16. Oxley, J., Welsh, D.: Chromatic, flow and reliability polynomials: the complexity of their coefficients. Comb. Probab. Comput. **11**(4), 403–426 (2002)
17. Pérez-Rosés, H.: Sixty years of network reliability. Maths. Comp. Sci. **12**(3), 275–293 (2018)
18. Provan, J.S.: Bounds on the reliability of networks. IEEE Trans. Reliab. **R–35**(3), 260–268 (1986)
19. Provan, J.S., Ball, M.O.: The complexity of counting cuts and of computing the probability that a graph is connected. SIAM J. Comput. **12**(4), 777–788 (1983)
20. Valiant, L.G.: The complexity of enumeration and reliability problems. SIAM J. Comput. **8**(3), 410–421 (1979)
21. von Neumann, J.: Probabilistic logics and the synthesis of reliable organisms from unreliable components. In: Shannon, C.E., McCarthy, J. (eds.) Automata Studies, pp. 43–98. Princeton University Press, Princeton (1956)

Reliability of Two-Terminal Networks Equivalent to Small Optimal Sorting Nets

Simon R. Cowell, Mariana Nagy, and Valeriu Beiu(✉)

"Aurel Vlaicu" University of Arad, 310330 Arad, Romania
{simon.cowell,mariana.nagy,valeriu.beiu}@uav.ro

Abstract. Sorting networks are a special case of "oblivious" sorting algorithms that can be implemented directly in hardware. Their underlying non-plane connectivity graph representations can be mapped onto a certain class of minimal two-terminal networks, allowing us to associate a two-terminal reliability polynomial to any (optimal) sorting network connectivity graph. This class of networks is interesting in that it intersects the class of "matchstick minimal" two-terminal networks (which includes the planar Moore-Shannon hammocks), yet neither of these two classes contains the other. We compare the two-terminal reliability polynomials associated in this manner to small optimal sorting network connectivity graphs, with the reliability polynomials of Moore-Shannon hammock networks of equivalent dimensions.

Keywords: Sorting networks · Two-terminal networks · Minimal networks · Reliability polynomials

1 Minimal Two-Terminal Networks

A two-terminal network N is essentially a graph, possibly having multiple edges (with their own identity), in which two vertices are nominated as the "terminals," normally denoted by S (the source) and T (the terminus). The "reliability" of such a network is the probability $Rel(N; p)$ that the two terminals (S and T) belong in the same component of a random sub-graph, each of the original n edges being present with probability p, independently of the other edges [1]. This turns out to be a polynomial of degree n in p [2, 3]. Alternately, $Rel(N; p)$ can be written in a unique way as a homogeneous polynomial of degree n in p and q (denoted as $Rel(N; p, q)$), where $q = 1 - p$. The latter form $Rel(N; p, q)$ is particularly convenient, since its coefficients count the numbers of a certain family of subsets of the original edges, and also since this form turns out to be numerically stable when evaluated on a computer (which is due to the fact that $Rel(N; p, q)$ is in Bernstein form).

A "minimal" two-terminal network, of "width" w and "length" l (a term coined by Moore and Shannon [4]), is a two-terminal network $N_{w,l}$ with minimally many edges, with respect to having those given width and length, i.e., $n = w \times l$. The width w and length l (non-negative integers) are parameters of the graph relating to its two-terminal st-connectivity. Moore and Shannon characterized minimal two-terminal networks as

I. Dzitac et al. (Eds.): ICCCC 2020, AISC 1243, pp. 45–56, 2021.
https://doi.org/10.1007/978-3-030-53651-0_4

being precisely those that are isomorphic to any one of the following graphs. Start with w copies of a path graph of l edges, and join w of the degree-1 vertices as a single vertex, and similarly for the remaining w degree-1 vertices, resulting in a "parallel-of-series" (PoS) graph. Nominate the two newly-made vertices as the terminals S and T. Let P be the partition of the remaining vertices according to their path-distance from a fixed one of the two terminals. Arbitrarily partition each part of P itself, uniting vertices in a common part.

In this paper, among the many minimal two-terminal networks we consider a special class, introduced in [4], known as the "hammock" networks. These are planar, having a brick-wall structure as shown in Fig. 1, and have recently been investigated and analyzed in [5–9]. We denote a hammock of width w and length l by $H_{w,l}$, or by $H^{+}_{w,l}$, accordingly as the top "course of bricks" begins with a whole- or a half-brick, respectively. In fact, $H_{w,l}$ and $H^{+}_{w,l}$ are isomorphic, and hence have the same $Rel(H_{w,l}; p, q)$, if, and only if, at least one of w or l is odd.

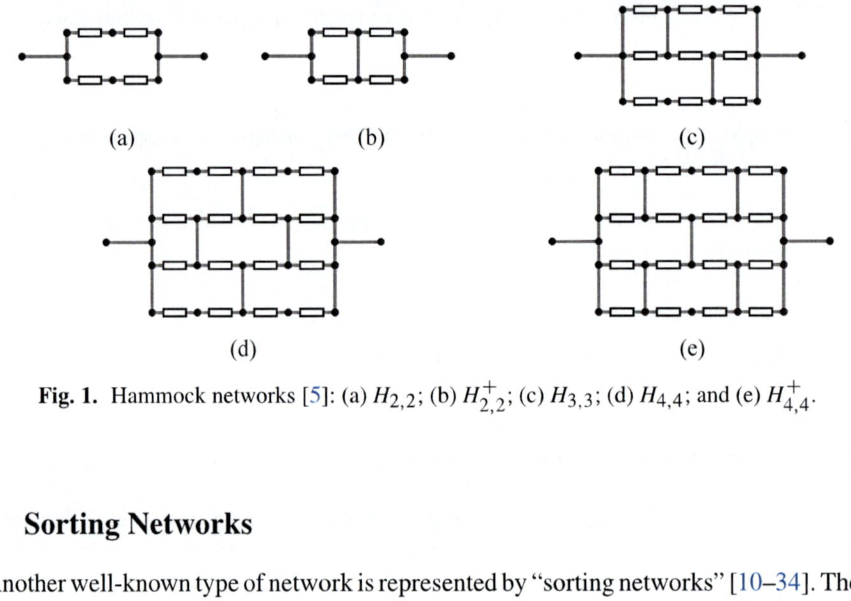

Fig. 1. Hammock networks [5]: (a) $H_{2,2}$; (b) $H^{+}_{2,2}$; (c) $H_{3,3}$; (d) $H_{4,4}$; and (e) $H^{+}_{4,4}$.

2 Sorting Networks

Another well-known type of network is represented by "sorting networks" [10–34]. They are a particular type of oblivious sorting algorithm that can be implemented directly in hardware [16], even exhibiting fault-tolerant properties [35–40]. They have w horizontal rails along which the w inputs start traveling from left to right, until they encounter a two-input comparator (i.e., the simplest two-input sorting network). A two-input comparator connects exactly two of the rails, and either preserves or exchanges the two incoming values, such that the two outgoing values are sorted in a consistent way (e.g., in non-decreasing order from top to bottom). The two-input comparators can be arranged in columns, such that in each column, each rail is connected to at most one two-input comparator. The two-input comparators in any given column operate independently of each other, hence are running in parallel [17]. The minimum number of such columns needed to sort the w inputs is known as the "depth" l' (or delay) of the sorting network (for many more details see [41]).

To deserve the modifier "sorting", such a $Sort_{w,l}$ network must be guaranteed to sort all possible inputs, which happens if, and only if, it correctly sorts all the 2^w possible binary inputs [12, 19]. Two notions of optimality have been considered in connection to $Sort_{w,l}$: (*i*) minimality with respect to number of comparators (smallest network) [42]; and (*ii*) minimality with respect to depth/delay (fastest network) [43].

3 Representations of Networks

Sorting networks $Sort_{w,l}$ are often represented either by a diagram of the structure of the network, or by a list of lists of pairs of integers, the j-th list of pairs corresponding to the j-th column, each pair in that list representing a two-input comparator, and the integers in that pair corresponding to the indices (counting w from top to bottom) of the horizontal rails connected by that two-input comparator [44]. Here we have chosen a few small sorting networks that are optimal in both senses simultaneously. We denote each one by $Sort_{w,l}$, where w is the number of inputs and $l = l' + 1$:

It is worth mentioning that many networks might be non-plane (of drawings of graphs), meaning being drawn in a way that is not a planar embedding, i.e., in a way that has edge crossings (see https://en.wikipedia.org/wiki/Planar_graph). Minimal networks that are not matchstick minimal (e.g., most of the sorting networks) may or may not be planar. The usual drawings of them are non-plane, but this does not guarantee that they have no planar embedding. Additionally, identifying optimal sorting networks for large w is far from trivial, and sustained efforts have been dedicated to this task with unexpectedly slow progress over time [12–34]. It is worth mentioning that when $w = 10$ we have two optimal sorting networks, one which is the smallest, i.e., achieving the minimum number of comparators (29 comparators in $depth = 8$), and another one which is the fastest, i.e., achieving the smallest delay ($depth = 7$ with 31 comparators), see [12, 18, 19, 26, 29, 33, 34].

A common way to represent a minimal two-terminal network (Fig. 1), is by a diagram very much like those used to represent sorting networks (Fig. 2). The horizontal rails correspond to the original w path graphs of l edges, and the comparators correspond to 2-element parts of partitions of the set of vertices at a common distance from a given terminal. Thus, any sorting network of depth l' on w inputs is graph-equivalent to a certain minimal two-terminal network of width w and length $l = l' + 1$. Note that the vertex partitions of that minimal two-terminal network necessarily have only parts of size 1 or 2, and conversely, any minimal two-terminal network of that kind is graph-equivalent to a *potential* sorting network of depth l' on w inputs.

Let us consider a network sorting w inputs. For each $i \in \{1, \ldots, w\}$, let k_i be the number of distinct paths of length l from S to T that start on the i-th input. Considering that a sorting network must correctly sort any of the $w!$ permutations of the input vector $(1, 2, \ldots, w)$, the number of distinct sequences of paths $(p_1, \ldots, p_w)$, where p_i is a path of length l from S to T starting at the i-th input, must be at least $w!$, therefore

$$w! \leq \prod_{i=1}^{w} k_i. \tag{1}$$

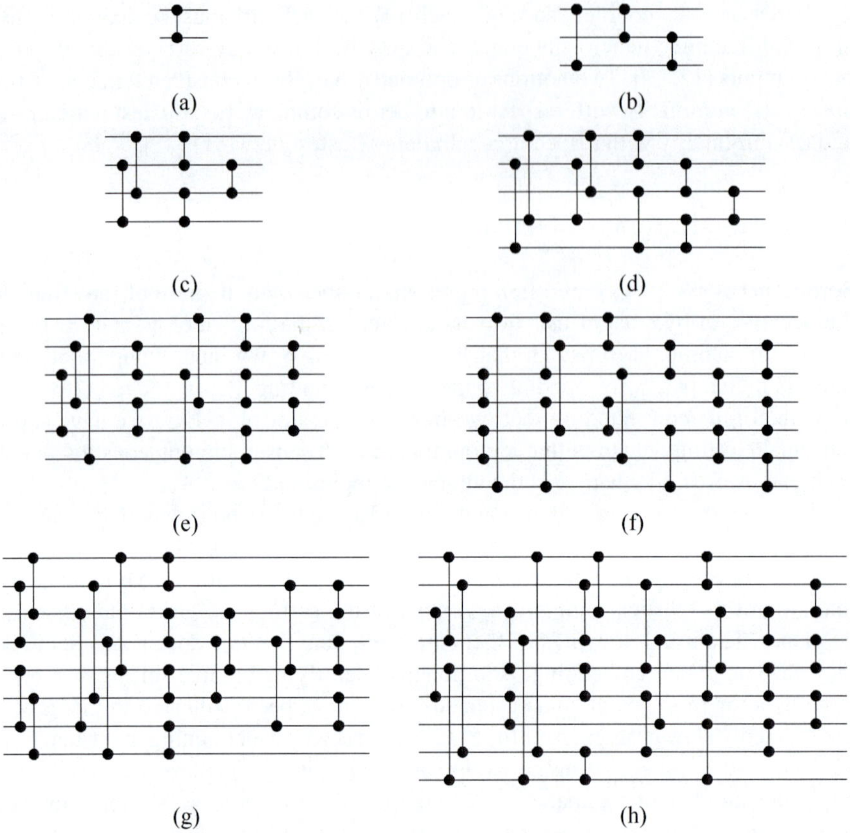

Fig. 2. Optimal sorting networks [44] for $w = 2, 3, \ldots, 9$: (a) $Sort_{2,2}$; (b) $Sort_{3,4}$; (c) $Sort_{4,4}$; (d) $Sort_{5,6}$; (e) $Sort_{6,6}$; (f) $Sort_{7,7}$; (g) $Sort_{8,7}$; and (h) $Sort_{9,8}$.

However, since the i-th input must be sorted into any of the w outputs (as necessary), it follows that

$$k_i \geq w \tag{2}$$

and

$$\prod_{i=1}^{w} k_i \geq w^w. \tag{3}$$

Since $w^w \geq w!$ (the inequality being strict precisely when $w > 1$), Eq. (3) gives a better bound on $\prod_{i=1}^{w} k_i$ than does Eq. (1).

From the reliability point of view, the first coefficient N_l of a reliability polynomial (in Bernstein form) counts the number of distinct length-l paths from S to T, and is therefore equal to $\sum_{i=1}^{w} k_i$. Therefore, from Eq. (2) we have

$$N_l \geq w^2. \tag{4}$$

Unfortunately, by applying the inequalities of the arithmetic and geometric means we do not improve on this, as simply recovering Eq. (4) as follows

$$N_l = \sum\nolimits_{i=1}^{w} k_i \geq w\left(\sqrt[w]{\prod\nolimits_{i=1}^{w} k_i}\right) \geq w\left(\sqrt[w]{w^w}\right) = w^2. \quad (5)$$

The main reason for the next steps of this investigation is to draw attention to the unexpected eminence optimal sorting nets seem to play for the wholly different field of highly reliable minimal two-terminal networks envisioned for binary computations. That is why in the following we will study the reliability of the four smallest optimal sorting networks for $w = 2, 3, 4, 5$, i.e., $Sort_{2,2}$, $Sort_{3,4}$, $Sort_{4,4}$, and $Sort_{5,6}$. These are all optimal with respect to both size (number of comparators) and depth (number of layers). We will compare them to Moore-Shannon hammocks $H_{w,l}$ (or $H^+_{w,l}$) of corresponding w and l. We will also compare their exact reliability polynomials with two other polynomials that approximate $Rel(H_{w,l}; p, q)$, namely: (*i*) Hermite polynomials introduced in [7]; and (*ii*) regularized incomplete beta functions studied in [9].

4 Simulations and Comparisons

In this section we present detailed simulations for $Sort_{2,2}$, $Sort_{3,4}$, $Sort_{4,4}$, $Sort_{5,6}$, and the equivalent $H^+_{2,2}$, $H_{3,4}$, $H^+_{4,4}$, $H_{5,6}$ (exact reliability polynomials are determined in [5]). The polynomials we have determined are as follows:

$$Rel(Sort_{2,2}; p, q) = 4p^2q^2 + 4p^3q + p^4 \quad (6)$$

$$Rel(H^+_{2,2}; p, q) = 4p^2q^2 + 4p^3q + p^4 \quad (7)$$

$$Hermite_{2,2}(p) = 3p^2 - 4p^3 \quad (8)$$

$$Rel(Sort_{3,4}; p, q) = 13p^4q^8 + 94p^5q^7 + 284p^6q^6 + 450p^7q^5 + 397p^8q^4 + 208p^9q^3 + 66p^{10}q^2 + 12p^{11}q + p^{12} \quad (9)$$

$$Rel(H_{3,4}; p, q) = 13p^4q^8 + 94p^5q^7 + 284p^6q^6 + 450p^7q^5 + 397p^8q^4 + 208p^9q^3 + 66p^{10}q^2 + 12p^{11}q + p^{12} \quad (10)$$

$$Hermite_{3,4}(p) = 15p^4 - 24p^5 + 10p^6 \quad (11)$$

$$Rel(Sort_{4,4}; p, q) = 24p^4q^{12} + 268p^5q^{11} + 1352p^6q^{10} + 3944p^7q^9 + 7182p^8q^8 + 8566p^9q^7 + 7068p^{10}q^6 + 4186p^{11}q^5 + 1804p^{12}q^4 + 560p^{13}q^3 + 120p^{14}q^2 + 16p^{15}q + p^{16}. \quad (12)$$

$$Rel(H^+_{4,4}; p, q) = 24p^4q^{12} + 264p^5q^{11} + 1302p^6q^{10} + 3740p^7q^9 + 6848p^8q^8 + 8312p^9q^7 + 6966p^{10}q^6 + 4164p^{11}q^5 + 1802p^{12}q^4 + 560p^{13}q^3 + 120p^{14}q^2 + 16p^{15}q + p^{16}. \quad (13)$$

$$Hermite_{4,4}(p) = 35p^4 - 84p^5 + 70p^6 - 20p^7 \quad (14)$$

$$Rel(Sort_{5,6}; p, q) = 70p^6q^{24} + 1629p^7q^{23} + 18178p^8q^{22} + 129110p^9q^{21} + 653213p^{10}q^{20} + 2495886p^{11}q^{19} + 7449982p^{12}q^{18} + 17728470p^{13}q^{17} + 34048682p^{14}q^{16} + 53174882p^{15}q^{15} + 67879409p^{16}q^{14} + 71161389p^{17}q^{13} + 61600916p^{18}q^{12} + 44303696p^{19}q^{11} + 26624123p^{20}q^{10} + 13420296p^{21}q^9 + 5679138p^{22}q^8 + 2011533p^{23}q^7 + 591618p^{24}q^6 + 142414p^{25}q^5 + 27405p^{26}q^4 + 4060p^{27}q^3 + 435p^{28}q^2 + 30p^{29}q + p^{30}. \tag{15}$$

$$Rel(H_{5,6}; p, q) = 94p^6q^{24} + 2172p^7q^{23} + 23988p^8q^{22} + 168010p^9q^{21} + 835135p^{10}q^{20} + 3124592p^{11}q^{19} + 9107514p^{12}q^{18} + 21122378p^{13}q^{17} + 39492093p^{14}q^{16} + 60026678p^{15}q^{15} + 74637417p^{16}q^{14} + 76380752p^{17}q^{13} + 64766145p^{18}q^{12} + 45813826p^{19}q^{11} + 27188520p^{20}q^{10} + 13583104p^{21}q^9 + 5714293p^{22}q^8 + 2016902p^{23}q^7 + 592137p^{24}q^6 + 142438p^{25}q^5 + 27405p^{26}q^4 + 4060p^{27}q^3 + 435p^{28}q^2 + 30p^{29}q + p^{30}. \tag{16}$$

$$Hermite_{5,6}(p) = 210p^6 - 720p^7 + 945p^8 - 560p^9 + 126p^{10} \tag{17}$$

For each of the four optimal sorting networks considered ($Sort_{2,2}$, $Sort_{3,4}$, $Sort_{4,4}$, $Sort_{5,6}$), we plot in Fig. 3 the reliability of a minimal two-terminal network having the same topology, together with the reliability of a Moore-Shannon hammock of equivalent dimensions, as well as the approximations to the hammock reliability polynomial by a Hermite polynomial [8] and by a regularized incomplete beta function (with positive integer parameters) [9]. In Fig. 4 we show the absolute and relative errors between each of these polynomials and the reliability polynomials of the equivalent hammocks.

As can be seen, the reliability polynomials corresponding to $Sort_{2,2}$ and $Sort_{3,4}$ coincide with those of $H^+_{2,2}$ and $H_{3,4}$ (the underlying graphs being isomorphic). For $Sort_{4,4}$ and $Sort_{5,6}$ we see that their corresponding reliability polynomials closely follow $H^+_{4,4}$ and $H_{5,6}$, even more closely than two approximations to $Rel(H_{w,l}; p, q)$. Unfortunately, because we know of no efficient algorithms for computing the reliability polynomials corresponding to optimal sorting networks this has no immediate practical application to approximating $Rel(H_{w,l}; p, q)$. On the other hand, it is of theoretical interest, since it *suggests that certain underlying graph qualities that make for an optimal sorting network also make for a highly reliable two-terminal network.*

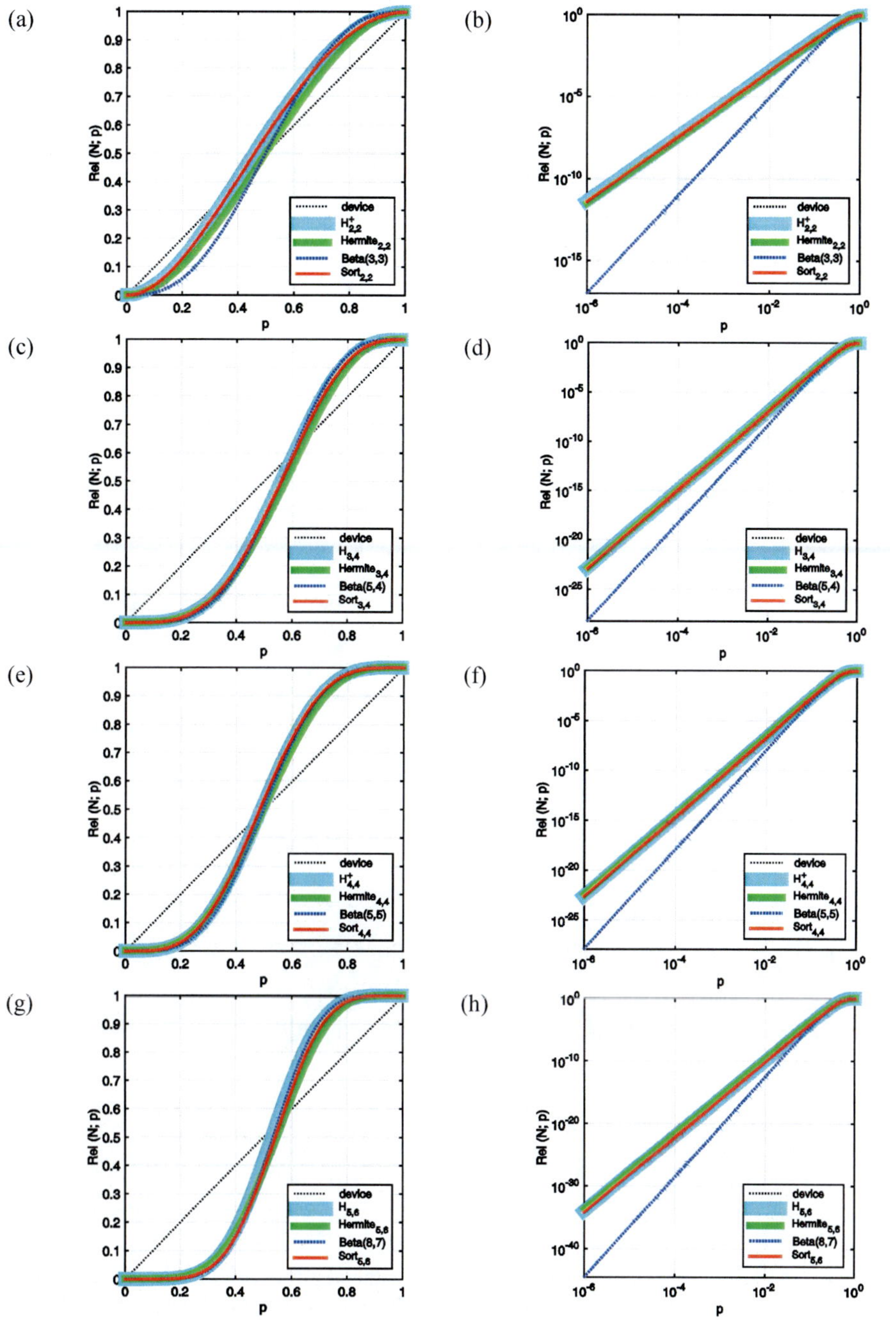

Fig. 3. Reliability of graph-equivalent optimal sorting networks ($Sort_{2,2}$, $Sort_{3,4}$, $Sort_{4,4}$, $Sort_{5,6}$), hammocks ($H^+_{2,2}$, $H_{3,4}$, $H^+_{4,4}$, $H_{5,6}$) as well as two approximations (Hermite and beta distribution), versus p in both linear (left column) and logarithmic (right column) scales.

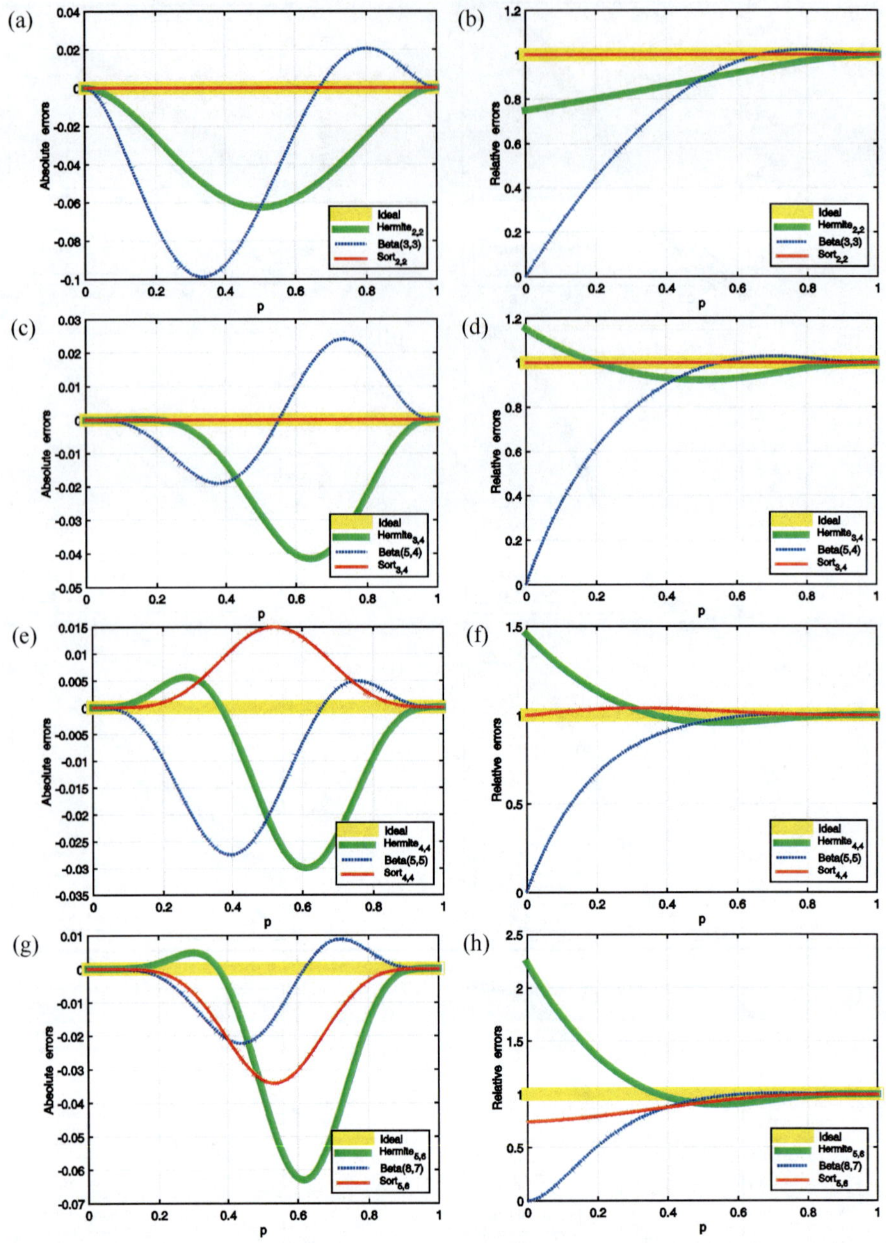

Fig. 4. Absolute (left column) and relative (right column) errors between the reliability of hammocks ($H^{+}_{2,2}$, $H_{3,4}$, $H^{+}_{4,4}$, $H_{5,6}$) and the reliability of graph-equivalent optimal sorting networks ($Sort_{2,2}$, $Sort_{3,4}$, $Sort_{4,4}$, $Sort_{5,6}$), as well as of two approximations (Hermite and beta distribution), versus p.

5 Conclusions

We observe that Moore's and Shannon's motivation for designing hammock networks [4] was to obtain a network that is very likely to admit a connection between its terminals S and T, provided that the edges are reasonably likely to be present, and at the same time very likely not to admit a connection between its terminals, provided that the edges are reasonably likely to be absent. All our studies to date [5–7] have shown that hammocks are optimal in this regard, although we (still) have no theoretical explanation for this phenomenon, as yet.

The current investigation was started from one particular sorting network, namely the one based on odd-even transposition sort [19]. This type of sorting network leads to a brick-wall graph identical to hammocks. We were curious to see if (some) optimal sorting networks might do better, i.e., be more reliable than hammocks—as odd-even transposition sort is not an optimal sorting network. Obviously, this investigation is in its early stages, arguing through simulations in support of such empirical observations and conjectures, while theoretical proofs are still pending. The aim is to draw the attention of more researchers to look into the unexplored issues presented here, as the associated problems are known to be far from trivial (computing exactly the reliability polynomials for $n = wl > 64$ is difficult, becoming daunting for $n > 128$, and basically impossible for $n > 256$, as the problems are #P-complete). Still, the preliminary simulation results reported here show that *the smallest optimal sorting networks do not surpass the hammocks with respect to reliability*, hence reinforcing our previous conjecture that hammocks might be the best.

Acknowledgements. This research was supported by the EU through the European Regional Development Fund under the Competitiveness Operational Program (*BioCell-NanoART = Novel Bio-inspired Cellular Nano-Architectures*, POC-A1.1.4-E-2015 nr. 30/01.09.2016).

References

1. Colbourn, C.J.: The Combinatorics of Network Reliability. Oxford University Press, New York & Oxford (1987)
2. Chari, M., Colbourn, C.J.: Reliability polynomials: a survey. J. Comb. Info. Syst. Sci. **22**(3–4), 177–193 (1997)
3. Pérez-Rosés, H.: Sixty years of network reliability. Maths. Comput. Sci. **12**(3), 275–293 (2018). https://doi.org/10.1007/s11786-018-0345-5
4. Moore, E.F., Shannon, C.E.: Reliable circuits using less reliable relays. Part I. J. Frankl. Inst. **262**(3), 191–208 (1956). https://doi.org/10.1016/0016-0032(56)90559-2. Preliminary version as: Reliable circuits using crummy relays. Tech. Rep. Memo. 54-114-42, Bell Labs (Nov. 1954)
5. Cowell, S.R., Beiu, V., Dăuş, L., Poulin, P.: On the exact reliability enhancements of small hammock networks. IEEE Access **6**, 25411–25426 (2018). https://doi.org/10.1109/ACCESS.2018.2828036
6. Dăuş, L., Beiu, V., Cowell, S.R., Poulin, P.: Brick-wall lattice paths and applications. Technical report. arXiv:1804.05277v1 [math.CO], 14 April 2018. https://arxiv.org/abs/1804.05277
7. Dăuş, L., Jianu, M.: The shape of the reliability polynomial of a hammock network. Technical report. arXiv:1901.04036v1 [math.CO], 13 January 2019. https://arxiv.org/abs/1901.04036

8. Dăuş, L., Jianu, M.: Full Hermite interpolation of the reliability of hammock network. Appl. Anal. Discrete Math. **14**(1), 198–220 (2020). https://doi.org/10.2298/AADM190805017D
9. Cowell, S.R., Hoară, S., Beiu, V.: Experimenting with beta distributions for approximating hammocks' reliability (Best Paper Award). In: Proceedings of International Conference on Computers Communications and Control (ICCCC 2020), Oradea, Romania, 11–15 May 2020, (in press)
10. O'Connor, D.G., Nelson, R.J.: Sorting system with n-line sorting switch. US Patent 3,029,413, 10 April 1962. https://patents.google.com/patent/US3029413A/en
11. Batcher, K.E.: Sorting networks and their applications. In: Proceedings of the Spring Joint Computer Conference (AFIPS 1968), Atlantic City, NJ, USA, pp. 307–314, 30 April–2 May 1968. https://doi.org/10.1145/1468075.1468121
12. Floyd, R.W., Knuth, D.E.: The Bose-Nelson sorting problem. Technical report. STAN-CS-70-177, CS Department, Stanford University, Stanford, CA, USA, November 1970. https://searchworks.stanford.edu/view/4591549. Also as Chap. 15 in: Srivastava, J.N. (ed.), A Survey of Combinatorial Theory, pp. 163-172 (1973). https://doi.org/10.1016/b978-0-7204-2262-7.50020-x
13. van Voorhis, D.C.: A generalization of the divide-sort-merge strategy for sorting networks. Technical report. STAN-CS-71-237, CS Department, Stanford University, Stanford, CA, USA, August 1971. http://i.stanford.edu/pub/cstr/reports/cs/tr/71/237/CS-TR-71-237.pdf
14. Green, M.W.: Letter to N. J. A. Sloane. Stanford Research Institute, Stanford, CA, USA, 31 May 1973. https://oeis.org/A003075/a003075.pdf
15. Dowd, M., Perl, Y., Saks, M., Rudolph, L.: The balanced sorting network. In: Proceedings of the Thirteenth Annual ACM Symposium on Principles of Distributed Computing (PODC 1983), Montreal, Quebec, Canada, pp. 161–172, 17–19 August 1983. https://doi.org/10.1145/800221.806719
16. Thompson, C.D.: The VLSI complexity of sorting. IEEE Trans. Comput. **C-32**(12), 1171–1184 (1983). https://doi.org/10.1109/TC.1983.1676178
17. Parberry, I.: Parallel Complexity Theory. Pitman, London (1987)
18. Parberry, I.: A computer assisted optimal depth lower bound for nine-input sorting networks. Math. Syst. Theor. **24**(1), 101–116 (1991). https://doi.org/10.1007/BF02090393
19. Knuth, D.E.: The Art of Computer Programming - Vol. 3: Sorting and Searching. 2nd edn. Addison-Wesley, Reading (1998)
20. Knuth, D.E.: Robert W. Floyd, in memoriam. ACM SIGACT News **34**(4), 3–13 (2003). https://doi.org/10.1145/954092.954488
21. Baddar, S.W.A.-H.: Finding better sorting networks. Ph.D. dissertation, Kent State University, May 2009. http://rave.ohiolink.edu/etdc/view?acc_num=kent1239814529
22. Valsalam, V.K.: Utilizing symmetry in evolutionary design. Ph.D. dissertation, University of Texas at Austin, August 2010. http://hdl.handle.net/2152/ETD-UT-2010-08-2021
23. Valsalam, V.K., Miikkulainen, R.: Evolving minimal-size sorting networks (Annual "Humies" Awards). In: Genetic & Evolutionary Computation Conference (GECCO 2011), Dublin, Ireland, 12–16 July 2011. http://www.human-competitive.org/sites/default/files/valsalam-slides.pdf
24. Valsalam, V.K., Miikkulainen, R.: Utilizing symmetry and evolutionary search to minimize sorting networks. J. Mach. Learn. Res. **14**(1), 303–331 (2013). http://www.cs.utexas.edu/users/ai-lab/pubs/valsalam.utcstr11.pdf
25. Bundala, D., Závodný, J.: Optimal sorting networks. In: Dediu, A.-H., Martín-Vide, C., Sierra-Rodríguez, J.-L., Truthe, B. (eds.) LATA 2014. LNCS, vol. 8370, pp. 236–247. Springer, Cham (2014). https://doi.org/10.1007/978-3-319-04921-2_19. Also as arXiv:1310.6271v2 [cs.DM] (22 Dec. 2013). https://arxiv.org/abs/1310.6271

26. Codish, M., Cruz-Filipe, L., Frank, M., Schneider-Kamp, P.: Twenty-five comparators is optimal when sorting nine inputs (and twenty-nine for ten). In: Proceedings of the IEEE International Conference on Tools with Artificial Intelligence (ICTAI 2014), Limassol, Cyprus, pp. 187–193, 10–12 November 2014. https://doi.org/10.1109/ictai.2014.36. Also as arXiv: 1405.5754v3 [cs.DM] (24 Jun. 2014). https://arxiv.org/abs/1405.5754
27. Codish, M., Cruz-Filipe, L., Schneider-Kamp, P.: The quest for optimal sorting networks: efficient generation of two-layer prefixes. In: Proceedings of the IEEE International Symposium on Symbolic & Numeric Algorithms Scientific Computing (SYNASC 2014), Timisoara, Romania, pp. 359–366, 22–25 September 2014. https://doi.org/10.1109/synasc.2014.55
28. Ehlers, T., Müller, M.: Faster sorting networks for 17, 19 and 20 inputs. Technical report. arXiv:1410.2736v1 [cs.DS], 10 October 2014. https://arxiv.org/abs/1410.2736
29. Codish, M., Cruz-Filipe, L., Frank, M., Schneider-Kamp, P.: Sorting nine inputs requires twenty-five comparisons. J. Comput. Syst. Sci. **82**(3), 551–563 (2016). https://doi.org/10.1016/j.jcss.2015.11.014
30. Ehlers, T.: Merging almost sorted sequences yields a 24-sorter. Info. Proc. Lett. **118**(C), 17–20 (2017). https://doi.org/10.1016/j.ipl.2016.08.005
31. Bundala, D., Codish, M., Cruz-Filipe, L., Schneider-Kamp, P., Závodný, J.: Optimal-depth sorting networks. J. Comp. Syst. Sci. **84**, 185–204 (2017). https://doi.org/10.1016/j.jcss.2016.09.004. Also as arXiv:1412.5302v1 [cs.DS] (17 Dec. 2014). https://arxiv.org/abs/1412.5302
32. López-Ramírez, B.C., Cruz-Cortés, N.: Designing minimal sorting networks using a bio-inspired technique. Computación y Sistemas **18**(4), 731–739 (2014). https://doi.org/10.13053/CyS-18-4-1959
33. Fonollosa, J.A.R.: Joint size and depth optimization of sorting networks. Technical report. arXiv:1806.00305v1 [cs.DS], 1 June 2018. https://arxiv.org/abs/1806.00305
34. Codish, M., Cruz-Filipe, L., Ehlers, T., Müller, M., Schneider-Kamp, P.: Sorting networks: to the end and back again. J. Comput. Syst. Sci. **104**, 184–201 (2019). https://doi.org/10.1016/j.jcss.2016.04.004
35. Yao, A.C., Yao, F.F.: On fault-tolerant networks for sorting. SIAM J. Comput. **14**(1), 120–128 (1985). https://doi.org/10.1137/0214009. Preliminary version as Tech. Rep. STAN-CS-79-721, CS Dept., Stanford Univ., Stanford, CA, USA, pp. 1-20 (Feb. 1979). http://i.stanford.edu/pub/cstr/reports/cs/tr/79/721/CS-TR-79-721.pdf
36. Sun, J., Cerny, E., Gecsei, J.: Fault tolerance in a class of sorting netwoks. IEEE Trans. Comput. **43**(7), 827–837 (1994). https://doi.org/10.1109/12.293261
37. Sharma, N.K., Tagle, P.U.: Reliability analysis of a fault-tolerant sorting network. In: Proceedings of the IEEE Global Telecommunication Conference (GLOBECOM 1997), Phoenix, AZ, USA, pp. 1403–1407, 3–8 November 1997. https://doi.org/10.1109/glocom.1997.644366
38. Salloum, S.N., Perrie, A.L.: Fault tolerance analysis of odd-even transportation sorting netwoks. In: Proceedings of the IEEE Pacific Rim Conference on Communications, Computer & Signal Processings (PACRIM 1999), Victoria, BC, Canada, pp. 155–157, 22–24 August 1999. https://doi.org/10.1109/pacrim.1999.799500
39. Salloum, S.N., Wang, D.-H.: Fault tolerance analysis of odd-even transposition sorting networks with single pass and multiple passes. In: Proceedings of the IEEE Pacific Rim Conference on Communications, Computer & Signal Processings (PACRIM 2003), Victoria, BC, Canada, pp. 193–196, 28–30 August 2003. https://doi.org/10.1109/pacrim.2003.1235750
40. Kung, T.-L., Hung, C.-N.: Estimating the subsystem reliability of bubblesort networks. Theor. Comput. Sci. **670**, 45–55 (2017). https://doi.org/10.1016/j.tcs.2017.01.021
41. Rivest, R.L., Knuth, D.E.: Bibliography 26. Computer sorting. Comput. Rev. **13**(6), 283–289 (1972). https://people.csail.mit.edu/rivest/pubs/RK72.pdf
42. The On-Line Encyclopedia of Integer Sequences®: Minimal number of comparisons needed for n-element sorting network. https://oeis.org/A003075

43. The On-Line Encyclopedia of Integer Sequences®: Minimal delay time for an n-element sorting network. https://oeis.org/A067782
44. Dobbelaere, B.O.: Smallest and fastest sorting networks for a given number of inputs. http://users.telenet.be/bertdobbelaere/SorterHunter/sorting_networks.html

Investigating Hammock Networks on IBM Q

Sorin Hoară, Roxana-Mariana Beiu, and Valeriu Beiu(✉)

"Aurel Vlaicu" University of Arad, 310330 Arad, Romania
{sorin.hoara,roxana.beiu,valeriu.beiu}@uav.ro

Abstract. IBM Q (https://www.ibm.com/quantum-computing/) represents a great opportunity offered by IBM to the quantum research community allowing running experiments, through a web interface, on several of their quantum systems on the cloud. One of the great technical challenges to making viable quantum computers is their qubit fidelity (quality/reliability) together with a plethora of error correction techniques—which, obviously, link to reliability theory. Hammock networks (a device-level alternative to gate-level reliability schemes) have shown outstanding reliability enhancements in the classical digital domain (e.g., about two-orders of magnitude better than gate-level von Neumann multiplexing). In spite of such performances, device-level reliability schemes in general, and hammock networks in particular, have never been assessed for quantum computations. A likely explanation is that device-level reliability seems much more akin to topological quantum computing concepts. That is why we have decided to test if and how much hammock networks might help in the quantum realm. Instead of theoretical analyses we have decided to perform simulations on IBM Q (unfortunately still gate-level constrained), and we report our preliminary findings in this paper.

Keywords: Reliability · Hammock networks · Quantum computing · IBM Q

1 Introduction

Today, quantum computing is certainly an extremely hot topic, which had reasonably modest beginnings [1–4], of interest only to a handful of physicists. Nowadays, it is taught [5], along many early possible implementations, e.g., [6, 7]. Since the very beginning, a major foreseeable obstacle was represented by errors [4, 8], an aspect which is still haunting us even now. In spite of such concerns, clever algorithms have been invented [9, 10], which were able to show advantages making top-level research worth pursuing and financing. The field is now experiencing a Cambrian explosion, and the relatively long list of references [11–86] (which was kept chronological) is in fact just a small selective sample of what it is being published lately.

Because IBM Q has made it possible to perform simulations on quantum circuits we have decided to investigate how a particular type of network known as hammock (see [87–90]) could help, and started this preliminary investigation.

I. Dzitac et al. (Eds.): ICCCC 2020, AISC 1243, pp. 57–69, 2021.
https://doi.org/10.1007/978-3-030-53651-0_5

2 Quantum Simulations on IBM Q

We have started by simulating very simple circuits with one or two gates, which were placed in series or/and in parallel. To these we have added either a swap (SW) or a CNOT gate. Obviously the SW gate got transpiled in 3 CNOT gates, which has to be properly accounted for (Fig. 1, Fig. 2, Fig. 3, Fig. 4 and Fig. 5).

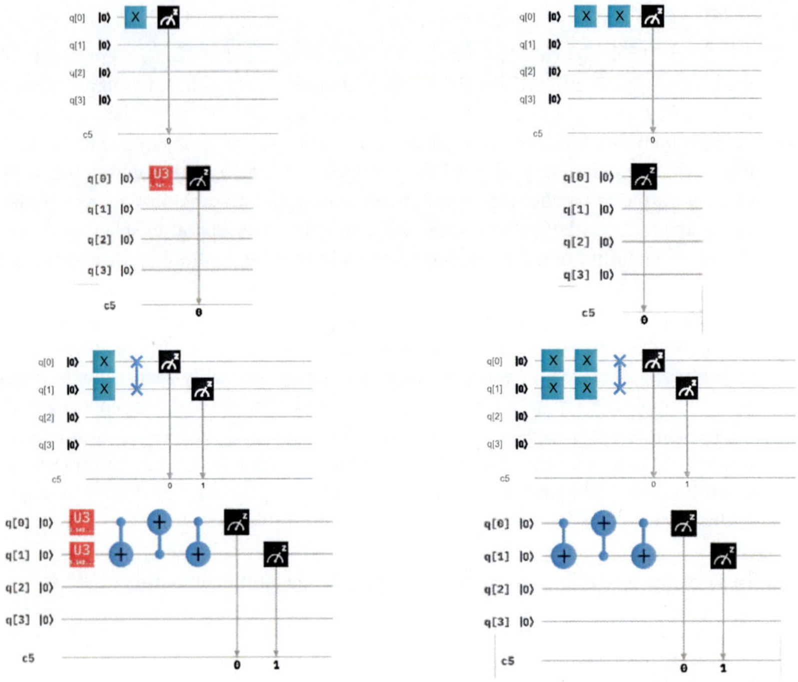

Fig. 1. Circuits 1X, 2X, X-SW, XX-SW as well as transpiled X-SW and XX-SW.

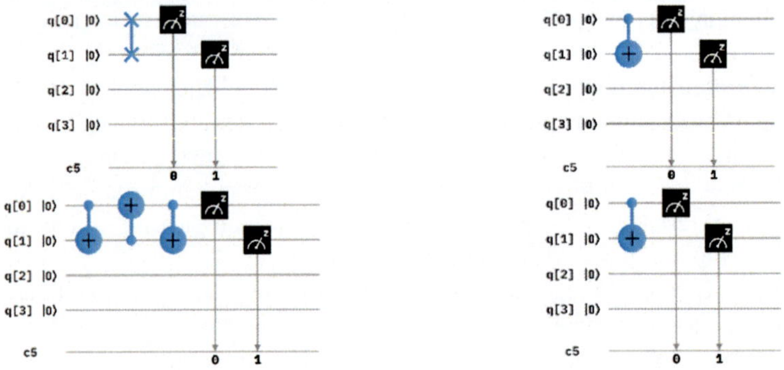

Fig. 2. Circuits for SW and CNOT as well as transpiled SW and CNOT.

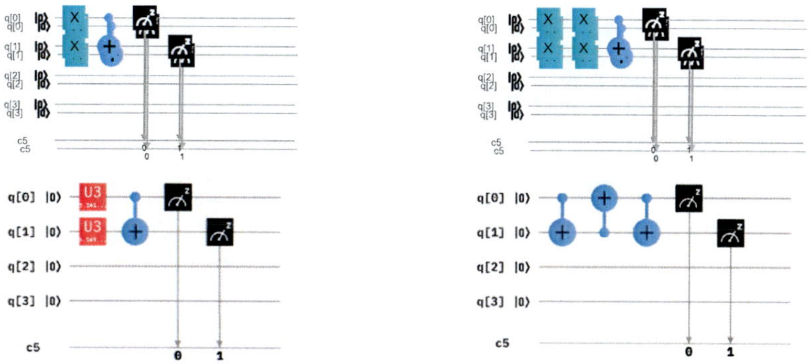

Fig. 3. Circuits for X-CNOT and XX-CNOT as well as their transpiled versions.

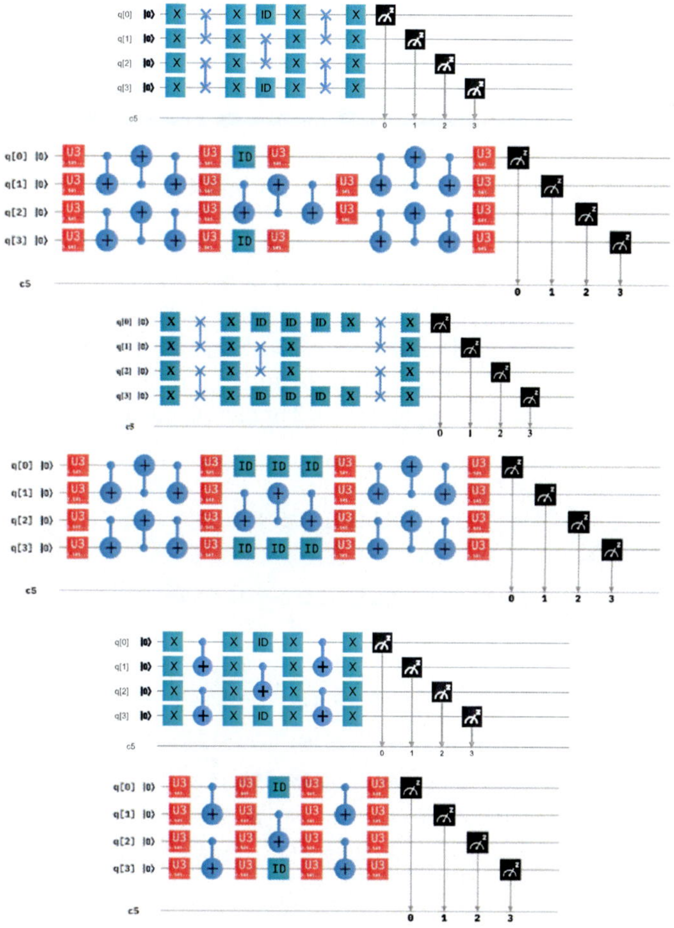

Fig. 4. Repetitive circuits: 4x4 X-SW (two versions) and 4x4 X-CNOT, together with their transpiled versions.

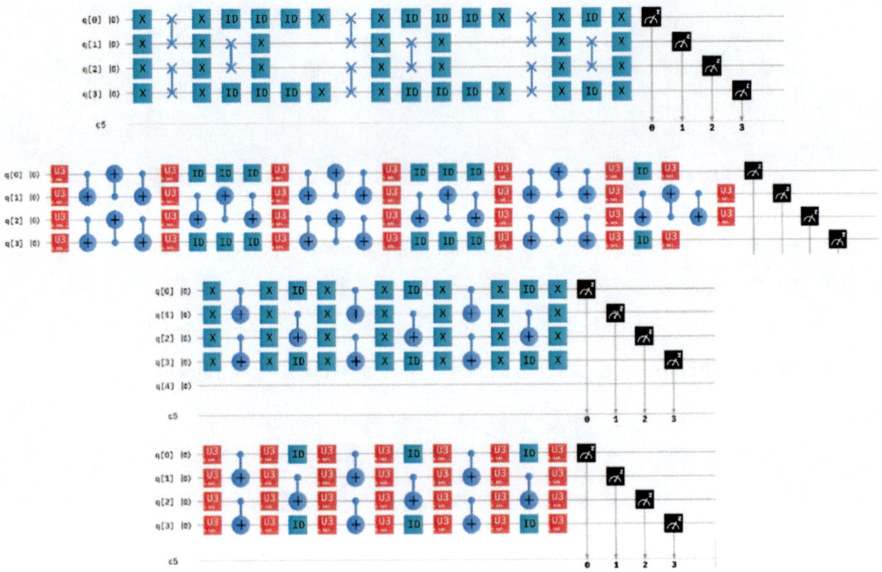

Fig. 5. Slightly larger repetitive circuits: 4x8 X-SW and 4x8 X-CNOT, together with their transpiled versions.

3 Results

All the (test) circuits presented above were run on the same quantum circuit (Melbourne) and each run was in fact 8192 Monte Carlo runs. In the following we present only a sample of the results we have gathered, as more than 2 million simulations have been performed (Tables 1, 2 and 3).

Table 1. X & XX.

		X	XX
0	#1	4.150	99.548
	#2	5.884	99.121
	#3	5.676	99.646
	#4	6.360	99.512
	Avg.	**5.518**	**99.457**
1	#1	95.850	0.452
	#2	94.116	0.879
	#3	94.324	0.354
	#4	93.640	0.488
	Avg.	**94.483**	**0.543**

Table 2. SW, CNOT, X-SW-X, X-CNOT-X, XX-SW-XX, and XX-CNOT-XX.

		SW	CNOT	X-SW-X	X-CNOT-X	XX-SW-XX	XX-CNOT-XX
00	#1	92.383	96.960	1.843	6.934	94.702	97.180
	#2	92.603	97.107	2.136	7.117	92.273	96.960
	#3	92.371	97.839	1.953	3.809	93.835	97.253
	#4	92.480	97.375	1.514	3.320	93.762	97.266
Avg. 01		**92.459**	**97.320**	**1.861**	**5.295**	**93.643**	**97.165**
	#1	3.223	0.354	5.640	90.625	2.832	0.256
	#2	2.600	0.366	7.422	89.795	4.259	0.305
	#3	3.088	0.281	7.263	93.640	3.321	0.208
	#4	2.991	0.256	6.75	94.055	3.308	0.171
Avg. 10		**2.976**	**0.314**	**6.768**	**92.029**	**3.497**	**0.235**
	#1	3.320	2.087	8.813	0.879	1.904	2.222
	#2	3.625	1.941	10.425	0.952	2.209	2.087
	#3	3.442	1.367	5.993	0.781	1.733	2.039
	#4	3.308	2.075	6.543	0.781	1.892	2.002
Avg. 11		**3.424**	**1.868**	**7.928**	**0.848**	**1.934**	**2.088**
	#1	1.074	0.598	83.704	1.563	0.562	0.342
	#2	1.172	0.586	80.017	2.136	0.989	0.647
	#3	1.099	0.513	84.851	1.770	1.111	0.500
	#4	1.221	0.293	85.193	1.843	1.038	0.562
Avg.		**1.142**	**0.498**	**83.441**	**1.828**	**0.903**	**0.513**

The following tables present for each circuit 4 successive runs (#1,#2, #3, and #4) of 8192 Monte Carlo each, as well as their average.

Table 3. a) 4x4-X: +SW, +CNOT, +SW+3ID; and b) 4x4-XX: +SW, +CNOT, +SW+3ID.

		SW[a)]	CNOT[a)]	SW+3ID[a)]	SW[b)]	CNOT[b)]	SW+3ID[b)]
0000	#1	66.638	2.319	58.911	0.330	1.536	1.331
	#2	71.484	2.454	56.848	1.038	1.624	1.318
	#3	72.900	2.209	60.437	1.624	1.575	0.940
	#4	69.920	2.319	58.472	1.580	1.868	1.367
Avg. 0001		**70.236**	**2.325**	**58.667**	**1.143**	**1.651**	**1.239**
	#1	5.151	0.610	5.627	0.854	2.759	2.429
	#2	4.773	0.867	5.884	0.330	3.015	2.393
	#3	4.138	0.739	4.919	0.598	2.759	2.173
	#4	4.773	0.806	5.176	0.366	2.454	2.271
Avg. 0010		**4.709**	**0.756**	**5.402**	**0.537**	**2.747**	**2.317**
	#1	7.629	1.746	6.250	0.830	2.588	2.405
	#2	5.713	2.014	6.372	3.357	2.869	2.441
	#3	7.336	2.124	6.360	3.174	2.917	2.380
	#4	5.627	2.002	6.360	3.040	2.734	2.380
Avg. 0011		**7.629**	**1.972**	**6.336**	**2.600**	**2.777**	**2.402**
	#1	1.648	0.037	1.184	2.661	4.419	4.318
	#2	1.160	0.085	1.355	1.001	4.272	3.894
	#3	1.257	0.098	1.697	1.111	4.956	3.748
	#4	1.160	0.061	1.501	0.977	4.284	4.456
Avg. 0100		**1.306**	**0.070**	**1.434**	**1.438**	**4.483**	**4.059**
	#1	4.810	11.646	4.504	0.732	1.782	1.648
	#2	4.834	11.047	5.090	3.564	1.929	1.587
	#3	4.590	10.693	4.150	4.553	1.807	1.624
	#4	6.384	11.426	4.321	3.882	2.063	1.648
Avg. 0101		**5.155**	**11.203**	**4.516**	**3.183**	**1.895**	**1.627**
	#1	2.637	0.049	3.027	2.930	4.541	4.333
	#2	2.490	0.061	3.235	0.562	4.517	4.297
	#3	1.294	0.061	2.893	1.306	4.309	4.626
	#4	1.782	0.061	2.759	0.903	4.504	4.443
Avg. 0110		**2.051**	**0.058**	**2.979**	**1.425**	**4.468**	**4.425**
	#1	0.537	0.574	0.488	2.808	3.257	3.625
	#2	0.574	0.464	0.708	14.709	3.137	2.771

(*continued*)

Table 3. (*continued*)

		SW[a)]	CNOT[a)]	SW+3ID[a)]	SW[b)]	CNOT[b)]	SW+3ID[b)]
	#3	0.623	0.562	0.537	11.462	3.748	3.650
	#4	0.562	0.586	0.464	11.365	3.320	3.589
Avg. 0111		**0.574**	**0.547**	**0.549**	**10.086**	**3.366**	**3.409**
	#1	0.281	0.159	0.415	10.913	9.045	9.436
	#2	0.439	0.110	0.526	1.990	8.728	10.205
	#3	0.208	0.232	0.476	2.734	8.862	9.912
	#4	0.293	0.244	0.330	2.368	9.192	9.143
Avg. 1000		**0.305**	**0.186**	**0.437**	**4.501**	**8.957**	**9.674**
	#1	4.346	5.493	9.363	0.818	2.527	2.222
	#2	3.345	5.627	8.972	2.673	2.747	2.075
	#3	3.821	5.408	8.948	3.894	3.210	2.295
	#4	4.309	5.017	10.571	3.845	2.832	2.307
Avg. 1001		**3.955**	**5.386**	**9.464**	**2.808**	**2.829**	**2.225**
	#1	1.013	0.061	2.002	2.222	4.858	5.066
	#2	0.842	0.037	2.283	0.574	5.176	4.907
	#3	0.769	0.085	2.405	1.196	4.932	2.431
	#4	1.501	0.049	2.356	0.879	5.200	4.675
Avg. 1010		**1.031**	**0.058**	**2.262**	**1.218**	**5.042**	**4.270**
	#1	1.512	0.378	2.063	2.563	5.481	5.042
	#2	0.928	0.354	1.831	7.092	4.968	4.321
	#3	1.208	0.525	1.672	7.886	5.847	4.956
	#4	1.208	0.354	1.978	8.508	5.542	4.382
Avg. 1011		**1.211**	**0.403**	**1.886**	**6.512**	**5.460**	**4.675**
	#1	0.281	0.049	1.392	7.544	12.024	10.278
	#2	0.244	0.098	1.318	2.734	11.475	10.583
	#3	0.195	0.049	1.147	2.979	12.292	10.901
	#4	0.269	0.085	1.331	2.600	12.024	11.316
Avg. 1100		**0.247**	**0.070**	**1.297**	**3.964**	**11.954**	**10.770**
	#1	2.612	72.412	3.503	2.222	3.394	2.893
	#2	2.429	72.278	3.711	6.445	3.540	3.052
	#3	1.282	72.620	2.893	9.229	3.223	3.333
	#4	1.587	72.217	2.905	7.629	3.894	3.162

(*continued*)

Table 3. (*continued*)

		SW[a)]	CNOT[a)]	SW+3ID[a)]	SW[b)]	CNOT[b)]	SW+3ID[b)]
Avg. 1101		**1.978**	**72.382**	**3.253**	**6.381**	**3.513**	**3.110**
	#1	0.427	0.439	0.903	6.433	7.385	7.751
	#2	0.281	0.269	0.977	1.358	7.751	7.678
	#3	0.208	0.256	0.635	3.149	7.190	7.593
	#4	0.256	0.220	0.745	2.161	7.336	7.617
Avg. 1110		**0.293**	**0.296**	**0.815**	**3.320**	**7.416**	**7.660**
	#1	0.378	3.003	0.256	8.020	6.494	5.542
	#2	0.305	3.162	0.562	47.75	6.689	6.421
	#3	0.134	3.247	0.586	39.136	6.982	6.494
	#4	0.208	3.467	0.464	43.689	7.104	6.482
Avg. 1111		**0.256**	**3.220**	**0.467**	**34.638**	**6.817**	**6.235**
	#1	0.110	1.025	0.110	48.120	27.893	31.860
	#2	0.159	1.074	0.293	4.688	27.563	32.056
	#3	0.037	1.099	0.244	5.969	25.391	30.945
	#4	0.159	1.086	0.269	6.201	25.684	30.762
Avg.		0.116	1.071	0.229	16.245	26.633	31.406

4 Conclusions

The Monte Carlo simulations of small and repetitive regular structures have revealed that:

- On IBM Q it is not currently possible to do single runs, a fact which would be helpful for better estimating variations and their associated distributions;
- The probabilities of failures of toy circuits made of sequentially connected elementary quantum gates is not following the classical equations for the probability of failure of serial/parallel connected elements;
- It is clear that correlations are playing a (significant) role, which explain the discrepancies mentioned above;
- It seems that the particular architecture of the quantum chip on which the designs were run, as well as the particular mapping onto the qubits, are affecting the overall reliability of the circuits;
- All the aspects mentioned/identified contribute to creating a better/clearer picture; which unfortunately looks like way more complex than expected.

The major missing element is represented by the spread/distribution of variations (which should be estimated). The other element of interest is represented by correlations with respect to different possible mappings, which will also have to be evaluated/estimated. This will require many simulations of the same (simple) circuits when mapped differently onto qubits, hence although this work has gathered about 2 million Monte Carlo simulations under its belt, way more are needed.

Acknowledgements. This research was supported by the EU through the European Regional Development Fund under the Competitiveness Operational Program (*BioCell-NanoART = Novel Bio-inspired Cellular Nano-Architectures*, POC-A1.1.4-E-2015 nr. 30/01.09.2016).

References

1. Bell, J.S.: Bertlmann's socks and the nature of reality. J. Physique Colloques **42**(C2), C2-41–C2-62 (1981). https://doi.org/10.1051/jphyscol:1981202
2. Feynman, R.P.: Simulating physics with computers. Intl. J. Theor. Phys. **21**(6–7), 467–488 (1982). https://doi.org/10.1007/BF02650179
3. Tonomura, A., Endo, J., Matsuda, T., Kawasaki, T., Ezawa, H.: Demonstration of single electron buildup of an interference pattern. Am. J. Phys. **57**(2), 117–120 (1989). https://doi.org/10.1119/1.16104
4. Schumacher, B.: Quantum coding. Phys. Rev. A **51**(4), 2738–2747 (1995). https://doi.org/10.1103/PhysRevA.51.2738
5. Preskill, J.: Lecture Notes for Physics 219/Computer Science 219: Quantum Computation. California Institute of Technology (2019–2020). http://www.theory.caltech.edu/~preskill/ph219/ph219_2019-20
6. Freedman, M.H., Kitaev, A., Larsen, M.J., Wang, Z.: Topological quantum computation. Bull. Am. Soc. **40**(1), 31–38 (2002). https://doi.org/10.1090/S0273-0979-02-00964-3
7. White, A.G.: Quantum information and optics. Aust. Optical Soc. News, March 2002. http://quantum.info/andrew/publications/2002/aos2.pdf
8. Szkopek, T., Boykin, P.O., Fan, H., Roychowdhury, V.P., Yablonovitch, E., Simms, G., Gyure, M., Fong, B.: Threshold error penalty for fault tolerant quantum computation with nearest neighbour communication. IEEE Trans. Nanotechn. **5**(1), 42–49 (2006). https://doi.org/10.1109/TNANO.2005.861402
9. Dasgupta, S., Papadimitriou, C.H., Vazirani, U.V.: Algorithms, Chapter 10 Quantum Algorithms. http://algorithmics.lsi.upc.edu/docs/Dasgupta-Papadimitriou-Vazirani.pdf
10. Strubell, E.: Introduction to quantum alGorithms. COS 498-Spring (2011). https://people.cs.umass.edu/~strubell/doc/quantum_tutorial.pdf
11. Szopek, T., Roychowdhury, V.P., Antoniadis, D.A., Damoulakis, J.N.: Physical fault tolerance of nanoelectronics. Phys. Rev. Lett. **106**(17), 176801 (2011). https://doi.org/10.1103/PhysRevLett.106.176801
12. Williams, C.P.: Explorations in Quantum Computing, 2nd edn. Springer, London Limited (2011)
13. Fowler, A.G.: Surface codes: towards practical large-scale quantum computation. Phys. Rev. A **86**, 032324 (2012). https://doi.org/10.1103/PhysRevA.86.032324
14. Bach, R., Pope, D., Liou, S.-H., Batelaan, H.: Controlled doubled-slit electron diffraction. New J. Phys. **15**, 033018 (2013). https://doi.org/10.1088/1367-2630/15/3/033018
15. Paetznick, A.: Resource optimization for fault-tolerant quantum computing. Ph.D. thesis, University of Waterloo, Waterloo, Ontario, Canada (2013). https://arxiv.org/abs/1410.5124

16. Yu, N., Ying, M.: Optimal simulation of three-qubit gates. Technical report, January 2013. https://arxiv.org/abs/1301.3727
17. Barends, R., Kelly, J., Megrant, A., Veitia, A., Sank, D., Jeffrey, E., White, T.C., Mutus, J., Fowler, A.G., Campbell, B., Chen, Y., Chen, Z., Chiaro, B., Dunsworth, A., Neill, C., O'Malley, P., Roushan, P., Vainsencher, A., Wenner, J., Korotkov, A.N., Cleland, A.N., Martinis, J.M.: Superconducting quantum circuits at the surface code threshold for fault tolerance. Nature **508**, 500–503 (2014). https://doi.org/10.1038/nature13171
18. O'Donnell, R.: Introduction to the quantum circuit model. Lecture Notes, September 2015. https://www.cs.cmu.edu/~odonnell/quantum15/lecture01.pdf
19. Monroe, C.R., Schoelkopf, R.J., Lukin, M.D.: Quantum connections. Sci. Am. **314**(5), 50–57 (2016). https://doi.org/10.1038/scientificamerican0516-50
20. Montanaro, A.: Quantum algorithms: an overview. npj Quantum Inf. **2**, 15023 (2016). https://doi.org/10.1038/npjqi.2015.23
21. Patel, R.B., Ho, J., Ferreyrol, F., Ralph, T.C., Pryde, G.J.: A quantum Fredkin gate. Sci. Adv. **2**(3), e1501531 (2016). https://doi.org/10.1126/sciadv.1501531
22. Gottesman, D.: Quantum fault tolerance in small experiments. Technical report, October 2016. https://arxiv.org/abs/1610.03507
23. Wendin, G.: Quantum information processing with superconducting circuits: a review. Reports Progress Phys. **80**(10), 106001 (2016). https://doi.org/10.1088/1361-6633/aa7e1a
24. Martinis, J., Ding, N., Babbush, R., Isakov, S.V., Neven, H., Smelyanskiy, V., Boixo, S.: Fidelity estimation for quantum computing systems. WO2017200536A1 (2016). https://patents.google.com/patent/WO2017200536A1/
25. Linke, N.M., Maslov, D., Roetteler, M., Debnath, S., Figgatt, C., Landsman, K.A., Wright, K., Monroe, C.: Experimental comparison of two quantum computing architectures. PNAS **114**(13), 3305–3310 (2017). https://doi.org/10.1073/pnas.1618020114
26. Gibney, E.: Billion-euro quantum project takes shape. Nature **545**(7652), 16 (2017). https://doi.org/10.1038/545016a
27. Riedel, M.F., Binosi, D., Roetteler, M., Debnath, S., Figgatta, C., Landsman, K.A., Wright, K., Monroe, C.: The European quantum technologies flagship programme. Quantum Sci. Technol. **2**(3), 030501 (2017). https://doi.org/10.1088/2058-9565/aa6aca
28. Campbell, E., Terhal, B., Vuillot, C.: Roads towards fault-tolerant universal quantum computation. Nature **549**, 172–179 (2017). https://doi.org/10.1038/nature23460
29. Loyd, S., Englund, D.: Future directions of quantum information processing. In: Workshop on the Emerging Science and Technology of Quantum, Computation, Communication, and Measurement, Virginia Tech Applied Research Corporation – VT-ARC.org, September 2017. https://community.apan.org/wg/afosr/w/researchareas/17803/future-directions-of-quantum-information-processing-workshop/
30. Calude, C.S., Calude, E.: The road to quantum computational supremacy. Technical report, December 2017. https://arxiv.org/abs/1712.01356
31. Veldhorst, M., Eenink, H.G.J., Yang, C.H., Dzurak, A.S.: Silicon CMOS architecture for a spin-based quantum computer. Nature Commun. **8**(1), 1766 (2017). https://doi.org/10.1038/s41467-017-01905-6
32. Patra, B., Incandela, R.M., van Dijk, J.P.G., Homulle, H.A.R., Song, L., Shahmohammadi, M., Staszewski, R.B., Vladimirescu, A., Babaie, M., Sebastiano, F., Charbon, E.: Cryo-CMOS circuits and systems for quantum computing applications. IEEE J. Solid State Circ. **53**(1), 309–321 (2018). https://doi.org/10.1109/JSSC.2017.2737549
33. Preskill, J.: Quantum computing in the NISQ era and beyond. Quantum **2**, 79 (2018). https://doi.org/10.22331/q-2018-08-06-79
34. Santagati, R., Wang, J., Gentile, A.A., Paesani, S., Wiebe, N., McClean, J.R., Morley-Short, S., Shadbolt, P.J., Bonneau, D., Silverstone, J.W., Tew, D.P., Zhou, X., O'Brien, J.L., Thompson,

M.G.: Witnessing eigenstates for quantum simulation of Hamiltonian spectra. Sci. Adv. **4**(1), eaap9646 (2018). https://doi.org/10.1126/sciadv.aap9646

35. Reagor, M., et al.: Demonstration of universal parametric entangling gates on a multi-qubit lattice. Sci. Adv. **4**(2), eaao3603 (2018). https://doi.org/10.1126/sciadv.aao3603
36. Boixo, S., Isakov, S.V., Smelyanskiy, V.N., Babbush, R., Ding, N., Jiang, Z., Bremner, M.J., Martinis, J.M., Neven, H.: Characterizing quantum supremacy in near-term devices. Nature Phys. **14**, 595–600 (2018). https://doi.org/10.1038/s41567-018-0124-x
37. Coles, P.J., Eidenbenz, S., Pakin, S., Adedoyin, A., Ambrosiano, J., Anisimov, P., Casper, W., Chennupati, G., Coffrin, C., Djidjev, H., Gunter, D., Karra, S., Lemons, N., Lin, S., Lokhov, A., Malyzhenkov, A., Mascarenas, D., Mniszewski, S., Nadiga, B., O'Malley, D., Oyen, D., Prasad, L., Roberts, R., Romero, P., Santhi, N., Sinitsyn, N., Swart, P., Vuffray, M., Wendelberger, J., Yoon, B., Zamora, R., Zhu, W.: Quantum algorithm implementations for beginners. Technical report, April 2018. https://arxiv.org/abs/1804.03719
38. Li, Y., Benjamin, S.C.: One-dimensional quantum computing with a 'segmented chain' is feasible with today's gate fidelities. npj Quantum Inf. **4**, Article no. 25 (2018). https://doi.org/10.1038/s41534-018-0074-2
39. Chamberland, C., Ronagh, P.: Deep neural decoders for near term fault-tolerant experiments. Quantum Sci. Technol. **3**(4), 044002 (2018). https://doi.org/10.1088/2058-9565/aad1f7
40. Acín, A., Bloch, I., Buhrman, H., Calarco, T., Eichler, C., Eisert, J., Esteve, D., Gisin, N., Glaser, S.J., Jelezko, F., Kuhr, S., Lewenstein, M., Riedel, M.F., Schmidt, P.O., Thew, R., Wallraff, A., Walmsley, I., Wilhelm, F.K.: The quantum technologies roadmap: a European community view. New J. Phys. **20**(8), 080201 (2018). https://doi.org/10.1088/1367-2630/aad1ea
41. Geller, M.R.: Sampling and scrambling on a chain of superconducting qubits. Phys. Rev. Appl. **10**(2), 024052 (2018). https://doi.org/10.1103/PhysRevApplied.10.024052
42. Caleffi, M., Cacciapuoti, A.S., Bianchi, G.: Quantum Internet: from communication to distributed computing! In: Proceedings of the ACM International Conference on Nanoscale Computing and Communication (NANOCOM), art. 3, pp. 1–4, September 2018. https://doi.org/10.1145/3233188.3233224
43. Chao, R., Reichardt, B.W.: Fault-tolerant quantum computation with few qubits. npj Quantum Inf. **4**, 42 (2018). https://doi.org/10.1038/s41534-018-0085-z
44. Lao, L., van Wee, B., Ashraf, I., van Someren, J., Khammassi, N., Bertels, K., Almudever, C.C.: Mapping of lattice surgery-based quantum circuits on surface code architectures. Quantum Sci. Technol. **4**(1), 015005 (2018). https://doi.org/10.1088/2058-9565/aadd1a
45. National Science & Technology Council USA, National Strategic Overview for Quantum Information Science, September 2018. https://www.whitehouse.gov/wp-content/uploads/2018/09/National-Strategic-Overview-for-Quantum-Information-Science.pdf
46. Alvarez-Rodriguez, U., Sanz, M., Lamata, L., Solano, E.: Quantum artificial life in an IBM quantum computer. Sci. Rep. **8**(1), 14793 (2018). https://doi.org/10.1038/s41598-018-33125-3
47. Haah, J., Hastings, M.B., Kothari, R., Low, G.H.: Quantum algorithm for simulating real time evolution of lattice Hamiltonians. In: IEEE Annual Symposium on Foundations of Computer Science (FOCS), pp. 350–360, Paris, France, October 2018. https://doi.org/10.1109/focs.2018.00041
48. Flamini, F., Spagnolo, N., Sciarino, F.: Photonic quantum information processing: a review. Rep. Progress Phys. **82**(1), 016001 (2018). https://doi.org/10.1088/1361-6633/aad5b2
49. Willsch, D., Willsch, M., Jin, F., De Raedt, H., Michielsen, K.: Testing quantum fault tolerance on small systems. Phys. Rev. A **98**(5), 052348 (2018). https://doi.org/10.1103/PhysRevA.98.052348
50. Humble, T.: Consumer applications of quantum computing. IEEE Consum. Electron. Mag. **7**(6), 8–14 (2018). https://doi.org/10.1109/MCE.2017.2755298

51. DeBenedictis, E.P., Frank, M.P.: The national quantum initiative will also benefit classical computers. Computer **51**(12), 69–73 (2018). https://doi.org/10.1109/MC.2018.2880025
52. Maslov, D., Nam, Y., Kim, J.: An outlook for quantum computing. Proc. IEEE **107**(1), 5–10 (2019). https://doi.org/10.1109/JPROC.2018.2884353
53. Stojanović, V.M.: Feasibility of single-shot realizations of conditional three-qubit gates in exchange-coupled qubit arrays with local control. Phys. Rev. A **99**(1), 012345 (2019). https://doi.org/10.1103/PhysRevA.99.012345
54. Harper, R., Flammia, S.T.: Fault-tolerant logical gates in the IBM quantum experience. Phys. Rev. Lett. **122**(8), 080504 (2019). https://doi.org/10.1103/PhysRevLett.122.080504
55. Koch, M., Keizer, J.G., Pakkiam, P., Keith, D., House, M.G., Peretz, E., Simmons, M.Y.: Spin read-out in atomic qubits in an all-epitaxial three-dimensional transistor. Nat. Nanotech. **14**(2), 137–140 (2019). https://doi.org/10.1038/s41565-018-0338-1
56. Litinski, D.: A game of surface codes: large-scale quantum computing with lattice surgery. Quantum **3**, 128 (2019). https://doi.org/10.22331/q-2019-03-05-128
57. Raymer, M.G., Monroe, C.: The US national quantum initiative. Quant. Sci. Technol. **4**(2), 020504 (2019). https://doi.org/10.1088/2058-9565/ab0441
58. Riedel, M., Kovacs, M., Zoller, P., Mlynek, J., Calarco, T.: Europe's quantum flagship initiative. Quant. Sci. Technol. **4**(2), 020501 (2019). https://doi.org/10.1088/2058-9565/ab042d
59. Roberson, T.M., White, A.G.: Charting the Australian quantum landscape. Quant. Sci. Technol. **4**(2), 020505 (2019). https://doi.org/10.1088/2058-9565/ab02b4
60. Sussman, B., Corkum, P., Blais, A., Cory, D., Damascelli, A.: Quantum Canada. Quant. Sci. Technol. **4**(2), 020503 (2019). https://doi.org/10.1088/2058-9565/ab029d
61. Wen, X.-G.: Choreographed entanglement dances: topological states of quantum matter. Science **363**(6429): eaal3099 (2019). https://doi.org/10.1126/science.aal3099
62. Kandala, A., Temme, K., Córcoles, A.D., Mezzacapo, A., Chow, J.M., Gambetta, J.M.: Error mitigation extends the computational reach of a noisy quantum processor. Nature **567**(7749), 491–495 (2019). https://doi.org/10.1038/s41586-019-1040-7
63. Wills, S.: Quantum effects in biology. Optics Photonics News **30**(4), 42–49 (2019). https://doi.org/10.1364/OPN.30.4.000042
64. Calafell, I.A., Cox, J.D., Radonjić, M., Saavedra, J.R.M., García de Abajo, F.J., Rozema, L.A., Walther, P.: Quantum computing with graphene plasmons. npj Quantum Inf **5**, 37 (2019). https://doi.org/10.1038/s41534-019-0150-2
65. Humble, T.S., DeBenedictis, E.P. (eds.) Special issue on quantum realism. Computer **52**(6) (2019). https://doi.org/10.1109/mc.2019.2908512
66. McGeoch, C.C., Harris, R., Reinhardt, S.P., Bunyk, P.I.: Practical annealing-based quantum computing. Computer **52**(6), 38–46 (2019). https://doi.org/10.1109/MC.2019.2908836
67. Paler, A., Herr, D., Devitt, S.J.: Really small shoe boxes: on realistic quantum resource estimation. Computer **52**(6), 27–37 (2019). https://doi.org/10.1109/MC.2019.2908621
68. Biamonte, J.D., Dorozhkin, P., Zacharov, I.: Keep quantum computing global and open. Nature **573**(7773), 190–191 (2019). https://doi.org/10.1038/d41586-019-02675-5
69. Slussarenko, S., Pryde, G.J.: Photonic quantum information processing: a concise review. Appl. Phys. Rev. **6**(4), 041303 (2019). https://doi.org/10.1063/1.5115814
70. Song, C., Cui, J., Wang, H., Hao, J., Feng, H., Li,Y.: Quantum computation with universal error mitigation on a superconducting quantum processor. Sci. Adv. **5**(9), eaaw568606 (2019). https://doi.org/10.1126/sciadv.aaw5686
71. Arute, F., et al.: Quantum supremacy using a programmable superconducting processor. Nature **574**(7779), 505–510 (2019). https://doi.org/10.1038/s41586-019-1666-5
72. Brooks, M.: Beyond quantum supremacy: the hunt for useful quantum computers. Nature **574**(7776), 19–21 (2019). https://doi.org/10.1038/d41586-019-02936-3

73. Forbes, A., Nape, I.: Quantum mechanics with patterns of light: progress in high dimensional and multidimensional entanglement with structured light. AVS Quantum Sci. **1**(1), 011701 (2019). https://doi.org/10.1116/1.5112027
74. Gibney, E.: The quantum gold rush. Nature **574**(7776), 22–24 (2019). https://doi.org/10.1038/d41586-019-02935-4
75. Chao, R., Reichardt, B.W.: Flag fault-tolerant error correction for any stabilizer code. Technical report, December 2019. https://arxiv.org/abs/1912.09549
76. Pauka, S.J., Das, K., Kalra, R., Moini, A., Yang, Y., Trainer, M., Bousquet, A., Cantaloube, C., Dick, N., Gardner, G.C., Manfra, M.J., Reilly, D.J.: A cryogenic interface for controlling many qubits. Technical report, December 2019. https://arxiv.org/abs/1912.01299
77. CB Insights: What is quantum computing? Technical report (2019). https://www.cbinsights.com/research/report/quantum-computing/
78. European Commission: Strategic Research Agenda. European Quantum Flagship, Brussels, February 2020. https://ec.europa.eu/newsroom/dae/document.cfm?doc_id=65402
79. Vasconcelos, F.: Quantum computing @ MIT: the past, present, and future of the second revolution in computing. Technical report, February 2020. https://arxiv.org/abs/2002.05559
80. Yu, Y., Ma, F., Luo, X.-Y., Jing, B., Sun, P.-F., Fang, R.-Z., Yang, C.-W., Liu, H., Zheng, M.-Y., Xie, X.-P., Zhang, W.-J., You, L.-X., Wang, Z., Chen, T.-Y., Zhang, Q., Bao, X.-H., Pan, J.-W.: Entanglement of two quantum memories via fibres over dozens of kilometres. Nature **578**(7794), 240–245 (2020). https://doi.org/10.1038/s41586-020-1976-7
81. Asaad, S., Mourik, V., Joecker, B., Johnson, M.A.I., Baczewski, A.D., Firgau, H.R., Mądzik, M.T., Schmitt, V., Pla, J.J., Hudson, F.E., Itoh, K.M., McCallum, J.C., Dzurak, A.S., Laucht, A., Morello, A.: Coherent electrical control of a single high-spin nucleus in silicon. Nature **579**(7798), 205–209 (2020). https://doi.org/10.1038/s41586-020-2057-7
82. Bultink, C.C., O'Brien, T.E., Vollmer, R., Muthusubramanian, N., Beekman, M.W., Rol, M.A., Fu, X., Tarasinski, B., Ostroukh, V., Varbanov, B., Bruno, A., DiCarlo, L.: Protecting quantum entanglement from leakage and qubit errors via repetitive parity measurements. Sci. Adv. **6**(12), eaay3050 (2020). https://doi.org/10.1126/sciadv.aay3050
83. Cartlidge, E.: Supreme or unproven? Optics Photonics News **31**(3), 42–49 (2020). https://doi.org/10.1364/OPN.31.3.000042
84. Parappurath, N., Alpeggiani, F., Kuipers, L., Verhagen, E.: Direct observation of topological edge states in silicon photonic crystals: spin, dispersion, and chiral routing. Sci. Adv. **6**(10), eaaw4137 (2020). https://doi.org/10.1126/sciadv.aaw4137
85. Kim, J.-H., Aghaeimeibodi, S., Carolan, J., Englund, D., Waks, E.: Hybrid integration methods for on-chip quantum photonics. Optica **7**(4), 291–308 (2020). https://doi.org/10.1364/OPTICA.384118
86. Versluis, R.: Quantum computers scale up. IEEE Spectr. **57**(4), 25–29 (2020). https://doi.org/10.1109/MSPEC.2020.9055969
87. von Neumann, J.: Probabilistic logics and the synthesis of reliable organisms from unreliable components. In: Shannon, C.E., McCarthy, J. (eds.) Automata Studies (AM-34), pp. 43–98. Princeton University Press, Princeton, April 1956. https://doi.org/10.1515/9781400882618-003
88. Moore, E.F., Shannon, C.E.: Reliable circuits using less reliable relays. Part I. J. Frankl. Inst. **262**(3), 191–208 (1956). https://doi.org/10.1016/0016-0032(56)90559-2
89. Pérez-Rosés, H.: Sixty years of network reliability. Maths. Comp. Sci. **12**(3), 275–293 (2018). https://doi.org/10.1007/s11786-018-0345-5
90. Cowell, S.R., Beiu, V., Dăuş, L., Poulin, P.: On the exact reliability enhancements of small hammock networks. IEEE Access **6**, 25411–25426 (2018). https://doi.org/10.1109/ACCESS.2018.2828036

Experimenting with Beta Distributions for Approximating Hammocks' Reliability

Simon R. Cowell, Sorin Hoară, and Valeriu Beiu

"Aurel Vlaicu" University of Arad, Arad, Romania
{simon.cowell,valeriu.beiu}@uav.ro, sorin.hoara@uav.roc

Abstract. It is a well-known fact that, in general, the combinatorial problem of finding the reliability polynomial of a two-terminal network belongs to the class of #P-complete problems. In particular, hammock (aka brick-wall) networks are particular two-terminal networks introduced by Moore and Shannon in 1956. Rather unexpectedly, hammock networks seem to be ubiquitous, spanning from biology (neural cytoskeleton) to quantum computing (layout of quantum gates). Because computing exactly the reliability of large hammock networks seems unlikely (even in the long term), the alternatives we are facing fall under approximation techniques using: (*i*) simpler 'equivalent' networks; (*ii*) lower and upper bounds; (*iii*) estimates of (some of) the coefficients; (*iv*) interpolation (e.g., Bézier, Hermite, Lagrange, splines, etc.); and (*v*) combinations of (some of) the approaches mentioned above. In this paper we shall advocate—for the first time ever—for an approximation based on ***an 'equivalent' statistical distribution***. In particular, we shall argue that as counting (lattice paths) is at the heart of the problem of estimating reliability for such networks, the ***binomial distribution*** might be a (very) good starting point. As the number of alternatives (lattice paths) gets larger and larger, a continuous approximation like the ***normal distribution*** naturally comes to mind. Still, as the number of alternatives (lattice paths) becomes humongous very quickly, more accurate and flexible approximations might be needed. That is why we put forward the ***beta distribution*** (as it can match the binomial distribution), and we use it in conjunction with a few exact coefficients (which help fitting the tails) to approximate the reliability of hammock networks.

Keywords: Network reliability · Hammock networks · Reliability polynomial · Approximations · Probability distributions

1 Introduction

The unrelenting pace of technological advances makes it possible to envisage 3 nm processors in the not too distant future [1]. In fact, a 3 nm Gate-All-Around (GAA) technology process development kit was announced in May 2019 [2], while a first prototype has been reported in January 2020 [3]. There is a large body of evidence [4] that the development of such minuscule devices leads to beneficial improvements of system level performances (e.g., lower power consumption), but has challenging counter effects (e.g., how to tame the statistical uncertainties unavoidable at the atomic scale) [5–7].

I. Dzitac et al. (Eds.): ICCCC 2020, AISC 1243, pp. 70–81, 2021.
https://doi.org/10.1007/978-3-030-53651-0_6

Fundamentally, although at the system level we are treading on information measured in bits (a word put forward around late 1940s by John W. Tukey, cf. [8]), which are either 0 or 1, scaling is hampered by physical imperfections at the device and circuit levels [9]. That is why the latest and mightiest chips (like, e.g., Cerebras Wafer Scale Engine, having 1.2 trillion transistors [10]) had to address this challenge, and are incorporating various types of redundancy (a trend started discreetly a long time ago with the parity bit). A much less mature but aspiring contender is represented by quantum computing, which lately has also been making big strides [11, 12]. In this case we are talking about qubits (a word coined by Benjamin Schumacher in 1995 [13]), and statistics is now the name of the game. Before going further it is worth quoting from John W. Tuckey: *"Far better an approximate answer to the right question, which is often vague, than an exact answer to the wrong question, which can always be made precise"* [14]. Although pertaining to data analysis, this statement bears a lot of meaning for both software and hardware, as advocated by this paper.

The solution to develop large scale reliable (hardware) systems in the face of statistical uncertainties is to design systems using schemes which can enhance reliability. The majority-voting and multiplexing schemes (both at the *gate-level*) were proposed by John von Neumann in five seminal lectures presented at Caltech in January 1952, https://sites.google.com/site/michaeldgodfrey/vonneumann/vN_Caltech_Lecture.pdf. A printed version of those lectures was published in April 1956 [15], followed in September 1956 by the pioneering hammock networks concept of Moore and Shannon [16] (the first *device-level* based reliability scheme). The network reliability field was born, and has been evolving ever since [17–20].

The basic questions in network reliability are related to two-terminal, k-terminal, and all-terminal reliability of a given network. All of them are known to be $\#P$-complete in general [21–23], hence exact algorithms are way too slow even for small networks. That is why lower and upper bounds were investigated as time-efficient alternatives [24], with bounding or approximating the coefficients of the reliability polynomial as another possible option [25–27].

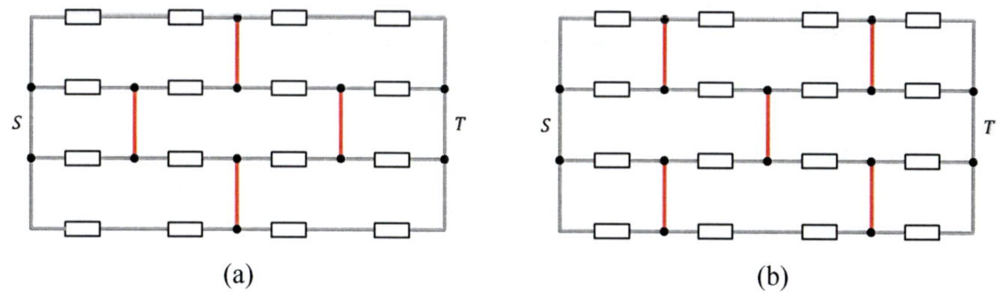

Fig. 1. Hammock networks for $w = l = 4$ (see [16]): (a) $H_{4,4}$ (S and T and 4 other nodes); and (b) $H^{+}_{4,4}$ (S and T and 5 other nodes).

In this paper the focus will be on hammock networks [16], which have started to be investigated in details only lately [28–30]. Hammocks are two-terminal networks $H_{w,l}$, of width w and length l. Reliability (or probabilistic st-connectedness) is defined as the probability that two nodes S (the source) and T (the terminus) can communicate (are connected). In fact, $H_{w,l}$ are modeled as graphs in which nodes communicate through n $(= wl)$ edges (see Fig. 1, where edges are represented as resistors), edges which have statistically independent probabilities of failing $q = 1 - p$ (nodes are assumed to be perfectly reliable).

2 A Fresh Take on Approximating the Reliability Polynomial

We have realized (about six years ago) that there are obvious similarities between reliability polynomials $h(p)$ (also $Rel(p)$ or $Rel(G; p)$) and cumulative distribution functions (cdfs) of random variables (rvs). This was not difficult to fathom as both are sigmoid (S-shaped) functions embodying probabilities. Following on that observation, we did a few preliminary simulations but have started only very recently to properly experiment with approximating the two-terminal reliability of Moore-Shannon hammocks by the cdf of various probability distributions.

We tried $\Phi\left(\frac{p-p_0}{\sigma}\right)$ first, being the cdf of *the normal distribution* $N\left(p_0, \sigma^2\right)$, where the median p_0 was found by solving numerically $h(p_0) = 1/2$, and σ was chosen by hand to produce a good fit. The results were quite promising. Looking for an explanation why $h(p)$ should resemble the cdf, we turned to *the Binomial distribution*

$$X \sim Bin(n, p_1), \tag{1}$$

n being the number of edges in the graph underlying a Moore-Shannon hammock network, reasoning that, in some sense, the connectivity of a random sub-graph of a given graph correlates with the number of edges that are present, which follows a binomial distribution, due to the customary assumption of independent, identically distributed (i.i.d.) Boolean edges. All of these suggested that $h(p)$ might have some connection, at least, with $X \sim Bin(n, p_1)$ for some p_1.

One disadvantage of $N\left(p_0, \sigma^2\right)$ is that it is supported on the whole real line, whereas $h(p)$, and hence $h'(p)$, are defined only for $0 \le p \le 1$ (with $h'(p)$ vanishing at the endpoints of that interval). Another advantage of a binomial distribution, therefore, is that its probability mass function (pmf) is compactly supported. Yet another reason for trying a binomial distribution is that these are, notoriously, well approximated by normal distributions, which we had already found to give promising results, in spite of their disadvantages.

Our first attempt with $X \sim Bin(n, p_1)$ was to scale its cdf horizontally, approximating $h(p)$ by the piecewise constant function

$$f(p) = P(X \le pn). \tag{2}$$

In order that $f(p_0)$ should approximately equal $1/2$, we needed X to have median approximately equal to $p_0 n$. However, any median of such an X lies between $\lfloor np_1 \rfloor$ and $\lceil np_1 \rceil$, so we set $p_1 = p_0$, and $X \sim Bin(n, p_0)$. Having thus forced $f(p)$ to have a median similar to that of $h(p)$, we found that the approximation was otherwise poor, the rate of change of f at p_0 being quite different from that of h at p_0. Replacing $X \sim Bin(n, p_0)$ by $X \sim Bin(N, p_0)$ and $f(p) = P(X \le pn)$ by $f(p) = P(X \le pN)$, we found that, choosing N by hand, we could greatly improve this. However, the optimal values of

N being relatively low, the piecewise constant function $f(p)$ was a relatively coarse approximation to the continuous function $h(p)$.

To solve this problem, we replaced *the scaled Binomial distribution* cdf $f(p) = P(X \leq pN)$, where $X \sim Bin(N, p_0)$, by *the regularised incomplete beta function*

$$I_x(a,b) = \frac{B(x;a,b)}{B(a,b)} \tag{3}$$

where

$$B(a,b) = \int_0^1 t^{a-1}(1-t)^{b-1}dt \tag{4}$$

is the beta function (see [31] for a data fitting/analysis application), and

$$B(x;a,b) = \int_0^x t^{a-1}(1-t)^{b-1}dt \tag{5}$$

is the incomplete beta function. The reason being that, as is well known, if $Y \sim Bin(m, r)$, then for $k \in \{0, 1, 2, \ldots, m\}$

$$P(Y \leq k) = I_{1-r}(m-k, k+1), \tag{6}$$

hence $I_{1-p_0}(N - pN, pN + 1)$ smoothly interpolates our piecewise constant function $f(p)$. We found that by using the optimal values of N that we had previously determined by hand, $I_{1-p_0}(N - pN, pN + 1)$ gave very good approximations to $h(p)$.

At this point, we realized that in associating the cdf of a binomial distribution with $h(p)$, we had perhaps not made the most logical choice of parameter. Rather than fixing p_0 in $X \sim Bin(N, p_0)$ and varying p in $f(p) = P(X \leq pN)$, we decided instead to fix a and let $g(p) = P(Z \geq a)$, while varying p in $Z \sim Bin(N, p)$. This would be a more natural way to assert some correlation between the event that the two terminals (S and T) of a hammock network are connected, and the event that sufficiently many edges are present in a random sub-graph of the graph underlying the hammock. This idea was partly inspired by the work of Erdős, Rényi and Gilbert [32–34] in what is now known as percolation theory in random graphs.

Since

$$P(Z \geq a) = P(N - Z \leq N - a) = P(W \leq k), \tag{7}$$

where $W = N - Z$, $k = N - a$, and $W \sim Bin(N, 1-p)$, and since, as noted above, we can write

$$P(W \leq k) = I_{1-(1-p)}(N-k, k+1) = I_p(N-k, k+1), \tag{8}$$

therefore, substituting $b - 1$ for k and $a + b - 1$ for N, we have

$$P(Z \geq a) = I_p(a, b), \tag{9}$$

where $Z \sim Bin(a + b - 1, p)$.

Equations (3)–(5) show that, in line with our initial ideas, the new approximation $I_p(a, b)$ for $h(p)$ is indeed a cdf, not of a binomial distribution, but of the so-called beta distribution whose probability density function (pdf) is

$$beta(a, b)(p) = \frac{(a+b-1)!}{(a-1)!(b-1)!} p^{a-1}(1-p)^{b-1}. \tag{10}$$

Indeed, for positive integers a and b, integrating by parts and then solving the resulting recurrence relation, one can demonstrate directly that $beta(a, b)(p)$ has cdf

$$\int_0^p beta(a, b)(t)dt = \sum_{k=a}^{a+b-1} \binom{a+b-1}{k} p^k (1-p)^{a+b-1-k} = P(Z \geq a). \tag{11}$$

where $Z \sim Bin(a + b - 1, p)$. This is one way to prove Eq. (6).

Bearing in mind that in future we may generalize to non-integer a and b by replacing $P(Z \geq a)$ by $I_p(a, b)$, our strategy in this paper was to choose integer parameters a and b so that our approximation $P(Z \geq a)$ to $h(p)$, where $Z \sim Bin(a + b - 1, p)$, is a homogeneous polynomial of degree $a + b - 1$ in p and q, where $q = 1 - p$. Provided $a + b - 1 \leq n$, we can then multiply this polynomial by $1 = (p + q)^{n-a-b+1}$ to obtain a homogeneous polynomial of degree n in p and q, so that we may easily compare the coefficients of the latter with those of $h(p, q)$ one by one—the expression of $h(p)$ in the scaled Bernstein basis of degree n. The resulting formula is:

$$P(Z \geq a) = \sum_{k=a}^{n} \sum_{j=\max(0,k-a-b+1)}^{\min(n-a-b+1,k-a)} \binom{n-a-b+1}{j} \binom{a+b-1}{k-j} p^k q^{n-k}. \tag{12}$$

In case we are dealing with a self-dual hammock (see [28, 30]), we have $h(p) = 1 - h(1 - p)$, hence $h'(p) = h'(1 - p)$, i.e., $h'(p)$ is symmetric about the line $p = 1/2$. Therefore our approximation $beta(a, b)$ to $h'(p)$ should also be symmetric about that line. For this we have to set $a = b$, which means, in this case, that we need only choose a single integer parameter (a) rather than two (a and b). In particular, all square hammocks of odd w and l are self-dual (see [28, 30]), and therefore allow us this simplification.

3 Simulations and Comparisons

In this section we present detailed simulations for $H_{5,5}$, starting with the exact reliability polynomial determined in [28]

$$\begin{aligned} h_{5,5}(p, q) = {} & 52p^5q^{20} + 994p^6q^{19} + 8983p^7q^{18} + 50796p^8q^{17} + 200559p^9q^{16} + \\ & 584302p^{10}q^{15} + 1294750p^{11}q^{14} + 2220298p^{12}q^{13} + \\ & 2980002p^{13}q^{12} + 3162650p^{14}q^{11} + 2684458p^{15}q^{10} + \\ & 1842416p^{16}q^9 + 1030779p^{17}q^8 + 471717p^{18}q^7 + 176106p^{19}q^6 + \\ & 53078p^{20}q^5 + 12650p^{21}q^4 + 2300p^{22}q^3 + 300p^{23}q^2 + 25p^{24}q + \\ & p^{25}. \end{aligned} \tag{13}$$

Table 1. The reliability polynomial for $H_{5,5}$ as well as several approximations. For simulations we have not used the three approximations which are shaded.

Hammock / approximation	Reliability polynomial
Exact $h_{5,5}(p,q)$ (reference) [28]	$52p^5q^{20} + 994p^6q^{19} + 8983p^7q^{18} + 50796p^8q^{17} + 200559p^9q^{16} + 584302p^{10}q^{15} + 1294750p^{11}q^{14} + 2220298p^{12}q^{13} + 2980002p^{13}q^{12} + 3162650p^{14}q^{11} + 2684458p^{15}q^{10} + 1842416p^{16}q^9 + 1030779p^{17}q^8 + 471717p^{18}q^7 + 176106p^{19}q^6 + 53078p^{20}q^5 + 12650p^{21}q^4 + 2300p^{22}q^3 + 300p^{23}q^2 + 25p^{24}q + p^{25}$
Hermite$_9$ (ver. 1) [35]	$126p^5 - 420p^6 + 540p^7 - 315p^8 + 70p^9$
Hermite$_9$ (ver. 2) [35]	$p^5 + 5p^5q + 15p^5q^2 + 35p^5q^3 + 70p^5q^4$
Hermite$_{25}$ (ver. 3)	$126p^5q^{20} + 2100p^6q^{19} + 16500p^7q^{18} + 81225p^8q^{17} + 280825p^9q^{16} + 724504p^{10}q^{15} + 1446600p^{11}q^{14} + 2288300p^{12}q^{13} + 2912000p^{13}q^{12} + 3010800p^{14}q^{11} + 2544256p^{15}q^{10} + 1762150p^{16}q^9 + 1000350p^{17}q^8 + 464200p^{18}q^7 + 175000p^{19}q^6 + 53004p^{20}q^5 + 12650p^{21}q^4 + 2300p^{22}q^3 + 300p^{23}q^2 + 25p^{24}q + p^{25}$
Hermite$_{25}$ adj (ver. 4)	$52p^5q^{20} + 994p^6q^{19} + 16500p^7q^{18} + 81225p^8q^{17} + 280825p^9q^{16} + 724504p^{10}q^{15} + 1446600p^{11}q^{14} + 2288300p^{12}q^{13} + 2912000p^{13}q^{12} + 3010800p^{14}q^{11} + 2544256p^{15}q^{10} + 1762150p^{16}q^9 + 1000350p^{17}q^8 + 464200p^{18}q^7 + 176106p^{19}q^6 + 53078p^{20}q^5 + 12650p^{21}q^4 + 2300p^{22}q^3 + 300p^{23}q^2 + 25p^{24}q + p^{25}$
Cubic spline [36]	$52p^5q^{20} + 994p^6q^{19} + 16500p^7q^{18} + 81225p^8q^{17} + 280825p^9q^{16} + 724504p^{10}q^{15} + 1446600p^{11}q^{14} + 2288300p^{12}q^{13} + 2912000p^{13}q^{12} + 3010800p^{14}q^{11} + 2544256p^{15}q^{10} + 1762150p^{16}q^9 + 1000350p^{17}q^8 + 464200p^{18}q^7 + 176106p^{19}q^6 + 53078p^{20}q^5 + 12650p^{21}q^4 + 2300p^{22}q^3 + 300p^{23}q^2 + 25p^{24}q + p^{25}$
Beta$_{25}$(6,6) (ver. 2) This paper	$462p^6q^{19} + 6798p^7q^{18} + 46827p^8q^{17} + 200563p^9q^{16} + 598378p^{10}q^{15} + 1320474p^{11}q^{14} + 2234246p^{12}q^{13} + 2966054p^{13}q^{12} + 3136926p^{14}q^{11} + 2670382p^{15}q^{10} + 1842412p^{16}q^9 + 1034748p^{17}q^8 + 473902p^{18}q^7 + 176638p^{19}q^6 + 53130p^{20}q^5 + 12650p^{21}q^4 + 2300p^{22}q^3 + 300p^{23}q^2 + 25p^{24}q + p^{25}$
Beta$_{25}$(6,6) adj (ver. 3) This paper	$52p^5q^{20} + 994p^6q^{19} + 6798p^7q^{18} + 46827p^8q^{17} + 200563p^9q^{16} + 598378p^{10}q^{15} + 1320474p^{11}q^{14} + 2234246p^{12}q^{13} + 2966054p^{13}q^{12} + 3136926p^{14}q^{11} + 2670382p^{15}q^{10} + 1842412p^{16}q^9 + 1034748p^{17}q^8 + 473902p^{18}q^7 + 176106p^{19}q^6 + 53078p^{20}q^5 + 12650p^{21}q^4 + 2300p^{22}q^3 + 300p^{23}q^2 + 25p^{24}q + p^{25}$

Promising approximation techniques are relying on:

- simpler 'equivalent' networks;
- upper/lower bounds;
- estimating/bounding the coefficients;
- interpolation (e.g., Hermite, Lagrange, splines, etc.); and
- combinations of these methods.

Looking into the literature published on hammock networks we have found a very fresh Hermite interpolation [35]

$$Hermite_{w,l}(p,q) = p^l\left[\sum\nolimits_{i=0}^{w-1}\binom{i+l-1}{l-1}q^i\right] \tag{14}$$

as well as a novel cubic spline interpolation [36].

The reliability polynomial together with the approximations we have used in this paper are presented in Table 1. This table includes several versions of the same type of approximation, where the first and second non-zero coefficients (red), as well as the associated symmetric coefficients (red), were forced to match the exact values determined in [29, 35]. For beta, forcing these coefficients allow us to fit the tails.

Detailed simulations are presented as follows. In Fig. 2 we can see five of the polynomials detailed in Table 1, together with the continuous beta function, plotted in both linear and logarithmic scale. From Fig. 2(a) it would seem that, with the exception of the cubic spline, all other approximations are performing quite well. Still, from Fig. 2(b), the cubic spline approximation is outstanding for $p < 0.1$, while beta$_{25}$(6,6) is clearly not that good. By adjusting/forcing the first and second non-zero coefficients for both the Hermite$_{25}$ and the beta$_{25}$ approximation they are becoming excellent themselves (for $p < 0.1$). An even more detailed understanding can be grasped from Fig. 3, where the absolute and relative errors are plotted in Figs. 3(a) and 3(b) respectively, with zooms on the regions of interest in Figs. 3(c) and 3(d). All of these unambiguously show that beta$_{25}$(6,6) adjusted (red line) is outperforming all other approximations.

Another perspective on the approximations from Table 1 has been obtained by plotting all the 26 coefficients of the polynomials under test, allowing for a one-on-one comparison. Figure 4 presents stair-step plots of the coefficients of the different polynomials together with the corresponding binomial coefficients in cyan (as polynomials are in Bernstein form). All approximations seem to behave quite similarly with the exception of the cubic spline, an assessment which is pairing that from Fig. 2. For completeness, Fig. 5 presents the absolute and relative errors on each and every one of the 26 coefficients. It can be seen that these stair-step plots are very closely resembling Fig. 3 (they look in fact like discretized versions of Fig. 3). These confirm once again that beta$_{25}$(6,6) adjusted (red line) gives by far the best match not only for the reliability polynomial but in fact for each and every coefficient.

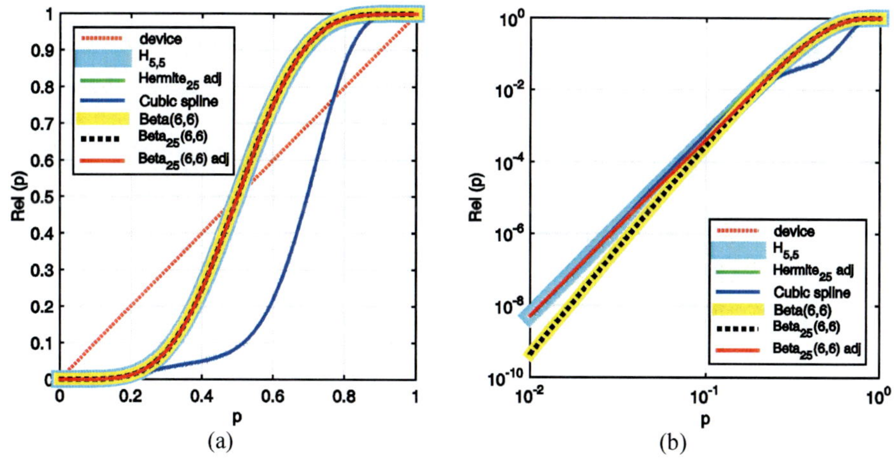

Fig. 2. Reliability of hammock $H_{5,5}$ as well as several approximations (see Table 1) versus p: (a) linear scale; and (b) logarithmic scale.

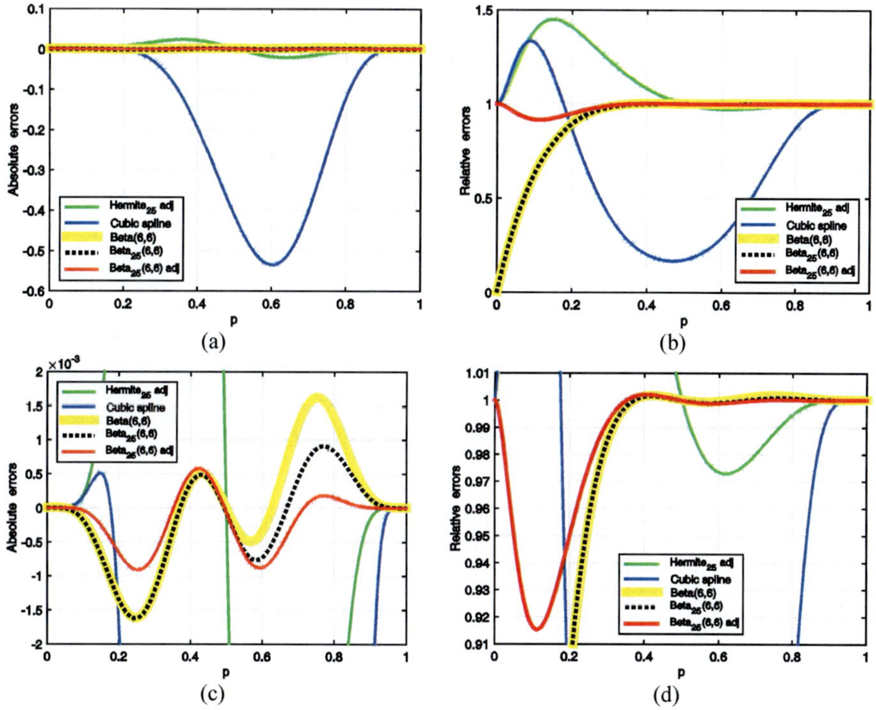

Fig. 3. The errors between the reliability approximations considered (see Table 1) and reliability of hammock $H_{5,5}$ versus p: (a) absolute errors; (b) relative errors; (c) zoom on absolute errors, and (d) zoom on relative errors.

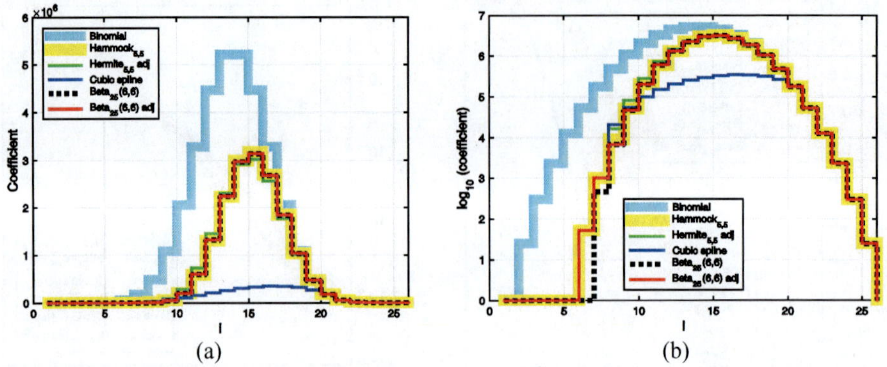

Fig. 4. The binomial coefficients $\binom{25}{i}$ for $i = 0, 1, \ldots, 25$ (cyan), together with the corresponding coefficients of the reliability polynomial for $H_{5,5}$ (yellow), as well as those of several approximations (see Table 1): (a) linear scale; and (b) logarithmic scale.

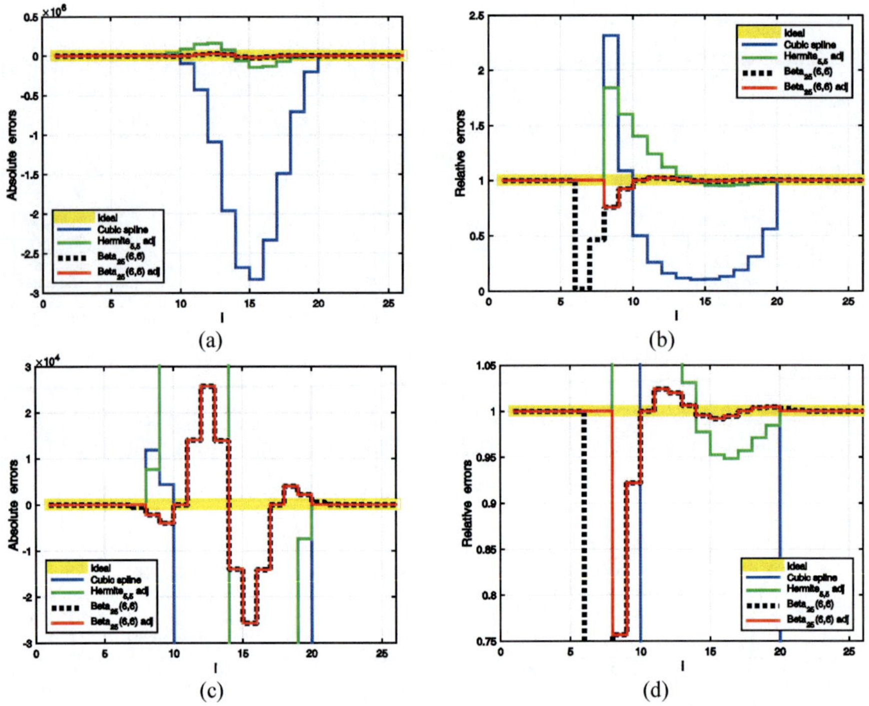

Fig. 5. The absolute and relative errors of the 26 coefficients of several approximations (see Table 1) versus the corresponding coefficients of the reliability polynomial of $H_{5,5}$: (a) absolute errors; (b) relative errors; (c) zoom on absolute errors, and (d) zoom on relative errors.

4 Conclusions

This paper has suggested a fresh approach for approximating the reliability of hammock networks $H_{w,l}$ [16, 28]. The unorthodox approximation method proposed relies on: (*i*) using the beta distribution; (*ii*) translating it into a Bernstein polynomial (of degree $n = wl$); and finally (*iii*) adjusting/correcting the tails of this cdf based on four known coefficients (for which exact computations are known and straightforward/simple [29, 35]).

We have reported detailed simulations and have compared the results obtained using our approach with those of other approximations. All the simulations performed are very encouraging, and support the claim that *beta approximation is much more accurate than any other known approximation.* We plan to continue this line of research by:

- trying to find both theoretical proofs and accurate bounds for the approximation method we have introduced here; and
- investigating other distributions (we plan to start with Poisson distributions of the coefficients, as suggested by $\text{beta}_{25}(6,6)$ adjusted (red line) in Fig. 4(a)).

Acknowledgements. This research was supported by the EU through the European Regional Development Fund under the Competitiveness Operational Program (*BioCell-NanoART = Novel Bio-inspired Cellular Nano-Architectures*, POC-A1.1.4-E-2015 nr. 30/01.09.2016).

References

1. Bae, G., Bae, D.-I., Kang, M., Hwang, S.M., Kim, S.S., Seo, B., Kwon, T.Y., Lee, T.J., Moon, C., Choi, Y.M., Oikawa, K., Masuoka, S., Chun, K.Y., Park, S.H., Shin, H.J., Kim, J.C., Bhuwalka, K.K., Kim, D.H., Kim, W.J., Yoo, J., Jeon, H.Y., Yang, M.S., Chung, S.-J., Kim, D., Ham, B.H., Park, K.J., Kim, W.D., Park, S.H., Song, G., Kim, Y.H.: 3 nm GAA technology featuring multi-bridge-channel FET for low power and high performance applications. In: IEEE International Electron Device Meeting (IEDM 2018), San Francisco, CA, 1–5 December 2018, pp. 656–659 (2018). https://doi.org/10.1109/iedm.2018.8614629
2. Samsung Electronics: Press Release, 15 May 2019. https://news.samsung.com/global/samsung-electronics-leadership-in-advanced-foundry-technology-showcased-with-latest-silicon-innovations-and-ecosystem-platform
3. Lee, J.Y.: Samsung develops world's first 3-nm processor. Korean Maeil Economy, 2 January 2020. https://n.news.naver.com/article/009/0004493648
4. International Roadmap for Devices and Systems (IRDS™), 2018 Edition. https://irds.ieee.org/editions/2018
5. Lapedus, M., Sperling, E.: Making chips at 3 nm and beyond. Semicon. Eng., 16 April 2020. https://semiengineering.com/making-chips-at-3nm-and-beyond/
6. Sperling, E.: Big changes in tiny interconnects. Semicon. Eng., 16 April 2020. https://semiengineering.com/big-changes-in-tiny-interconnects/
7. Bae, D.-I., Choi, B.-D.: Short channels and mobility control of GAA multi stacked nanosheets through the perfect removal of SiGe and post treatment. Electr. Lett. **56**(8), 400–402 (2020). https://doi.org/10.1049/el.2019.3459

8. Shannon, C.E.: A mathematical theory of communication. Bell System Tech. J. **27**(3), 379–423 (1948). https://doi.org/10.1002/j.1538-7305.1948.tb01338.x
9. Yeric, G.: IC design after Moore's law. In: IEEE Custom Integrated Circuit Conference (CICC 2019), Austin, TX, 14–17 April 2019. https://doi.org/10.1109/cicc.2019.8780343
10. Moore, S.K.: Huge chip smashes deep learning's speed barrier. IEEE Spectr. **57**(1), 24–27 (2020). https://doi.org/10.1109/MSPEC.2020.8946303
11. Brooks, M.: Beyond quantum supremacy: the hunt for useful quantum computers. Nature **574**(7776), 19–21 (2019). https://doi.org/10.1038/d41586-019-02936-3
12. Gibney, E.: The quantum gold rush. Nature **574**(7776), 22–24 (2019). https://doi.org/10.1038/d41586-019-02935-4
13. Schumacher, B.: Quantum coding. Phys. Rev. A **51**(4), 2738–2747 (1995). https://doi.org/10.1103/PhysRevA.51.2738
14. Tukey, J.W.: The future of data analysis. Annals Math. Stat. **33**(1), 1–67 (1962). https://doi.org/10.1214/aoms/1177704711. https://www.jstor.org/stable/2237638
15. von Neumann, J.: Probabilistic logics and the synthesis of reliable organisms from unreliable components. In: Shannon, C.E., McCarthy, J. (eds.) Automata Studies (AM-34), pp. 43 – 98. Princeton University Press, Princeton, April 1956. https://doi.org/10.1515/9781400882618-003
16. Moore, E.F., Shannon, C.E.: Reliable circuits using less reliable relays, part I. J. Frankl. Inst. **262**(3), 191–208 (1956). https://doi.org/10.1016/0016-0032(56)90559-2
17. Colbourn, C.J.: The Combinatorics of Network Reliability. Oxford University Press, Oxford (1987)
18. Chari, M., Colbourn, C.J.: Reliability polynomials: a survey. J. Combin. Inform. Syst. Sci. **22**(3–4), 177–193 (1997)
19. Stanley, R.P.: Enumerative Combinatorics, vol. 1, 2nd edn. Cambridge University Press, Cambridge (2012). https://math.mit.edu/~rstan/ec/ec1/
20. Pérez-Rosés, H.: Sixty years of network reliability. Maths. Comp. Sci. **12**(3), 275–293 (2018). https://doi.org/10.1007/s11786-018-0345-5
21. Valiant, L.G.: The complexity of computing the permanent. Theor. Comp. Sci. **8**(2), 189–201 (1979). https://doi.org/10.1016/0304-3975(79)90044-6
22. Valiant, L.G.: The complexity of enumeration and reliability problems. SIAM J. Comput. **8**(3), 410–421 (1979). https://doi.org/10.1137/0208032
23. Provan, J.S., Ball, M.O.: The complexity of counting cuts and computing the probability that a graph is connected. SIAM J. Comput. **12**(4), 777–788 (1983). https://doi.org/10.1137/0212053
24. Provan, J.S.: Bounds on the reliability of networks. IEEE Trans. Reliab. **R-35**(3), 260–268 (1986). https://doi.org/10.1109/tr.1986.4335429
25. Camarda, P.: Bounds evaluation of coefficients in the reliability polynomial. Microelectron. Reliab. **30**(6), 1099–1110 (1990). https://doi.org/10.1016/0026-2714(90)90288-X
26. Oxley, J., Welsh, D.: Chromatic, flow and reliability polynomials: the complexity of their coefficients. Comb. Prob. Comput. **11**(4), 403–426 (2002). https://doi.org/10.1017/s0963548302005175
27. Beichl, I., Cloteaux, B., Sullivan, F.: An approximation algorithm for the coefficients of the reliability polynomial. Congr. Numer. **197**, 143–151 (2009). https://tsapps.nist.gov/publication/get_pdf.cfm?pub_id=902705
28. Cowell, S.R., Beiu, V., Dăuş, L., Poulin, P.: On the exact reliability enhancements of small hammock networks. IEEE Access **6**, 25411–25426 (2018). https://doi.org/10.1109/ACCESS.2018.2828036
29. Dăuş, L., Beiu, V., Cowell, S.R., Poulin, P.: Brick-wall lattice paths and applications. arXiv, art. 1804.05277 [math.CO], 14 April 2018. http://arxiv.org/abs/1804.05277

30. Dăuş, L., Jianu, M.: The shape of the reliability polynomial of a hammock network. In: International Conference on Computers Communication & Control (ICCCC 2020), online, Agora Univ. Oradea (host), Romania, 11–15 May 2020, Springer, in press. Early version in arXiv, art. 1901.0436 [math.CO], 13 January 2019. http://arxiv.org/abs/1901.0436
31. Kipping, D.M.: Parametrizing the exoplanet eccentricity distribution with the beta distribution. Month. Notices Royal Astro. Soc. **434**(1), L51–L55 (2013). https://doi.org/10.1093/mnrasl/slt075
32. Erdős, P., Rényi, A.: On random graphs I. Publ. Math. Debrecen **6**, 290–297 (1959)
33. Gilbert, E.N.: Random graphs. Ann. Math. Statist. **30**(4), 1141–1144 (1959). https://doi.org/10.1214/aoms/1177706098
34. Erdős, P., Rényi, A.: On the evolution of random graphs. Publ. Math. Inst. Hung. Acad. Sci. **5**(1), 17–61 (1960)
35. Dăuş, L., Jianu, M.: Full Hermite interpolation of the reliability of a hammock network. Appl. Anal. Discrete Math. **14**(1), 198–220 (2020). https://doi.org/10.2298/aadm190805017d. Preliminary version in Proc. Maths. Info. & Edu. MIE 2019, pp. 59–64. Tech. Univ. Civil Eng., Bucharest, Romania, May 2019. http://civile-old.utcb.ro/mie/proceedings2019.pdf
36. Cristescu, G., Drăgoi, V.-F.: Cubic spline approximation of the reliability polynomials of two dual hammock networks. Transylvanian J. Math. Mech. **11**(1–2), 77–99 (2019). http://tjmm.edyropress.ro/journal/19111208.pdf

General Frameworks for Designing Arithmetic Components for Residue Number Systems

Ahmad Hiasat(✉)

Princess Sumaya University for Technology, Amman, Jordan
a.hiasat@psut.edu.jo

Abstract. In many previous works, researchers have proposed Residue-based arithmetic components for the two classical moduli sets $(2^p, 2^p - 1, 2^p + 1)$, and $(2^p, 2^p - 1, 2^{p-1} - 1)$, where p is a positive integer. These components included reverse converters, sign identifiers, and scalers. In this paper, we are widening the umbrella of these two sets to be $(2^k, 2^p - 1, 2^p + 1)$, and $(2^k, 2^p - 1, 2^{p-1} - 1)$, where k is a positive integers such that $0 < k \leq 2p$. The classical moduli sets are special cases of these expanded sets when $p = k$. This paper introduces multiplicative inverses for these expanded moduli sets. The introduced multiplicative inverses will ease the process of designing residue-based arithmetic components. This paper also proposes general frameworks for designing reverse converters, sign identifiers, comparators, and scalers. Additionally, this work expands the options available for a designer willing to design a RNS processor based on these enhanced moduli sets.

Keywords: Computer arithmetic · Multiplicative inverses · Residue number system · Reverse conversion · Scaling · Sign identification · Magnitude comparison

1 Introduction

Residue Number System (RNS) is an arithmetic representation which is used in designing application-specific high-speed processors. These processors are utilized in digital signal processing, cryptography, embedded systems, and modulation Techniques of telecommunication systems. The high-speed processors are considered integrated solutions for different control and communications systems [1–14]. This paper proposes general frameworks for designing high-speed residue-based processors, which consist basically of: reverse converters, sign identifiers, comparators, and scalers. This work expands the options available for a designer willing to design a RNS processor based on these enhanced moduli sets.

The RNS has the merit of being carry-free representation across different residue-based digits. Computations on different digits are conducted separately form each other. Separability of digits reduces complexity of specific components such as adders and multipliers.

I. Dzitac et al. (Eds.): ICCCC 2020, AISC 1243, pp. 82–92, 2021.
https://doi.org/10.1007/978-3-030-53651-0_7

On the other hand, RNS has a few major limitations that make its advantage limited to applications which require mainly arithmetic addition/multiplication. These limitations are basically: sign identification, comparison, scaling, division, and decoding to binary-based systems. As a matter of fact, these limitations have prevented RNS from being adopted in general purpose computers. Thus, building efficient RNS-based arithmetic architectures is a cornerstone process in implementing more effective computing processors for many applications.

Different researchers have considered, attentively, the above listed limitations, particularly, the residue-to-binary decoding process of the two well known three moduli sets: $(2^p, 2^p - 1, 2^p + 1)$, and $(2^p, 2^p - 1, 2^{p-1} - 1)$ [15–28]. Other residue-based operations are still very demanding, such as division, sign identification, and scaling, [29–42]. In this paper, we are introducing the multiplicative inverses of the more general and expanded moduli sets $(2^k, 2^p - 1, 2^p + 1)$, and $(2^k, 2^p - 1, 2^{p-1} - 1)$, where p and k are positive integers such that $0 < k \leq 2p$. Obviously, the former sets are special cases of the last two ones, where $p = k$.

In Sect. 2 of the paper, the notation is introduced, along with the multiplicative inverses for the sets $(2^p, 2^p - 1, 2^p + 1)$, and $(2^p, 2^p - 1, 2^{p-1} - 1)$, which were reported in earlier publications [15,16]. Section 2 also gives a justification for introducing the expanded moduli sets. The new multiplicative inverses for the 3-moduli sets $(2^k, 2^p - 1, 2^p + 1)$ and $(2^p, 2^p - 1, 2^{p-1} - 1)$ are presented and proved in Sect. 3. Section 4 introduces general frameworks for designing different arithmetic residue-based components through three case-studies: reverse conversion, sign identification, and magnitude comparison.

2 Notations and Work Justification

2.1 Notations

The notations and symbols used in this paper are:

- $\{p_1, p_2, \ldots\ldots, p_N\}$, is the set of N moduli where these moduli are positive integers and relatively prime.
- $P = \prod_{i=1}^{N} p_i$, is defined as the "Dynamic Range".
- Given a non-negative integer X such that $X \in [0, P)$, the residue-based representation of X is given by: $X \overset{RNS}{\rightarrow} (R_1, R_2, \ldots\ldots, R_N)$
- $R_i = \langle X \rangle_{p_i}$, is the least integer remainder of dividing X by p_i.
- $\hat{p}_i = \frac{P}{p_i}$, $\left\langle \frac{1}{\hat{p}_i} \right\rangle_{p_i}$, is multiplicative inverse of $\hat{p}_i$ (that is $\left\langle \langle \hat{p}_i \rangle_{p_i} \left\langle \frac{1}{\hat{p}_i} \right\rangle_{p_i} \right\rangle_{p_i} = 1$)
- $\lfloor . \rfloor$ is the largest integer less than or equal to (.).

2.2 Multiplicative Inverses of the Moduli Sets $(2^p, 2^p - 1, 2^p + 1)$, and $(2^p, 2^p - 1, 2^{p-1} - 1)$ and Work Justification

In previous works, a considerable number of authors have researched the issue of converting RNS notation into binary. The conversions relied on the use of the

Chinese Remainder Theorem (CRT), which utilizes the multiplicative inverses. Moduli sets of $(2^p, 2^p - 1, 2^p + 1)$, and $(2^p, 2^p - 1, 2^{p-1} - 1)$, and other similar moduli were of particular importance due to the easiness by which a designer can deal with building arithmetic components for the given moduli [2–14]. The relative simplicity of conversion from RNS to binary was another important factor. This simplicity was based on the fact that these moduli have compact forms for their multiplicative inverses. This enabled the design of converters using arithmetic components without the need for memory elements such as ROM or RAM. Most designs were restricted to the moduli sets $(2^p, 2^p - 1, 2^p + 1)$, $(2^p, 2^p - 1, 2^{p-1} - 1)$, $(2^{p+1} - 1, 2^p, 2^p - 1)$, $(2^{2p} - 1, 2^p, 2^{2p+1} - 1)$ and other slightly modified ones [15–28].

In [15], authors presented the multiplicative inverse for the moduli set $(2^p, 2^p - 1, 2^p + 1)$. It has been proved in [15] that:

$$\left\langle \frac{1}{\hat{p}_1} \right\rangle_{p_1} = 2^p - 1, \left\langle \frac{1}{\hat{p}_2} \right\rangle_{p_2} = 2^{p-1}, \left\langle \frac{1}{\hat{p}_3} \right\rangle_{p_3} = 2^{p-1} + 1 \tag{1}$$

In [16], authors also presented multiplicative inverses for the 3-moduli set $(2^p, 2^p - 1, 2^{p-1} - 1)$. It has been proved in [16] that:

$$\left\langle \frac{1}{\hat{p}_1} \right\rangle_{p_1} = \left\langle 2^{p-1} + 1 \right\rangle_{p_1}, \left\langle \frac{1}{\hat{p}_2} \right\rangle_{p_2} = 2^p - 3, \left\langle \frac{1}{\hat{p}_3} \right\rangle_{p_3} = 2^{p-2} \tag{2}$$

These multiplicative inverses were used in designing RNS to Binary converters for the moduli sets: $(2^p, 2^p - 1, 2^p + 1)$ and $(2^p, 2^p - 1, 2^{p-1} - 1)$.

When using RNS in different applications, it is highly wanted to build a system of a large dynamic range [1,2]. This can be achieved by having more moduli because the dynamic range is the arithmetic product of all moduli in any set. Increasing the number of moduli will increase the complexity of the conversion in terms of area needs, time delay, and energy/power consumed. The 3-moduli sets were recognized by many authors as a reasonable solution between time requirements and conversion-circuit complexity. Moreover, a three-moduli set of the classical type or its expanded form is easier for scaling, sign identification, and comparison than other moduli sets.

The well known $(2^p, 2^p - 1, 2^p + 1)$ moduli set has a $3p$ bits dynamic range. When designing a residue-based processor using this particular set, the processor would involve modulus $(2^p \pm 1)$ adder, subtracter, and multiplier. The delay required to perform modulo $(2^p \pm 1)$ arithmetic operation would need, approximately, the same delay needed to perform a modulo 2^{2p} arithmetic operation. A close look to the published articles in this field supports this result. This have raised a concern of the necessity of suggesting expanded 3-moduli sets $(2^k, 2^p - 1, 2^p + 1)$ and $(2^k, 2^p - 1, 2^{p-1} - 1)$, $k \leq 2p$, which expand the dynamic range by $(k - p)$ bits, if $k > p$, at no additional computational complexity.

In this paper, we are introducing the multiplicative inverses for the more general formulas of these sets, expressed as: $(2^k, 2^p - 1, 2^p + 1)$, and $(2^k, 2^p - 1, 2^{p-1} - 1)$. We are also showing that the multiplicative inverses introduced in

[15] and [16] are special cases of the more general multiplicative inverse formulas introduced in this paper.

3 New General Closed Forms of the Multiplicative Inverses

3.1 The Three-Moduli Set $(2^k, 2^p - 1, 2^p + 1)$

For the moduli set $(2^k, 2^p-1, 2^p+1)$, where p and k are positive integers, $p_1 = 2^k$, $p_2 = 2^p - 1$, and $p_3 = 2^p + 1$. Thus, $P = (2^{2p} - 1)2^k$, whereas, $p_2p_3 = (2^{2p} - 1)$. Moreover, $\hat{p}_1 = (2^{2p} - 1)$, $\hat{p}_2 = 2^k(2^p + 1)$, and $\hat{p}_3 = 2^p(2^p - 1)$. To avoid highly unbalanced moduli, we will assume that $0 < k \leq 2p$.

Theorem 1. *For the three-moduli set $(2^k, 2^p-1, 2^p+1)$, the general closed-forms of the multiplicative inverses are:*

$$\left\langle \frac{1}{\hat{p}_1} \right\rangle_{p_1} = \langle -1 \rangle_{p_1} \tag{3}$$

$$\left\langle \frac{1}{\hat{p}_2} \right\rangle_{p_2} = 2^{\langle p-k-1 \rangle_p} \tag{4}$$

$$\left\langle \frac{1}{\hat{p}_3} \right\rangle_{p_3} = \left\langle \pm 2^{\langle p-k-1 \rangle_p} \right\rangle_{p_3} \tag{5}$$

It can be easily verified that formulas in (1) are special cases of (3) through (5), respectively.

Proof: The intention is to prove that $\left\langle \langle \hat{p}_i \rangle_{p_i} \left\langle \frac{1}{\hat{p}_i} \right\rangle_{p_i} \right\rangle_{p_i} = 1.$

– Proof of (3)

$$\langle \hat{p}_1 \rangle_{2^k} = \left\langle 2^{2p} - 1 \right\rangle_{2^k} = \langle -1 \rangle_{2^k}$$

Therefore,

$$\left\langle \langle \hat{p}_1 \rangle_{p_1} \frac{1}{\hat{p}_1} \right\rangle_{2^k} = \langle (-1)(-1) \rangle_{2^k} = 1$$

– Proof of (4)

- First Case: $k < p$
  This case implies that $\left\langle 2^k \right\rangle_{2^p-1} = 2^k$, therefore,

$$\langle \hat{p}_2 \rangle_{2^p-1} = \left\langle 2^k(2^p + 1) \right\rangle_{2^p-1} = \left\langle 2^k(1 + 1) \right\rangle_{2^p-1} = \left\langle 2^{k+1} \right\rangle_{2^p-1}$$

The inverse in this particular case is 2^{p-k-1}, hence,

$$\left\langle \langle \hat{p}_2 \rangle_{p_2} \frac{1}{\hat{p}_2} \right\rangle_{2^p-1} = \left\langle 2^{k+1} \; 2^{p-k-1} \right\rangle_{2^p-1} = \langle 2^p \rangle_{2^p-1} = 1$$

- Second Case: $k \geq p$: This case implies that $\langle 2^k \rangle_{2^p-1} = 2^{k-p}$, therefore,

$$\langle \hat{p}_2 \rangle_{2^p-1} = \langle 2^k(2^p+1) \rangle_{2^p-1} = \langle 2^{k-p}(1+1) \rangle_{2^p-1} = \langle 2^{k-n+1} \rangle_{2^p-1}$$

The multiplicative inverse in this case would be 2^{2p-k-1}, therefore,

$$\left\langle \langle \hat{p}_2 \rangle_{p_2} \tfrac{1}{\hat{p}_2} \right\rangle_{2^p-1} = \langle 2^{k-n+1} \ 2^{2p-k-1} \rangle_{2^p-1} = \langle 2^p \rangle_{2^p-1} = 1$$

The formulas of the multiplicative inverses in both cases can be unified into one formula as: $\left\langle \frac{1}{\hat{p}_2} \right\rangle_{2^p-1} = 2^{\langle p-k-1 \rangle_p}$

– Proof of (5)

- First Case: $k < p$: This case implies that $\langle 2^k \rangle_{2^p+1} = 2^k$, therefore,

$$\langle \hat{p}_3 \rangle_{2^p+1} = \langle 2^k(2^p-1) \rangle_{2^p+1} = \langle 2^k(-2) \rangle_{2^p+1} = \langle -2^{k+1} \rangle_{2^p+1}$$

The inverse in this particular case is 2^{p-k-1}, hence,

$$\left\langle \langle \hat{p}_3 \rangle_{p_3} \tfrac{1}{\hat{p}_3} \right\rangle_{2^p+1} = \langle (-2^{k+1})\,(2^{p-k-1}) \rangle_{2^p+1} = \langle -2^p \rangle_{2^p+1} = 1$$

- Second Case: $k \geq p$: This case implies that $\langle 2^k \rangle_{2^p+1} = -2^{k-p}$, hence,

$$\langle \hat{p}_3 \rangle_{2^p+1} = \langle 2^k(2^p-1) \rangle_{2^p+1} = \langle -2^{k-p}(-2) \rangle_{2^p+1} = \langle 2^{k-p+1} \rangle_{2^p+1}$$

The multiplicative inverse in this case would be -2^{2p-k-1}, therefore,

$$\left\langle \langle \hat{p}_3 \rangle_{p_3} \tfrac{1}{\hat{p}_3} \right\rangle_{2^p+1} = \langle 2^{k-p+1}\,(-2^{2p-k-1}) \rangle_{2^p+1} = \langle -2^p \rangle_{2^p+1} = 1$$

The formulas of the multiplicative inverses in both cases can be unified into one formula as: $\left\langle \frac{1}{\hat{p}_3} \right\rangle_{2^p+1} = \left\langle \pm 2^{\langle p-k-1 \rangle_p} \right\rangle_{2^p+1}$, where the plus sign hold when $k < p$, and negative sign holds otherwise.

3.2 The Moduli Set $(2^k, 2^p-1, 2^{p-1}-1)$

For the moduli set $(2^k, 2^p-1, 2^{p-1}-1)$, where p and k are positive integers, $p_1 = 2^k$, $p_2 = 2^p-1$, and $p_3 = 2^{p-1}-1$. Thus, $\hat{p}_1 = (2^p-1)(2^{p-1}-1)$, $\hat{p}_2 = 2^k(2^{p-1}-1)$, and $\hat{p}_3 = 2^k(2^p-1)$. Similarly, to avoid highly unbalanced moduli, we will assume that $p_1 \leq p_2p_3$, which implies that $0 < k \leq (2p-2)$.

Theorem 2. *For the three-moduli set* $(2^k, 2^p-1, 2^{p-1}-1)$, *the general closed-forms of the multiplicative inverses are:*

$$\left\langle \frac{1}{\hat{p}_1} \right\rangle_{p_1} = 1 + \left\lfloor \frac{k}{p} \right\rfloor 2^{p-1} + \left\lfloor \frac{k}{p+1} \right\rfloor 2^p \tag{6}$$

$$\left\langle \frac{1}{\hat{p}_2} \right\rangle_{p_2} = \left\langle -2^{\langle p-k+1 \rangle_p} \right\rangle_{2^p-1} \tag{7}$$

$$\left\langle \frac{1}{\hat{p}_3} \right\rangle_{p_3} = 2^{\langle p-k-1 \rangle_{p-1}} \tag{8}$$

It can be easily verified that formulas in (2) are special cases of (6) through (8), respectively. It is intended to prove that $\left\langle \langle \hat{p}_i \rangle_{p_i} \left\langle \frac{1}{\hat{p}_i} \right\rangle_{p_i} \right\rangle_{p_i} = 1$

– Proof of (6)

- First Case: $k \leq (p-1)$: In this case,

$$\langle \hat{p}_1 \rangle_{2^k} = \left\langle (2^p - 1)(2^{p-1} - 1) \right\rangle_{2^k} = \langle 1 \rangle_{2^k}$$

The multiplicative inverse would then be: 1, thus,

$$\left\langle \langle \hat{p}_1 \rangle_{p_1} \frac{1}{\hat{p}_1} \right\rangle_{2^k} = \langle (1)(1) \rangle_{2^k} = 1$$

- Second Case: $k = p$: In this case,

$$\langle \hat{p}_1 \rangle_{2^k} = \left\langle (2^p - 1)(2^{p-1} - 1) \right\rangle_{2^k} = \left\langle -(2^{p-1} - 1) \right\rangle_{2^k}$$

The inverse would be: $(2^{p-1} + 1)$, hence,

$$\left\langle \langle \hat{p}_1 \rangle_{p_1} \frac{1}{\hat{p}_1} \right\rangle_{2^k} = \left\langle -(2^{p-1} - 1)(2^{p-1} + 1) \right\rangle_{2^k} = 1$$

- Third Case: $p < k < (2p-1)$: In this case,

$$\langle \hat{p}_1 \rangle_{2^k} = \left\langle (2^p - 1)(2^{p-1} - 1) \right\rangle_{2^k} = \left\langle 2^{2p-1} - 2^p - 2^{p-1} + 1 \right\rangle_{2^k} = \left\langle 1 - (3)2^{p-1} \right\rangle_{2^p}$$

The multiplicative inverse would then be: $(3)2^{p-1} + 1$, therefore,

$$\left\langle \langle \hat{p}_1 \rangle_{p_1} \frac{1}{\hat{p}_1} \right\rangle_{2^k} = \left\langle (1 - (3)2^{p-1})(1 + (3)2^{p-1}) \right\rangle_{2^k} = 1$$

Thus, for all the three cases, the unified formula for the multiplicative inverse is: $\left\langle \frac{1}{\hat{p}_1} \right\rangle_{p_1} = 1 + \left\lfloor \frac{k}{p} \right\rfloor 2^{p-1} + \left\lfloor \frac{k}{p+1} \right\rfloor 2^p$

– Proof of (7)

- First Case: $k < p$: This case implies that $\left\langle 2^k \right\rangle_{2^p - 1} = 2^k$, therefore,

$$\langle \hat{p}_2 \rangle_{2^p-1} = \left\langle 2^k (2^{p-1} - 1) \right\rangle_{2^p-1} = \left\langle 2^k (-2^{p-1}) \right\rangle_{2^p-1} = \left\langle -2^{k+p-1} \right\rangle_{2^p-1} = \left\langle -2^{k-1} \right\rangle_{2^p-1}$$

The inverse in this particular case is -2^{p-k+1}, hence,

$$\left\langle \langle \hat{p}_2 \rangle_{p_2} \frac{1}{\hat{p}_2} \right\rangle_{2^p-1} = \left\langle (-2^{k-1})(-2^{p-k+1}) \right\rangle_{2^p-1} = \langle 2^p \rangle_{2^p-1} = 1$$

- Second Case: $k \geq p$: This case implies that $\left\langle 2^k \right\rangle_{2^p-1} = 2^{k-p}$, therefore,

$$\langle \hat{p}_2 \rangle_{2^p-1} = \left\langle 2^k (2^{p-1} - 1) \right\rangle_{2^p-1} = \left\langle 2^{k-p} (-2^{p-1}) \right\rangle_{2^p-1} = \left\langle -2^{k-1} \right\rangle_{2^p-1} = \left\langle -2^{k-p-1} \right\rangle_{2^p-1}$$

The inverse in this particular case is -2^{2p-k+1}, therefore,

$$\left\langle \langle \hat{p}_2 \rangle_{p_2} \frac{1}{\hat{p}_2} \right\rangle_{2^p-1} = \left\langle (-2^{k-p-1})(-2^{2p-k+1}) \right\rangle_{2^p-1} = \langle 2^p \rangle_{2^p-1} = 1$$

The formulas of the multiplicative inverses in both cases can be unified as: $\left\langle \frac{1}{\hat{p}_2} \right\rangle_{2^p-1} = \left\langle -2^{\langle p-k+1 \rangle_p} \right\rangle_{2^p-1}$.

– Proof of (8)

- First Case: $k < (p-1)$: This case implies that $\langle 2^k \rangle_{2^{p-1}-1} = 2^k$, therefore,

$$\langle \hat{p}_3 \rangle_{2^{p-1}-1} = \langle 2^k(2^p-1) \rangle_{2^{p-1}-1} = \langle 2^k(2-1) \rangle_{2^{p-1}-1} = \langle 2^k \rangle_{2^{p-1}-1}$$

The multiplicative inverse in this case would be 2^{p-k-1}, therefore,

$$\left\langle \langle \hat{p}_3 \rangle_{p_3} \frac{1}{\hat{p}_3} \right\rangle_{2^{p-1}-1} = \langle 2^k \ 2^{p-k-1} \rangle_{2^{p-1}-1} = \langle 2^{p-1} \rangle_{2^{p-1}-1} = 1$$

- Second Case: $k \geq (p-1)$: This case implies that $\langle 2^k \rangle_{2^{p-1}-1} = 2^{k-n+1}$, therefore,

$$\langle \hat{p}_3 \rangle_{2^{p-1}-1} = \langle 2^k(2^p-1) \rangle_{2^{p-1}-1} = \langle 2^{k-p+1}(2-1) \rangle_{2^{p-1}-1} = \langle 2^{k-p+1} \rangle_{2^{p-1}-1}$$

The inverse in this specific case is 2^{2p-k-2}, therefore,

$$\begin{aligned} \left\langle \langle \hat{p}_3 \rangle_{p_3} \frac{1}{\hat{p}_3} \right\rangle_{2^{p-1}-1} &= \langle 2^{k-p+1} \ 2^{2p-k-2} \rangle_{2^{p-1}-1} \\ &= \langle 2^{p-1} \rangle_{2^{p-1}-1} \\ &= 1 \end{aligned}$$

The formulas of these multiplicative inverses in both cases can be unified into one formula as: $\left\langle \frac{1}{\hat{p}_3} \right\rangle_{2^{p-1}-1} = 2^{\langle p-k-1 \rangle_{p-1}}$

4 General Frameworks for Designing Residue-Based Arithmetic Components: Case Studies

This section introduces three case studies. The first and the second cases deal with a reverse conversion and scaling for the moduli set $\{2^k, 2^p-1, 2^p+1\}$. However, the third case study deals with sign identification and magnitude comparison for the set $\{2^k, 2^p-1, 2^{p-1}-1\}$.

4.1 Case Study 1: Residue to Binary Reverse Converter for $\{2^k, 2^p-1, 2^p+1\}$

The process of converting residue number representation into binary weighted representation is realized using the CRT, or one of its modified versions. For a 3-moduli set, the CRT is defined by [2]:

$$X = \hat{m}_1 \left\langle \frac{1}{\hat{m}_1} \right\rangle_{m_1} R_1 + \hat{m}_1 \left\langle \frac{1}{\hat{m}_2} \right\rangle_{m_2} R_2 + \hat{m}_3 \left\langle \frac{1}{\hat{m}_3} \right\rangle_{m_3} R_3 - IP \qquad (9)$$

where I is the number of multiplies of P in the previous formula.

Substituting the elements of the moduli set $(2^k, 2^p - 1, 2^p + 1)$ into (9) produces:

$$X = (2^{2p} - 1)\left\langle \frac{1}{\hat{p}_1} \right\rangle_{p_1} R_1 + 2^k(2^p + 1)\left\langle \frac{1}{\hat{p}_2} \right\rangle_{p_2} R_2 + 2^k(2^n - 1)\left\langle \frac{1}{\hat{p}_3} \right\rangle_{p_3} R_3 - IP \quad (10)$$

Dividing (10) by 2^k, applying the floor value $\lfloor . \rfloor$ operator, and taking modulo $(2^{2p} - 1)$ produces:

$$X_s = \left\langle -2^{2p-k} R_1 + (2^p + 1)2^{p-k-1} R_2 + (2^p - 1)2^{p-k-1} R_3 \right\rangle_{(2^{2p}-1)} \quad (11)$$

where $X_s = \left\lfloor \frac{X}{2^k} \right\rfloor$. Considering and manipulating the three terms of (11) and recalling the three terms as μ_1, μ_2, and μ_3, where: $\mu_1 = \left\langle -R_1 2^{p+1} \right\rangle_{(2^{2p}-1)}$, $\mu_2 = \langle R_2(2^p + 1) \rangle_{(2^{2p}-1)}$, and $\mu_3 = \langle R_3(2^p - 1) \rangle_{(2^{2p}-1)}$, (11) can be rephrased to produce:

$$X_s = \left\langle (\mu_1 + \mu_2 + \mu_3)\, 2^{p-k-1} \right\rangle_{(2^{2p}-1)} \quad (12)$$

Hence, X is expressed as $X = X_s \star R_1$, where $\star$ is the process of concatenating both integers. The values of μ_1, μ_2, and μ_3 are obtained by bit manipulation of R_1, R_2, and R_3, respectively, following the same approach adopted in [15,16]. The structure needed to implement (12) is simply a carry-save adder (CSA) followed by modulo $(2^{2p} - 1)$ adder.

4.2 Case Study 2: Scaling Unit for $\{2^k, 2^p - 1, 2^p + 1\}$

Scaling is an important arithmetic component that is needed to deal with overflow whenever it happens [1,2]. Selecting the modulus $p_1 = 2^k$ to be the scaling factor for the moduli set $\{2^k, 2^p - 1, 2^p + 1\}$, then for the residue-based representation (R_1, R_2, R_3), the scaled value of this representation is given by (R_{1s}, R_{2s}, R_{3s}), where $R_{1s} = \langle X_s \rangle_{2^k}$, $R_{2s} = \langle X_s \rangle_{2^p-1}$, and $R_{3s} = \langle X_s \rangle_{2^p+1}$.

Computing $R_{1s} = \langle X_s \rangle_{2^k}$ is simply obtained by obtaining the least significant k bits of X_s. Computing $R_{2s} = \langle X_s \rangle_{2^p-1}$ is obtained by applying modulo $(2^p - 1)$ to (11), which leads to $R_{2s} = \left\langle (R_2 - R_1)2^{p-k} \right\rangle_{2^p-1}$. Similarly, $R_{3s} = \langle X_s \rangle_{2^p+1}$ is also obtained by applying modulo $(2^p + 1)$ to (11), which leads to $R_{3s} = \left\langle (R_1 - R_3)2^{p-k} \right\rangle_{2^p+1}$.

In fact, the reverse converter is used as the p_1 channel of the scaling unit to produce R_{1s}. A modulo $(2^p - 1)$ adder and a modulo $(2^p + 1)$ adder are needed to establish the p_2 and p_3 channels, and produce R_{2s} and R_{3s}, respectively.

4.3 Case Study 3: Sign Identification Unit for $\{2^k, 2^p - 1, 2^{p-1} - 1\}$

A sign identification formula was proposed in [31]. Applying the formula to the three-moduli set $\{2^k, 2^p - 1, 2^{p-1} - 1\}$ produces:

$$\phi = \left\langle \left\langle \frac{1}{\hat{p}_1} \right\rangle_{p_1} R_1 + \frac{p_1 \left\langle \frac{1}{\hat{p}_2} \right\rangle_{p_2}}{p_2} R_2 + \frac{p_1 \left\langle \frac{1}{\hat{p}_3} \right\rangle_{p_3}}{p_3} R_3 - \frac{R_1}{p_2 p_3} \right\rangle_{p_1} \quad (13)$$

where the value ϕ consists of k bits such that $\phi = \overbrace{\phi_{k-1}\phi_{k-2}\cdots\cdots\phi_2\phi_1\phi_0}^{k}$. If the bit $\phi_{k-1} = 0$, value of X is non-negative, otherwise, it is negative.
Customizing (6)–(8) for the value $(k > p)$ produces:

$$\left\langle \frac{1}{\hat{p}_1} \right\rangle_{p_1} = \langle 1 + 2^{p-1} + 2^p \rangle_{2^k}, \left\langle \frac{1}{\hat{p}_2} \right\rangle_{p_2} = \left\langle -2^{p-k+1} \right\rangle_{(2^p-1)}, \left\langle \frac{1}{\hat{p}_3} \right\rangle_{p_3} = \left\langle 2^{p-k-1} \right\rangle_{(2^{p-1}-1)}. \tag{14}$$

Substituting (14) into (13) and then simplifying the resulting equation produces:

$$\phi = \left\langle (1 + 2^{p-1} + 2^p)R_1 - \frac{2^{p+1}}{2^p - 1}R_2 + \frac{2^{p-1}}{2^{p-1} - 1}R_3 - \frac{R_1}{(2^p - 1)(2^{p-1} - 1)} \right\rangle_{2^k} \tag{15}$$

The bit-manipulation approach depicted in earlier works, such as [31], can be used to simplify and implement (15).

Moreover, magnitude comparison in RNS is carried differently from other number systems. Assuming the residue-based representations of any two variables Y and $Z \in P$ are given by (Y_1, Y_2, Y_3) and (Z_1, Z_2, Z_3), respectively, then the comparison is carried by replacing (X_1, X_2, X_3) in (15) by $(Y_1 - Z_1, Y_2 - Z_2, Y_3 - Z_3)$, respectively. If the most significant bit of ϕ resulting from (15) is 0, then $Y \geq Z$.

5 Conclusions

This work has introduced the closed-form forms of the multiplicative inverses of the expanded moduli sets $(2^k, 2^p - 1, 2^p + 1)$, and $(2^k, 2^p - 1, 2^{p-1} - 1)$, where p and k are positive integers. Using these closed-form formulas, the design of residue-based to binary-based decoders would be an easier task for all values of $0 < k \leq 2p$. This work, in fact, is a generalization of many published converters that dealt with the specific moduli sets where $k = p$ or other modifications. Additionally, the introduced multiplicative inverses are also inevitable when considering the design of other RNS-based arithmetic elements. Therefore, this paper has also introduced abstractly case studies demonstrating the utilization of these multiplicative inverses in designing different arithmetic components that perform residue decoding, scaling, sign determination and comparison.

References

1. Szabo, N., Tanaka, R.: Residue Arithmetic and Its Applications to Computer Technology. McGraw Hill, New York, NY, USA (1967)
2. Soderstrand, M., Jenkins, W., Jullien, G., Taylor, F.: Residue Number System Arithmetic: Modern Applications in Digital Signal Processing. IEEE Press, Piscataway (1986)
3. Ashur, A., Ibrahim, M., Aggoun, A.: Novel RNS structures for the moduli set $(2^n - 1, 2^n, 2^n + 1)$ and their application to digital filter implementation. Signal Process. **46**, 331–343 (1995)

4. Bhardwaj, M., Ljusanin, B.: The renaissance-a residue number system based vector co-processor for DSP dominated embedded ASICs. In: Conference Record of Thirty-Second Asilomar Conference on Signals, Systems and Computers, vol. 1, pp. 202–207 (1998)
5. Chren, W.A.: Delta-sigma modulator with large osr using the one-hot residue number system. IEEE Trans. Circuits Syst. II Analog. Digit. Signal Process. **46**(8), 1002–1008 (1999)
6. Hiasat, A., Hasan, O.: Bit-serial architecture for rank order and stack filters. Integr. VLSI J. **36**(1–2), 3–12 (2003)
7. Hiasat, A., Al-Khateeb, A.: Efficient digital sweep oscillator with extremely low sweep rates. IEE Proc. Circuits Devices Syst. **145**(6), 409–414 (1998)
8. Hiasat, A., AL-Khateeb, A.: New high-resolution digital sinusoidal oscillator structure with extremely low frequency and sensitivity. Int. J. Electron. **86**(3), 287–296 (1999)
9. Keller, T., Liew, T., Hanzo, L.: Adaptive redundant residue number system coded multicarrier modulation. IEEE J. Sel. Areas Commun. **18**(11), 2292–2301 (2000)
10. Conway, R., Nelson, J.: Improved RNS fir filter architectures. IEEE Trans. Circuits Syst. II Express Briefs **51**(1), 26–28 (2004)
11. Dutta, C., Garai, P., Sinha, A.: Design of a reconfigurable DSP processor with bit efficient residue number system. Int. J. VLSI Des. Commun. Syst. (VLSICS) **3**(5), 175–189 (2012)
12. Esmaeildoust, M., Schinianakis, D., Javashi, H., Stouraitis, T., Navi, K.: Efficient RNS implementation of elliptic curve point multiplication over {gf}(p). IEEE Trans. Very Large Scale Integr. Syst. **21**(8), 1545–1549 (2013)
13. Zheng, X., Wang, B., Zhou, C., Wei, X., Zhang, Q.: Parallel DNA arithmetic operation with one error detection based on 3-moduli set. IEEE Trans. Nano Biosci. **15**(5), 499–507 (2016)
14. Sousa, L., Antao, S., Martins, P.: Combining residue arithmetic to design efficient cryptographic circuits and systems. IEEE Circuits Syst. Mag. **16**(4), 6–32 (2016)
15. Sweidan, A., Hiasat, A.: A new efficient memoryless residue to binary converter. IEEE Trans. Circuits Syst. **35**(11), 1441–1444 (1988)
16. Wang, W., Swamy, M.N.S., Ahmad, M.O., Wang, Y.: A parallel residue-to-binary converter. In: 1999 IEEE International Conference on Acoustics, Speech, and Signal Processing. Proceedings. ICASSP99, vol. 3, pp. 1541–1544 (1999)
17. Hiasat, A., Sousa, L.: On the design of RNS inter-modulo processing units for the arithmetic-friendly moduli sets $\{2^{n+k}, 2^n - 1, 2^{n+1} - 1\}$. Comput. J. **62**(2), 292–300 (2018)
18. Hiasat, A.: A residue-to-binary converter with an adjustable structure for an extended RNS three-moduli set. J. Circuits Syst. Comput. **28**(8), 1950, 126:1–1950, 126:24 (2019)
19. Hiasat, A.: A reverse converter for three-moduli set $(2^k, 2^n - 1, 2^n + 1), k < n$. In: 2019 IEEE Jordan International Joint Conference on Electrical Engineering and Information Technology (JEEIT) (2019)
20. Pettenghi, H., Chaves, R., Sousa, L.: RNS reverse converters for moduli sets with dynamic ranges up to \$(8n+1)\$ -bit. IEEE Trans. Circuits Syst. I Regul. Pap. **60**(6), 1487–1500 (2013)
21. Hiasat, A.: An efficient reverse converter for the three-moduli set $(2^{n+1} - 1, 2^n, 2^n - 1)$. IEEE Trans. Circuits Syst. II Express Briefs **64**(8), 962–966 (2017)
22. Hiasat, A.: A residue-to-binary converter for the extended four-moduli set $\{2^n - 1, 2^n + 1, 2^{2n} + 1, 2^{2n+p}\}$. IEEE Trans. Very Large Scale Integr. (VLSI) Syst. **25**(7), 2188–2192 (2017)

23. Hiasat, A., Sweidan, A.: Residue number system to binary converter for the moduli set $\{2^n - 1, 2^n - 1, 2^n + 1\}$. J. Syst. Arch. **49**(1), 53–58 (2003)
24. Hiasat, A., Sweidan, A.: Residue-to-binary decoder for an enhanced moduli set. IEE Proc. Comput. Digit. Tech. **151**(2), 127–130 (2004)
25. Hiasat, A.: A reverse converter and sign detectors for an extended RNS five-moduli set. IEEE Trans. Circuits Syst. I Regul. Pap. **64**(1), 111–121 (2017)
26. Hiasat, A., Abdel-Aty-Zohdy, H.: Design and implementation of an RNS division algorithm. In: Proceedings 13th IEEE Symposium on Computer Arithmetic, pp. 240–249 (1997)
27. Hiasat, A.: A sign detector for a group of three-moduli sets. IEEE Trans. Comput. **65**(12), 3580–3591 (2016)
28. Hiasat, A.: Sign detector for the extended four-moduli set $\{2^n - 1, 2^n + 1, 2^{2n} + 1, 2^{n+k}\}$. IET Comput. Digit. Tech. **12**(2), 39–43 (2018)
29. Niras, C., Kong, Y.: Fast sign-detection algorithm for residue number system moduli set $\{2^n - 1, 2^n, 2^{n+1} - 1\}$. IET Comput. Digit. Tech. **10**, 54–58 (2016)
30. Hiasat, A.: RNS arithmetic multiplier for medium and large moduli. IEEE Trans. Circuits Syst. II Analog. Digit. Signal Process. **47**(9), 937–940 (2000)
31. Xu, M., Bian, Z., Yao, R.: Fast sign detection algorithm for the RNS moduli set $\{2^{n+1} - 1, 2^n - 1, 2^n\}$. IEEE Trans. Very Large Scale Integr. (VLSI) Syst. **23**(2), 379–383 (2015)
32. Kumar, S., Chang, C.: A new fast and area-efficient adder-based sign detector for RNS $2^n - 1, 2^n, 2^n + 1$. IEEE Trans. Very Large Scale Integr. (VLSI) Syst. **24**(7), 2608–2612 (2016)
33. Sousa, L., Martins, P.: Sign detection and number comparison on RNS 3-moduli sets $\{2^n - 1, 2^{n+x}, 2^n + 1\}$ and $\{2n - 1, 2n + x, 2n + 1\}$. Circuits Syst. Signal Process. **36**(3), 1224–1246 (2017)
34. Hiasat, A., Sousa, L.: Sign identifier for the enhanced three moduli set $\{2^{n+k}, 2^n - 1, 2^{n+1} - 1\}$. J. Signal Process. Syst. **91**, 953–961 (2019)
35. Tomczak, T.: Fast sign detection for RNS $(2^n - 1, 2^n, 2^n + 1)$. IEEE Trans. Circuits Syst. I Regul. Pap. **55**(6), 1502–1511 (2008)
36. Y. Ye, S.M., Hu, J.: An efficient 2^n RNS scaler for moduli set $(2^n - 1, 2^n, 2^n + 1)$. In: 2008 International Symposium on Information Science and Engineering, pp. 511–515 (2008)
37. Chang, C., Low, J.: Simple, fast, and exact RNS scaler for the three-moduli set $\{2^n - 1, 2^n, 2^n + 1\}$. IEEE Trans. Circuits Syst. I Regul. Pap. **58**(11), 2686–2697 (2011)
38. Sousa, L.: 2^n RNS scalers for extended 4-moduli sets. IEEE Trans. Comput. **64**(12), 3322–3334 (2015)
39. Hiasat, A.: Efficient RNS scalers for the extended three-moduli set $(2^n - 1, 2^{n+p}, 2^n + 1)$. IEEE Trans. Comput. **66**(7), 1253–1260 (2017)
40. Hiasat, A.: New residue number system scaler for the three-moduli set $\{2^{n+1} - 1, 2^n, 2^n - 1\}$. Computers **7**(3), 1–7 (2018)
41. Hiasat, A.: General modular adder designs for residue number system applications. IET Circuits Devices Syst. **12**(4), 424–431 (2018)
42. Hiasat, A.: A suggestion for a fast residue multiplier for a family of moduli of the form $\{2^n - (2^p \pm 1)\}$. Comput. J. **47**(1), 93–102 (2004)

The Shape of the Reliability Polynomial of a Hammock Network

Leonard Dăuş and Marilena Jianu(✉)

Department of Mathematics and Computer Science,
Technical University of Civil Engineering of Bucharest, 020396 Bucharest, Romania
{leonard.daus,marilena.jianu}@utcb.ro

Abstract. Motivated by the study of hammock (aka brick-wall) networks, we introduce in this paper the notion of **X**-path, which generates all possible connections through the network. The new concept of **X**-path, together with the Jordan curve theorem for piecewise smooth curves, allow us to provide a direct proof of duality properties for hammock networks. Afterwards, we closely link the reliability polynomial of the hammock network having length l and width w to the reliability polynomial of the dual hammock network (of length w and width l). An important consequence is that the computations required for finding the reliability polynomials of all hammock networks are reduced by half.

Keywords: Networks · Reliability polynomial · Lattice path · Jordan curve theorem

1 Introduction

The field of reliability was properly established in 1956, when von Neumann [16] and Moore & Shannon [13], respectively, published their two prescient papers. Focusing on the second one, it should be emphasized that the original purpose of Moore & Shannon was to get an understanding of the reliability of electrical circuits/networks made of unreliable individual devices/components. In order to improve the reliability of such networks, they introduced a new type of reliability enhancement scheme called hammock (or brick-wall) network. In the last few years, the interest regarding the work of Moore & Shannon has been revived (see [4,9,14,15]), not only from its theoretical perspectives, but also as it can be applied in various fields ranging from biology/medicine to engineering and even social sciences.

The problem of finding the reliability polynomial of a network belongs to the class of $\#P$-complete problems, a class of computationally equivalent counting problems (introduced by Valiant in [22]) that are at least as difficult as the NP-complete problems [2,23]. Although hammock (brick-wall) networks were proposed more than sixty years ago, no general closed-form expressions have been reported yet for their associated reliability polynomials.

I. Dzitac et al. (Eds.): ICCCC 2020, AISC 1243, pp. 93–105, 2021.
https://doi.org/10.1007/978-3-030-53651-0_8

Recently [4], the reliability polynomials have been calculated exactly for a few particular cases of small size, more precisely for the 29 hammock networks presented by Moore & Shannon in their original paper [13]. For completing this task, the authors have developed an algorithm based on a recursive depth-first traversal of a binary tree. Lately, Drăgoi et al. [9] evaluated and compared compositions of series and parallel networks of two devices and hammock networks of the same size, showing that hammock networks constantly outperform such compositions in terms of reliability.

An important breakthrough was achieved by [7] where the first and second non-zero coefficients of the reliability polynomial of any hammock have been computed exactly. The methods used to prove the formulas for these leading coefficients involved the transition matrices of certain linear transformations, lattice paths and generating functions. It is worth mentioning that the algorithm behind the method presented in [7] is straightforward and valid in general, i.e., for any directed acyclic network. In particular, the algorithm was briefly described and used for calculating the first non-zero coefficient of cylindrical hammock networks in [3] (see also Figs. 3 and 4 in that paper). Very recently [8], those two non-zero coefficients were determined exactly using combinatorial methods, and were used to approximate the reliability polynomial by a full Hermite interpolation.

There are also intimate links to electrical circuits in general, and resistor networks in particular. In 1845 Gustav R. Kirchhoff established his two well-known electric circuit laws which became fundamental pillars of electrical engineering. One very practical application of those laws is that of *calculating the equivalent resistance of a given finite network of resistors*, such resistance problems being easy to define but amazingly difficult to solve [1]. Way later, Green's function was used for computing the two-point resistances of an infinite networks [6], followed by the Laplacian matrix for finite ones [24], while the Recursion-Transform method [19]—expressed using either currents or potentials – has made significant inroads (see also [20]). On one side, neither the Green's function (normally used for infinite networks) nor the Laplacian matrix (dependent on the eigenvalues of two matrices along two orthogonal directions) methods can tackle resistor networks with complex boundaries. On the other side, using the Recursion-Transform method it was possible to obtain exact results for the equivalent resistance of hammock networks [5, 10, 21]. Still, the equivalent resistances have been determined only for perfect cases, while failing resistors as well as resistors' variations still need to be considered [17].

Finally, another aspect of interest is graph duality which directly corresponds to network duality (by adding a link from S to T), and inherently links to the duality of electrical/electronic circuits. As integrated circuits are nothing else but large networks of transistors (variable resistors), it follows that results from graph theory (including duality) should be carefully weighted when aiming to design reliability-enhanced integrated circuits.

The main goal of this paper is to propose, in Theorem 1, a direct proof of the duality properties of hammock networks. A direct consequence is that the computations required for finding the reliability polynomials of all hammock networks are reduced by half. It should be noted that, while this paper studies an applied mathematical subject, it uses, as a key tool, the Jordan curve theorem, which is a pure mathematical result (being in fact the first theorem discovered in set-theoretic topology).

The concept of brick-wall lattice path, introduced in [7], has been proved to be a useful and versatile tool in the study of hammock (brick-wall) networks. In this paper studying hammock networks, we naturally define and to use the concept of **X**-path, which generates all possible connections through the network. We refer the reader to [12] and [18] for more details about lattice paths, and to [2] for definitions and results about network reliability.

2 The Reliability Polynomial of a Network

A network is a probabilistic graph [2], $\mathbf{N} = (V, E)$, where V is the set of nodes (vertices) and E is the set of (undirected) edges. The edges can be represented as independent identically distributed random variables: each edge operates (is closed) with probability p and fails (is open) with probability $q = 1 - p$. We assume that nodes do not fail, hence the failure of the network is always a consequence of edge failures.

Let K be a subset of V containing some special nodes (called terminals). The K-reliability of the network $\mathbf{N}$ is the probability that there exists a path (a sequence of adjacent edges) made of operational (closed) edges between any pair of nodes in K. This is a polynomial in p denoted by $h_K(p)$. If $K = V$ then $h_K(p)$ is called the all-terminal reliability of the network. If the subset K contains just two nodes: S (source/input) and T (terminus/output), then $h_K(p)$ is called two-terminal reliability. This paper studies only the two-terminal reliability which will be denoted $h(p)$.

A *pathset* of the network $\mathbf{N}$ is a subset of E which contains a path between the nodes S and T. A minimal pathset (*minpath*) is a pathset P such that, if any edge e of P is removed, then $P - \{e\}$ is no longer a pathset (the nodes S and T are disconnected). We denote by $\mathcal{P}$ the set of all the pathsets of $\mathbf{N}$.

A *cutset* of the network $\mathbf{N}$ is a subset of edges $C \subset E$, such that the complementary set, $E - C$, contains no path between S and T ($E - C$ is not a pathset). A minimal cutset (*mincut*) is a cutset C such that, if any edge e of C is removed, then $C - \{e\}$ is no longer a cutset ($E - C \bigcup \{e\}$ is a pathset). We denote by $\mathcal{C}$ the set of all the cutsets of $\mathbf{N}$.

If $n = |E|$ is the size of the graph, N_i is the number of pathsets with exactly i edges, and C_i is the number of cutsets with exactly i edges, then the two-terminal reliability of the network $\mathbf{N}$ can be expressed as (see [2])

$$h(p) = \sum_{P \in \mathcal{P}} p^{|P|} q^{n-|P|} = \sum_{i=1}^{n} N_i p^i (1-p)^{n-i}, \qquad (1)$$

or, in terms of cutsets, as

$$h(p) = 1 - \sum_{C \in \mathcal{C}} q^{|C|} p^{n-|C|} = 1 - \sum_{i=1}^{n} C_i (1-p)^i p^{n-i}. \tag{2}$$

3 Hammock Networks

A hammock (brick-wall) network is formed by $w \times l$ identical devices arranged on w lines (horizontal), each line consisting of l devices connected in series. Besides these horizontal connections, there exist also vertical connections (only between adjacent horizontal lines). Out of all $(l-1)(w-1)$ possible vertical connections, half are present and the other half are absent. The vertical connections are arranged regularly in an alternate way, giving rise to the "brick-wall" pattern shown in Fig. 1.

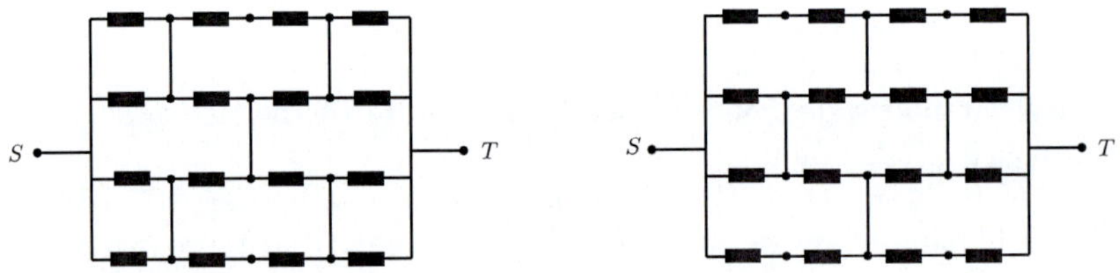

Fig. 1. Two brick-wall networks of dimensions $w = 4$, $l = 4$.

Such brick-wall networks were named hammock networks by Moore & Shannon, from their appearance when the nodes S and T are pulled apart and every vertical connection collapses into a node. In this case the rectangular "bricks" are reshaped into rhombuses. As can be seen from Fig. 2, the "hammock" representation perfectly fits the definition of a probabilistic graph, unlike the "brick-wall" representation, where all the vertical edges behave like having $p = 1$ (i.e., they are always closed).

In Fig. 2 a brick-wall network with $w = 3$, $l = 7$ and the equivalent hammock network are presented. Notice that, in order to preserve the regularity of the hammock network, the nodes S and T can be replaced by "fictitious" nodes, $S_1, S_2, \ldots, S_k$, and, respectively, $T_1, T_2, \ldots, T_h$, where $k, h \in \left\{\left\lfloor \frac{w}{2} \right\rfloor, \left\lfloor \frac{w}{2} \right\rfloor + 1\right\}$.

Definition 1. *Let $S \subset \mathbb{Z}^2$. A lattice path with steps in S is a sequence of lattice points, $v_0, v_1, \ldots, v_k \in \mathbb{Z}^2$, such that $v_i - v_{i-1} \in S$ for all $i = 1, 2, \ldots, k$.*

Definition 2. *An **X**-path is a lattice path $v_0, v_1, \ldots, v_k$ with steps in the set $S = \{(1,1), (-1,1), (1,-1), (-1,-1)\}$, such that $v_i \neq v_j$, $\forall i \neq j$. In particular, we consider an **X**-path to be the set of k edges connecting the points $v_0, v_1, \ldots, v_k$.*

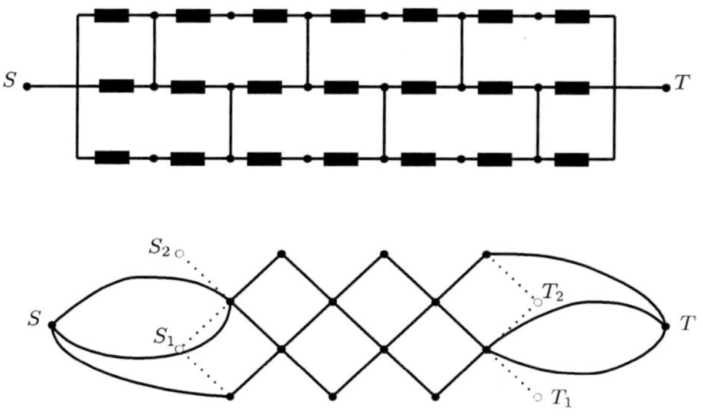

Fig. 2. Hammock network of dimensions $w = 3$, $l = 7$.

As can be understood, from a lattice point (x, y) it is allowed to move in four directions and reach one of the four neighboring points $(x + 1, y + 1)$, $(x - 1, y + 1)$, $(x+1, y-1)$ and $(x-1, y-1)$. If (x, y) is a starting point then any direction may be chosen, if not, we must take into account that $v_i \neq v_j$, $\forall i \neq j$.

We notice that the sum of the coordinates of any neighboring point has the same parity as $x + y$. We say that a lattice point (x, y) is even (odd) if $x + y$ is even (odd). An **X**-path is even (odd) if it contains even (respectively, odd) points. For example, the **X**-path represented in Fig. 3 contains only odd points.

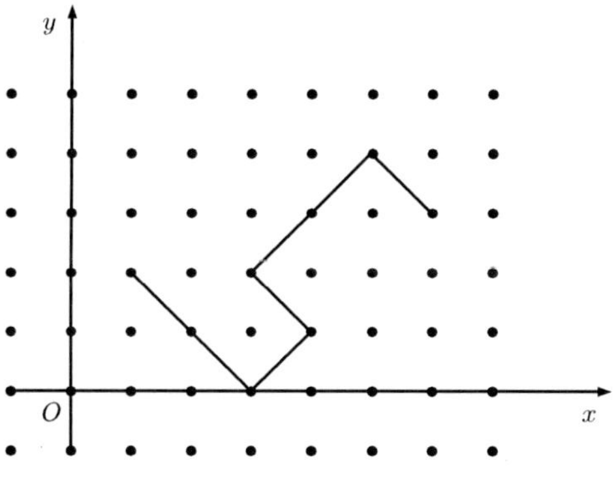

Fig. 3. The **X**-path $(1,2),(2,1),(3,0),(4,1),(3,2),(4,3),(5,4),(6,3)$

Let $\mathcal{V}_{l,w} = \{A_{x,y} = (x, y) \in \mathbb{Z}^2 : 0 \leq x \leq l,\ 0 \leq y \leq w\}$ be the set of all lattice points in the rectangle $[0, l] \times [0, w]$ and $V_{l,w}^{[1]} = \{A_{x,y} \in \mathcal{V}_{l,w} : x + y = \text{even}\}$, $V_{l,w}^{[2]} = \{A_{x,y} \in \mathcal{V}_{l,w} : x + y = \text{odd}\}$ be the subsets of even (respectively, odd) points in the rectangle $[0, l] \times [0, w]$.

We denote by $\mathcal{E}_{l,w} = \{A_{x,y}A_{x',y'} : A_{x,y}, A_{x',y'} \in \mathcal{V}_{l,w}, |x - x'| = |y - y'| = 1\}$ the set of all the edges that can be represented as line segments of length $\sqrt{2}$

connecting points of $\mathcal{V}_{l,w}$. Let $E_{l,w}^{[1]} = \{A_{x,y}A_{x',y'} \in \mathcal{E}_{l,w} : x + y = \text{even}\}$ be the subset of all even edges of $\mathcal{E}_{l,w}$, and let $E_{l,w}^{[2]} = \{A_{x,y}A_{x',y'} \in \mathcal{E}_{l,w} : x + y = \text{odd}\}$ be the subset of odd edges (the two disjoint subsets form a partition of $\mathcal{E}_{l,w}$).

A *hammock network of the first kind* of dimensions (l, w) is the probabilistic graph $H_{l,w}^{[1]} = \left(V_{l,w}^{[1]}, E_{l,w}^{[1]}\right)$, while a *hammock network of the second kind* is $H_{l,w}^{[2]} = \left(V_{l,w}^{[2]}, E_{l,w}^{[2]}\right)$. We assume that each edge is closed with probability p and open with probability $1 - p$. The input (source) nodes are $S_j = A_{0,y}$ (with y even, for the first kind, and odd for the second kind), and the output (terminus) nodes are $T_k = A_{l,z}$ (with $l + z$ even, respectively, odd).

A subset of even (respectively, odd) edges $P \subset E_{l,w}^{[i]}$ is a *pathset* in $H_{l,w}^{[i]}$ if it contains an **X**-path connecting a source node S_j with a target node T_k. Let $\mathcal{P}_{l,w}^{[i]}$ be the set of all pathsets in $H_{l,w}^{[i]}$. A subset $C \subset E_{l,w}^{[i]}$ is a *cutset* in $H_{l,w}^{[i]}$ if $E_{l,w}^{[i]} - C$ contains no **X**-path connecting a source node S_j with a terminus node T_k. Let $\mathcal{C}_{l,w}^{[i]}$ be the set of all cutsets in $H_{l,w}^{[i]}$. By using these notations in formulas (1) and (2), the reliability polynomials of hammock networks of the first and of second type, $h_{l,w}^{[1]}(p)$ and $h_{l,w}^{[2]}(p)$, can be written:

$$h_{l,w}^{[i]}(p) = \sum_{P \in \mathcal{P}_{l,w}^{[i]}} p^{|P|}(1-p)^{lw-|P|} = 1 - \sum_{C \in \mathcal{C}_{l,w}^{[i]}} (1-p)^{|C|}p^{lw-|C|}, \quad i = 1, 2 \qquad (3)$$

Remark 1. If either l or w is odd, the hammock networks $H_{l,w}^{[1]}$ and $H_{l,w}^{[2]}$ are isomorphic, and the associated reliability polynomials are identical: $h_{l,w}^{[1]} = h_{l,w}^{[2]}$. If both l and w are even, we have two different networks of dimensions (l, w): $h_{l,w}^{[1]} \neq h_{l,w}^{[2]}$.

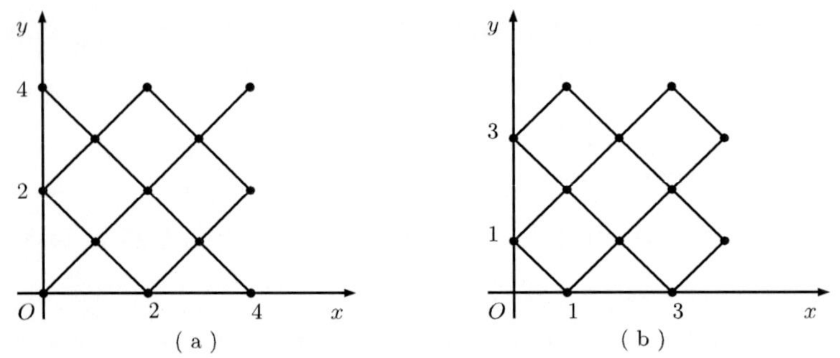

Fig. 4. Hammock networks of the first kind (a) and of the second kind (b).

For example, in Fig. 4 the hammock networks of the first and second type having dimensions $w = l = 4$ are presented.

4 Dual Networks

For every edge $e \in \mathcal{E}_{l,w}$, $e = A_{x,y}A_{x+1,y\pm1}$, we denote by $\bar{e} = A_{x+1,y}A_{x,y\pm1}$ its complementary edge (the edge that *cuts* e). It can be seen that the complementary edge of an even edge is odd, and the complementary edge of an odd edge is even. Thus, if $e \in E^{[i]}_{l,w}$, then $\bar{e} \in \overline{E^{[i]}_{l,w}} = \mathcal{E}_{l,w} - E^{[i]}_{l,w} = E^{[2/i]}_{l,w}$. By using the notation $\overline{V^{[i]}_{l,w}} = \mathcal{V}_{l,w} - V^{[i]}_{l,w} = V^{[2/i]}_{l,w}$, the dual network of $H^{[i]}_{l,w} = \left(V^{[i]}_{l,w}, E^{[i]}_{l,w}\right)$ is $\overline{H^{[i]}_{l,w}} = \left(\overline{V^{[i]}_{l,w}}, \overline{E^{[i]}_{l,w}}\right)$ with the source nodes $S'_j = A_{x,0} \in \overline{V^{[i]}_{l,w}}$ and the terminus nodes $T'_k = A_{z,w} \in \overline{V^{[i]}_{l,w}}$ (see Fig. 5). The probability of an edge $\bar{e} \in \overline{E^{[i]}_{l,w}}$ being closed is the probability of the edge $e \in E^{[i]}_{l,w}$ being open ("cut"): $q = 1 - p$.

Remark 2. Since for every edge $e \in \mathcal{E}_{l,w}$, $\bar{\bar{e}} = e$ it follows that $\overline{\overline{H^{[i]}_{l,w}}} = H^{[i]}_{l,w}$.

Remark 3. The networks $\overline{H^{[i]}_{l,w}}$ and $H^{[2/i]}_{w,l}$ are isomorphic (since they are symmetric with respect to the first bisectrix) and the reliability polynomial of the dual network $\overline{H^{[i]}_{l,w}}$ can be written

$$\overline{h^{[i]}_{l,w}}(p) = h^{[2/i]}_{w,l}(1 - p) \tag{4}$$

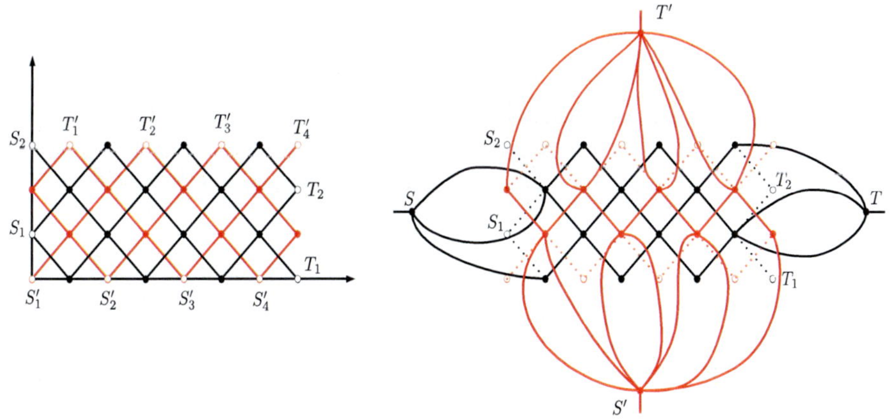

Fig. 5. Hammock dual networks.

Let $G^{[i]}_{l,w}$ be the graph obtained from $H^{[i]}_{l,w}$ by replacing the "fictitious" nodes $S_1, S_2, \ldots, S_k$ and $T_1, T_2, \ldots, T_h$ with the nodes S and T, respectively, and let $\overline{G^{[i]}_{l,w}}$ be the graph obtained from $\overline{H^{[i]}_{l,w}}$ by the same operation (the terminal nodes, in this case, are S' and T'). We notice that, if we consider the terminal nodes S and T as being placed to $\pm\infty$, then $\overline{G_{l,w}^{[i]}}$ as defined above corresponds to the definition of the dual graph of $G_{l,w}^{[i]}$. In Fig. 5 the hammock network of dimensions $w = 3$, $l = 7$ (from Fig. 2), and its dual network (red), are presented.

5 The Reliability Polynomial of a Hammock Network

The main result of this paper is represented by Theorem 1 whose corollaries detail the connection between the reliability polynomials of a hammock network and of that of its dual. The proof of this theorem relies on the *Jordan curve theorem* [11] which states that *every simple closed plane curve divides the plane into an "interior" region bounded by the curve and an "exterior" region, so that every continuous path connecting a point from one region to a point from the other intersects that curve somewhere.*

Theorem 1. *Let* $\Sigma = \{e_1, e_2, \ldots, e_n\} \subset E_{l,w}^{[i]}$ *be a subset of edges of the network* $H_{l,w}^{[i]}$ *and let* $\overline{\Sigma} = \{\bar{e}_1, \bar{e}_2, \ldots, \bar{e}_n\} \subset \overline{E_{l,w}^{[i]}}$ *be the set of complementary edges* $(i = 1, 2)$*. Then the following statements hold:*

i) If Σ *is a mincut in* $H_{l,w}^{[i]}$*, then* $\overline{\Sigma}$ *is a minpath in* $\overline{H_{l,w}^{[i]}}$*;*
ii) If Σ *is a minpath in* $H_{l,w}^{[i]}$*, then* $\overline{\Sigma}$ *is a mincut in* $\overline{H_{l,w}^{[i]}}$*.*

Proof. i) Since Σ is a mincut, for every $e_i \in \Sigma$ there exists an **X**-path which contains e_i and connects a source node (denoted by S_i) to a terminus node (denoted by T_i): $\xi_i = \sigma_i \cup e_i \cup \tau_i$, where σ_i is an **X**-path from S_i to e_i and τ_i is an **X**-path from e_i to T_i and $\sigma_i, \tau_i \subset E_{l,w}^{[i]} - \Sigma$. Let E_i, F_i be the end vertices of e_i, where E_i is reachable from the source node S_i and F_i from the target node T_i. We can see that $\sigma_i \cap \tau_j = \emptyset$ for $i \neq j$ (otherwise Σ would not be a cutset). As a consequence, $E_i \neq F_j$ for $i \neq j$. Obviously, in order to be a cutset, Σ must contain an edge with a vertex on the Ox axis. It can be proved that Σ cannot contain two such edges. Suppose $e_i, e_j \in \Sigma$ are two edges with a vertex on Ox, and e_i is closer to O then e_j. If $F_i \in Ox$, we consider the simple closed curve $\gamma = \tau_i \cup T_i A_{l+1,-1} \cup A_{l+1,-1} F_i$ (see Fig. 6), otherwise we take $\gamma = \tau_i \cup T_i A_{l+1,-1} \cup A_{l+1,-1} F_i' \cup F_i' F_i$, where F_i' is the projection of F_i on the Ox axis. We notice that E_j belongs to the interior domain region bounded by γ (otherwise, if E_j was on τ_i, we would have $\sigma_j \cap \tau_i \neq \emptyset$). Since E_j is an interior point and S_j is an exterior point of γ, it follows (by Jordan curve theorem) that the continuous path σ_j connecting the two points intersects γ somewhere, so $\sigma_j \cap \tau_i \neq \emptyset$, which is impossible (see Fig. 6). Therefore, Σ contains exactly one edge with a vertex on the Ox axis. Let e_I be this "initial" edge. Similarly, Σ contains exactly one edge with a vertex on the straight line $y = w$, and let e_F be this final edge.

We shall prove that any square with sides in $E_{l,w}^{[i]}$ has either two sides or none in Σ. Let $MNPQ$ be a square with sides in $E_{l,w}^{[i]}$ such that $MN = e_i \in \Sigma$. We know that one of the endpoints of e_i (suppose $M = E_i$) is connected to a

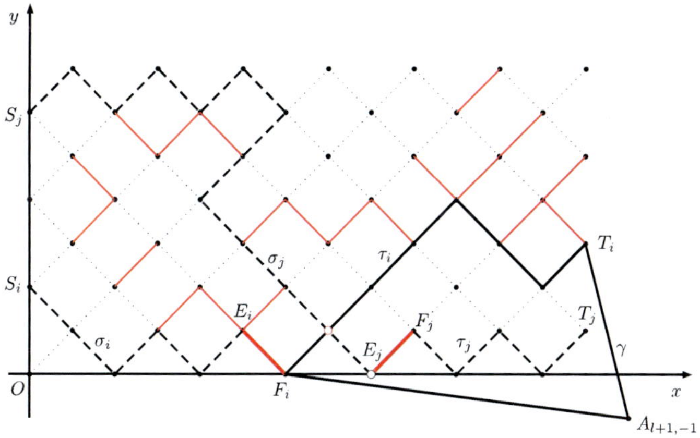

Fig. 6. A mincut Σ (red) cannot have two edges with a vertex on the Ox axis.

source node by an **X**-path σ_i, and that the other one, $N = F_i$, is connected to one of the terminus nodes by an **X**-path τ_i. If all the other sides of the square were in $E_{l,w}^{[i]} - \Sigma$, then the **X**-path $\sigma_i \cup MQ \cup QP \cup PN \cup \tau_i \subset E_{l,w}^{[i]} - \Sigma$ would connect a source node to a terminus node, so Σ would not be a cutset. On the other hand, if $MNPQ$ has at least 3 edges in Σ, $e_i = MN$, $e_j = MQ$, and $e_k = NP$, it follows that two opposite vertices are reachable from source nodes (suppose $M = E_i = E_j$, $P = E_k$) and the other two are reachable from terminus nodes ($N = F_i = F_k$, $Q = F_j$). If $\tau_i \cap \tau_j \neq \emptyset$, we denote by γ the simple loop formed by QN, τ_i and τ_j. Otherwise, $\gamma = QN \cup \tau_i \cup T_iT_j \cup \tau_j$. One of the points M and P is in the interior region bounded by γ. Let M be this point. Since M is connected to the source node S_i by σ_i and S_i is in the exterior of γ, it follows that $\sigma_i \cap \gamma \neq \emptyset$, so $\sigma_i \cap \tau_j \neq \emptyset$, which is impossible (see Fig. 7). Thus, the square $MNPQ$ has exactly two sides in Σ.

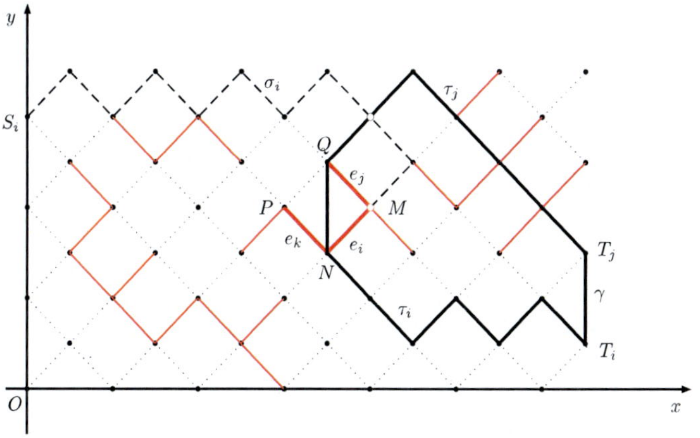

Fig. 7. A square of the network cannot have three sides in Σ.

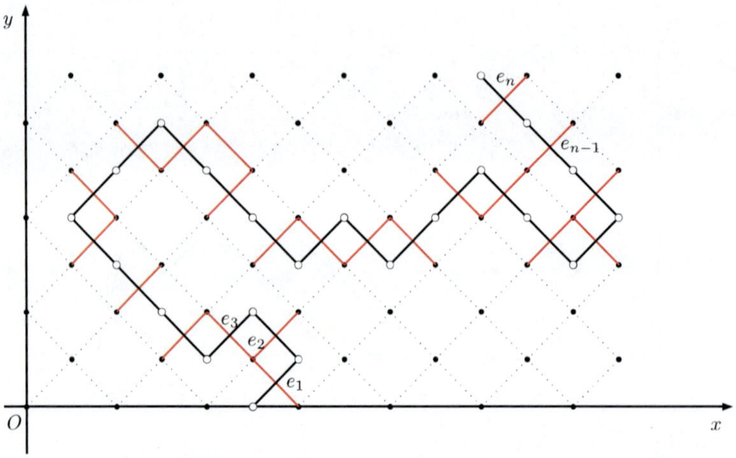

Fig. 8. If Σ is a mincut in $H_{l,w}^{[i]}$, then $\overline{\Sigma}$ is a minpath in $\overline{H_{l,w}^{[i]}}$.

We can change the indices of the edges of Σ such that $e_1 = e_I$, e_2 is the other edge in Σ (of the square with side e_1), and so on up to the final edge $e_n = e_F$. It follows that $\overline{\Sigma} = \{\bar{e}_1, \bar{e}_2, \ldots, \bar{e}_n\}$ is an **X**-path connecting a point on Ox to a point on the horizontal line $y = w$, so $\overline{\Sigma}$ is a minpath in $\overline{H_{l,w}^{[i]}}$ (see Fig. 8).

ii) Since Σ is a minpath in $H_{l,w}^{[i]}$, it follows that it is an **X**-path connecting a source node S_i (located on Oy) to a terminus node T_j (located on the vertical line $x = l$). We consider the simple closed curve $\gamma = \Sigma \cup T_j A_{l,-1} \cup A_{l,-1} A_{0,-1} \cup A_{0,-1} S_i$. If σ is an **X**-path in $\overline{H_{l,w}^{[i]}}$, connecting a source node $S_i' = A_{x,0}$ (an interior point of γ) to a target node $T_j' = A_{z,w}$ (an exterior point) then, by Jordan curve theorem, it follows that $\sigma \cap \gamma \neq \emptyset$. Hence σ contains at least one edge that "cuts" an edge of Σ (an edge of $\overline{\Sigma}$). Thus, any pathset in $\overline{H_{l,w}^{[i]}}$ contains at least one edge of $\overline{\Sigma}$, so $\overline{\Sigma}$ is a cutset. It remains to prove that $\overline{\Sigma}$ is a mincut. Suppose that some edges can be eliminated from $\overline{\Sigma}$ to obtain a mincut $\Sigma' \subset \overline{\Sigma}$. As it was shown above, it follows that $\overline{\Sigma'} \subset \Sigma$ is a minpath in $H_{l,w}^{[i]}$, which is impossible, because Σ is a minpath. □

Theorem 1 states that $\Sigma \subset E_{l,w}^{[i]}$ is a minpath in $H_{l,w}^{[i]}$ if and only if $\overline{\Sigma} \subset \overline{E_{l,w}^{[i]}}$ is a mincut in $\overline{H_{l,w}^{[i]}}$. The symmetric statement is also true, by Remark 2: Σ is a *mincut* in $H_{l,w}^{[i]}$ if and only if $\overline{\Sigma}$ is a *minpath* in $\overline{H_{l,w}^{[i]}}$. The following corollary gives a more general result, for any pathset and, respectively, cutset.

Corollary 1. *Let $\Sigma = \{e_1, e_2, \ldots, e_n\} \subset E_{l,w}^{[i]}$ be a subset of edges of the network $H_{l,w}^{[i]}$, and let $\overline{\Sigma} = \{\bar{e}_1, \bar{e}_2, \ldots, \bar{e}_n\} \subset \overline{E_{l,w}^{[i]}}$ be the set of complementary edges. Then Σ is a pathset in $H_{l,w}^{[i]}$ if and only if $\overline{\Sigma}$ is a cutset in $\overline{H_{l,w}^{[i]}}$.*

As a consequence, by Eq. (3) and Remark 3, Corollary 2 follows:

Corollary 2. *For any* $l, w \geq 1$ *and* $i = 1, 2$ *the following relation is true for all* $p \in [0, 1]$:

$$h_{l,w}^{[i]}(p) = 1 - h_{w,l}^{[2/i]}(1 - p). \tag{5}$$

By Remark 1, if either l or w is odd, then $h_{l,w}^{[1]} = h_{l,w}^{[2]} = h_{l,w}$.

Corollary 3. *If either* l *or* w *is odd, then the following relation is true for all* $p \in [0, 1]$:

$$h_{l,w}(p) = 1 - h_{w,l}(1 - p). \tag{6}$$

For $l \neq w$ this means that the plots of the reliability polynomials $h_{l,w}(p)$ and $h_{w,l}(p)$ are symmetric to one another with respect to the point $\left(\frac{1}{2}, \frac{1}{2}\right)$. For $l = w = 2k + 1$ this means that the point $\left(\frac{1}{2}, \frac{1}{2}\right)$ is a center of symmetry for the plot of the reliability polynomial $h_{2k+1,2k+1}(p)$ (see Fig. 9).

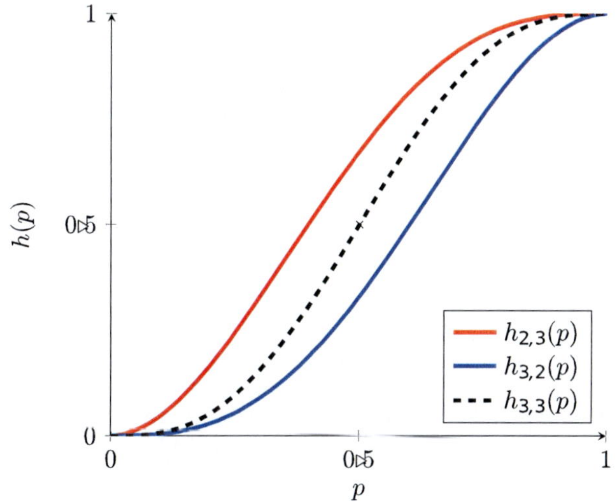

Fig. 9. Symmetry of $h_{l,w}(p)$ and $h_{w,l}(p)$ w.r.t $\left(\frac{1}{2}, \frac{1}{2}\right)$ when at least one dimension is odd; self-symmetry of $h_{2k+1,2k+1}(p)$ for $k = 1$.

Corollary 4. *Let* $h(p) = h_{l,w}^{[i]}(p)$ *be the reliability polynomial of a hammock network of dimensions* (l, w), *of either kind 1 or 2. The derivatives of* h *satisfy the following relations:*

$$h^{(k)}(0) = 0, \forall k = 0, 1, \ldots, l - 1 \tag{7}$$

$$h(1) = 1,\ h^{(k)}(1) = 0, \forall k = 1, 2, \ldots, w - 1. \tag{8}$$

Proof. Since any pathset of a hammock network has at least l edges, by Eq. (1), we obtain:

$$h(p) = \sum_{i=l}^{wl} N_i p^i (1-p)^{wl-i} = \sum_{i=l}^{wl} b_i p^i \qquad (9)$$

and relation (7) follows immediately.

Let $\overline{h}(p) = h_{w,l}^{[2/i]}(p)$ be the reliability polynomial of the dual network. Since w is the length of the dual network, it follows by (7) that $\overline{h}^{(k)}(0) = 0$, $\forall k = 0, 1, \ldots, w-1$. By Corollary 2 we have that $h(p) = 1 - \overline{h}(1-p)$, and it follows that $h(1) = 1$ and $h^{(k)}(p) = (-1)^{k+1}\overline{h}^{(k)}(1-p)$ for all $k \geq 1$. For $p = 1$ we obtain $h^{(k)}(1) = (-1)^{k+1}\overline{h}^{(k)}(0) = 0$, $\forall k = 1, 2, \ldots, w-1$. □

Acknowledgments. This research was funded by the European Union (EU) through the European Regional Development Fund (ERF) under the Competitiveness Operational Program (*BioCell-NanoART = Novel Bio-inspired Cellular Nano-Architectures*, POC-A1-A1.1.4-E nr. 30/2016).

References

1. Bouwkamp, C.J.: Solution to problem 63–14: a resistance problem. SIAM Rev. **7**(2), 286–290 (1965)
2. Colbourn, C.J.: The Combinatorics of Network Reliability. Oxford University Press, Oxford (1987)
3. Cowell, S.R., Beiu, V., Dăuş, L., Poulin, P.: On cylindrical hammock networks. In: Proceedings of the IEEE International Conference on Nanotechnology (IEEE-NANO 2017), Pittsburgh, KS, USA, 25–28 July 2017, pp. 185–188 (2017)
4. Cowell, S.R., Beiu, V., Dăuş, L., Poulin, P.: On the exact reliability enhancements of small hammock networks. IEEE Access **6**, 25411–25426 (2018)
5. Cowell, S.R., Drăgoi, V.-F., Rohatinovici, N.-C., Beiu, V.: Effective conductances of Moore-Shannon hammocks. In: Proceedings of the IEEE International Conference on Nanotechnology (IEEE-NANO 2018), Cork, Ireland, 23–26 July 2018, pp. 1–4 (2018). Article no. 8626295
6. Cserti, J.: Application of the lattice Green's function for calculating the resistance of an infinite network of resistors. Amer. J. Phys. **68**(10), 896–906 (2000)
7. Dăuş, L., Beiu, V., Cowell, S.R., Poulin, P.: Brick-wall lattice paths and applications, Technical report, arXiv (2018). http://arxiv.org/abs/222.7960
8. Dăuş, L., Jianu, M.: Full Hermite interpolation of the reliability of a hammock network. Appl. Anal. Discret. Math. **14**(1), 198–220 (2020)
9. Drăgoi, V.-F., Cowell, S.R., Beiu, V., Hoară, S., Gaşpar, P.: How reliable are compositions of series and parallel networks compared with hammocks? Int. J. Comput. Commun. Control. **13**(5), 772–791 (2018)
10. Essam, J.W., Izmailian, N.Sh., Kenna, R., Tan, Z.-Z.: Comparison of methods to determine point-to-point resistance in nearly rectangular networks with application to a hammock' network. R. Soc. Open Sci. **2**(4) (2015). 13 p. Article no. 140420
11. Hales, T.: The Jordan curve theorem, formally and informally. Am. Math. Mon. **114**(10), 882–894 (2007)

12. Humphreys, K.: A history and a survey of lattice path enumeration. J. Stat. Plan. Inference **140**, 2237–2254 (2010)
13. Moore, E.F., Shannon, C.E.: Reliable circuits using less reliable relays - Part I. J. Frankl. Inst. **262**(3), 191–208 (1956)
14. Nath, M., Ren, Y., Khorramzadeh, Y., Eubank, S.: Determining whether a class of random graphs is consistent with an observed contact network. J. Theor. Biol. **440**, 121–132 (2018)
15. Nath, M., Ren, Y., Eubank, S.: An approach to structural analysis using Moore-Shannon network reliability. In: Complex Networks and Their Applications VII. Studies in Computational Intelligence, vol. 812, pp. 537–549. Springer (2019)
16. von Neumann, J.: Probabilistic logics and the synthesis of reliable organisms from unreliable components. In: Automata Studies, pp. 43–98. Princeton University Press (1956)
17. Sevüktekin, N.C., Raginsky, M., Singer, A.C.: Linear noisy networks with stochastic components. In: Proceedings of the IEEE Conference on Decision and Control (CDC 2019), Nice, France, 11–13 December 2019, pp. 5386–5391 (2019)
18. Stanley, R.P.: Enumerative Combinatorics, vol. I, 2nd edn. Cambridge University Press, Cambridge (2012)
19. Tan, Z.-Z.: Recursion-transform approach to compute the resistance of a resistor network with an arbitrary boundary. Chin. Phys. B **24**(2) (2015). Article no. 020503, 7 p
20. Tan, Z.-Z., Tan, Z.: The basic principle of $m \times n$ resistor networks. Commun. Theor. Phys. **72**(5) (2020). Article no. 055001, 12 p
21. Tan, Z., Tan, Z.-Z., Zhou, L.: Electrical properties of an $m \times n$ hammock network. Commun. Theor. Phys. **69**(5), 610–616 (2018)
22. Valiant, L.: The complexity of computing the permanent. Theor. Comput. Sci. **8**(2), 189–201 (1979)
23. Valiant, L.: The complexity of enumeration and reliability problems. SIAM J. Comput. **8**(3), 410–421 (1979)
24. Wu, F.Y.: Theory of resistor networks: the two-point resistance. J. Phys. A Math. Gen. **37**(26), 6653–6673 (2004)

Integrated Solutions in Computer-Based Control

Queuing Theory Application on DTN Buffer Management

Corina-Ştefania Nănău(✉)

Transilvania University of Braşov, Iuliu Maniu, 50, 500091 Braşov, Romania
corina.nanau@unitbv.ro

Abstract. This paper aims to draw a parallel between the component elements of a queuing system and the buffer management used by the Delay Tolerant Network (DTN) nodes. Given that waiting in a queue is a widespread practice, many times it has been tried to optimize the time spent in such queues. The Introduction of the paper contains a briefly description for some elements of buffer management in DTN networks. The second section presents an initial information of queuing theory and several related works. The third section draws a parallel between buffer management in DTN and queuing systems by implementing a new drop policy as a part of buffer management. The obtained results will be illustrated with the help of a practical network context, using the ONE simulator.

Keywords: DTN · Queuing theory · Buffer management · Performance metrics · Priority queue · M/G/1 model · Simulation

1 Introduction

The Delay Tolerant Network (DTN) [1,2] is a communications network that supports long-term discontinuities and tolerates long delays of the data transmission. Unlike a traditional network, with a TCP/IP communication protocol that accepts a few seconds delay before being transmitted to the next hop in the path to destination node, the message transmission in a DTN network can take hours or even days, if circumstances require. These problems can occur over very long distances, such those in space, but also in smaller networks, where interference is very high and network resources are low and overloaded. Other challenges facing the DTN network are: high mobility of the nodes, intermittent connectivity, high error rate, limited buffer size, limited power capacity and so on. The notion of DTN was introduced in [3].

On the other hand, in real life situations, people are forced many times to wait in line to solve different tasks. They can stand in line at the supermarket, at a medical office, at various counters of an institution, etc. All these situations can be stochastically modeled within the mathematical framework known as queuing theory. The current situation, in which the messages in a DTN node's buffer wait in the queue to be forwarded to another node, fits perfectly to be

I. Dzitac et al. (Eds.): ICCCC 2020, AISC 1243, pp. 109–120, 2021.
https://doi.org/10.1007/978-3-030-53651-0_9

modeled in this way. Thus, we can consider each node as an individual waiting system, where the clients are the messages arriving in node's buffer, the servers are the nodes to which messages are forwarded and the waiting queue is the node buffer.

This paper aims to highlight the connection between DTN and certain elements of queuing theory. This connection is made by the buffer management in the DTN, where a queue management policy has to be implemented. Such a management policy, based on priority queues, will be simulated using the Opportunistic Network Simulator (ONE) [4]. The practical application consists in the implementation of this new technique on a couple of routing protocols that do not have a buffer management model. The simulation results will be analyzed by comparison.

2 Aspects of the Delay Tolerant Networks

In a DTN network there is no guarantee that there is an end-to-end path between nodes at a given time [5] because of the large discontinuities between the links. In this condition, the nodes have to store and forward the messages [6]. If a node is not the end destination of a message, it may store that message and forward it to another node when a connection opportunity appears. The forwarding is specified by the routing protocol and, if any, the buffer management.

There are many routing protocols specialized for different environments and they try to optimize different metric performance of the network, so it is difficult to compare them. The performance metrics for a DTN network are: message delivery rate, the average latency of the delivery, the average time spent in the buffer by the messages, the average number of copies of messages, etc.

The DTN routing protocols can be classified into single copy (like Direct Delivery) and multiple copy (like most of the routing protocols). Some of the basic multiple copy routing protocols are: Epidemic [7], Spray and Wait [8], PRoPHET [9], MaxProp [10]. A short description of them can be seen below.

Epidemic routing protocol floods the network with copies of the messages, until that messages reach the destination. Every time when two nodes meet, they change their missing messages. The main purpose of this protocol is to maximize delivery ratio.

Spray and Wait has two phases: spray phase and wait phase. During the spray phase the nodes spread into the network a finite number of message copies. When a node has one copy of a message, it enters the wait phase and it carries the message until it meets the destination. Thus, this is an extension of the Epidemic approach, in which a limitation on the number of copies is required.

PRoPHET (Probabilistic Routing Protocol) use a statistical approach that assigns to each node a delivery predictability. This predictability is calculated based on the probability of meeting between nodes. A node will forward a message only to another node with a higher delivery predictability than itself. This protocol uses historical information about the network and requires more processing. The scope of this algorithm is to minimize the delivery latency.

MaxProp is one of the few protocols that combines message routing with the management of buffer. The buffer is logically divided in two parts: messages that exceed a certain number of hops and messages that have a smaller number of hops. The messages with fewer hop counts receive a higher priority, because they are relatively new into the system and they have not traveled too far. The messages that are above the preset threshold are sorted in descending order of computed probability of being delivered. In case of buffer overload, the latter are removed.

In DTN, the selection of the next node in the message path plays a very important role in routing performance. But an efficient buffer management policy also has a major impact over network performance. This is an important challenge for DTN due to the limited space available in the nodes, but also because a part of DTN routing algorithms spread many copies of the messages, favoring buffer overload and network congestion. Thus, routing algorithms can be improved using an efficient buffer management strategy. A good buffer management can greatly increase the performance of routing protocols.

The management of the buffer is composed of two distinct policies. One of them is the Scheduling Policy, which determines the order of message delivery to the contact nodes. The other is the Drop Policy, which determines the order of the message drop from the buffer.

The main purpose of a buffer management policy is to improve the rate of message delivery over the network, but also to minimize the average delay of sending messages is a factor to be considered. The DTN networks are designed to tolerate large delays in message transmission, but if the delay is too high, there is a risk that the message may have lost relevance when it reaches the destination.

3 The Queuing Model

This paper highlights the process of buffer management in a DTN network, based on a mathematical model of queuing theory.

The proposed queuing model uses a DTN network that requires:

- a routing protocol that decides which messages may be accepted, which may be rejected and which are the nodes that will receive the stored messages
- a buffer management that organizes the buffer to avoid overloading

The routing protocols used to analyze the new buffer management technique are Epidemic protocol and PRoPHET protocol. They were chosen because they do not have a specific queuing management. Thus, they use a FIFO approach to remove messages if the queue became full. The proposed buffer management consists of a new technique of dropping messages and it is based on M/G/1 queuing model because it uses a priority queue.

3.1 Introduction in Queuing Theory

Queuing theory represents the study of queues or waiting lines, and queuing systems are simplified mathematical models to explain congestion [11]. The purpose of queuing systems study is to improve certain aspects of customers life, namely: the average time in the queue, the average time in the system, the expected queue length, the expected number of served customers, the minimization number of customers leaving the system, etc. There are several queuing models that solve different of these problems. The mathematical computation is sufficient for a simple system, but for more complex systems it is required a simulation to analyze them.

A queuing model has the following main entities:

- **clients** that are waiting in line to be served
- **servers** that provide the requested services
- **population** that is the source of potential clients, which can be finite or infinite
- in some case an **orbit**, that is a source of clients who re-interrogate the queue availability

The basic notation used for a simulation model are:

- n - the number of customers in the system
- λ - the arrival rate
- μ - the rate of service for one server
- ρ - the mean fraction of time for which server is busy (the server usage factor: $\rho = \lambda/\mu$)
- p_j - the probability of being j clients in the system in statistical steady state
- L_q - the mean number of clients waiting in the queue
- L_s - the mean number of clients waiting in the system (waiting + being served)
- W_q - the mean waiting time for a client in the queue
- W_s - the mean waiting time for a client in the system

In a queuing system, the client's arrival rate and the service rate are very important. The time between two consecutive arrivals generally occurs according to a Poisson distribution. The time required for serving a client can have a Poisson distribution, too, or a general distribution. The arriving and the serving process are statistically independent.

In the related works we can find various applications of models in queuing theory. In [12] the author presents a monitoring network for a tactical battle. There are used Markov processes to highlight the state transitions between the network nodes. The link between any two nodes can be connected or disconnected. The probability of transition from connected state into disconnected state is p_d with intensity parameter λ_d, and the probability of transition from disconnected state into connected state is p_c, with intensity parameter λ_c.

In [13], the authors create a Delay Tolerant Queue Management Model, consisting of three queues of messages. One of them have a FIFO discipline, and the others are priority queues (PQ). A stochastic M/G/1 model is applied to the system and there is a single-server system with PQs, which is feed by three client streams with Poisson distribution and three different arrival rates.

In [14] it is used a Jackson queuing network [15] and a M/M/n state independent queuing model. There is considered the mobility of the nodes and a Markov chain of state space. The purpose of [14] is to develop a new queuing model that can be used to approximate the average end-to-end delay in a DTN network.

3.2 M/G/1 Model with Priority Queues

An M/G/1 queue [16] is a queuing model where arrivals are *Markovian* (modulated by a Poisson process), service times have a *General distribution* and there is a single server. The model name is written in Kendall's notation, and it is an extension of the M/M/1 queue, where service times have to be exponentially distributed. In M/G/1, a transition from state i to state $i+1$ represents the arrival of a new client and the transition between state i to state $i-1$ represents a client being served.

We consider R to be the mean residual service time and S to be the second service moment.

In [17] there are the following formulas:

$$S = \frac{2}{\mu^2} \tag{1}$$

and

$$R = \frac{\lambda S}{2} \tag{2}$$

According to [13], the formulas for W_q and W_s are computed based on R, μ and ρ, as follows:

$$W_q = R + \frac{L_q}{\mu} \tag{3}$$

Little's formula:

$$L_q = \lambda W_q \tag{4}$$

$$(3),(4) \Rightarrow W_q = R + \frac{\lambda W_q}{\mu} \Rightarrow W_q\left(1 - \frac{\lambda}{\mu}\right) = R \tag{5}$$

$$W_q(1-\rho) = R \Rightarrow W_q = \frac{R}{1-\rho} \tag{6}$$

If it is considered $\frac{1}{\mu}$ be the message service time, then:

$$W_s = W_q + \frac{1}{\mu} = \frac{1}{\mu}(W_q\mu + 1) = \frac{1}{\mu}\left(\frac{R\mu}{1-\rho} + 1\right) \tag{7}$$

$$(2),(7) \Rightarrow W_s = \frac{1}{\mu}\left(\frac{\frac{\lambda S\mu}{2}}{1-\rho}+1\right) \tag{8}$$

$$W_s = \frac{1}{\mu}\left(\frac{\frac{\lambda}{\mu}}{1-\rho}+1\right) \tag{9}$$

$$W_s = \frac{1}{\mu}\left(\frac{\rho}{1-\rho}+1\right) \Rightarrow W_s = \frac{1}{\mu(1-\rho)} \tag{10}$$

4 New Buffer Management Approach

Evaluating such a queuing theory needs an actual space network and a complex simulation. The mathematical approach of queuing theory is materialized here by a buffer management model for a DTN network.

The goal of buffer management in DTN is to prioritize the messages from the queue. In this case, the prioritization is done through a message deletion policy. In order to achieve this, the information provided by the messages is used. Besides these, the node can take into account other network information.

4.1 The Utility Function Description

The work in this article is to complete Epidemic and PRoPHET routing methods with a message drop policy and to analyze the results. The new approach computes a utility function $f(x)$ for each message in the system. A priority queue based on M/G/1 model is used in addition to the following utility function formula:

$$f(x) = \frac{TTL}{Init\_TTL}(HC + MF)$$

- **The initial Time to Live (Init_TTL)** represents the time to live associated to the message when creating it.
- **The Time to Live (TTL)** represents the time remained until the message expires. When TTL value becomes 0, the message will be removed from the node buffer.
- **Hop Count (HC)** represents the number of nodes the message has crossed in its path to the current node. When a source node generates a message, the newly created message will have $HC = 0$. When the message is transmitted to another node, the HC value is incremented by 1.
- **My Forwarding (MF)** represents the number of message copies that the current node forwards to the contact nodes.

The rate $\frac{TTL}{Init\_TTL}$ favors the dropping of the newest messages, the opposite of SHLI (Shortest life First) [18] drop policy. The result of this rate is counterbalanced by the transmission count value of the message. The forwarding number is used by MOFO (Most Forwarded) drop policy in [18]. The proposed approach wants to balance the dropping choice for younger messages and for those who have already traveled far enough in the network.

In buffer overloading case, the messages with the higher utility are removed from the buffer. The idea is to drop the messages with a balanced value of the lowest life time and the highest forwarding number. These messages are supposed to have the least chance of reaching their destination.

In this situation, the nodes sort their buffers in ascending order of the message utility. When a new message is coming and the buffer is full, the node will remove the message with the highest utility, and so on until the new message can be buffered.

4.2 Validation Scenarios for the Utility Function

In the above formula, $HC + MF$ is denoted by TCV, which represents the Transmission Count Value. The $Init\_TTL$ is considered constant for all the messages in the system.

In case of buffer overload there are considered two potential messages to be dropped, M_1 and M_2, with the associated utility functions, $f(M_1)$ and $f(M_2)$.

$$f(M_1) = \frac{TTL_1}{Init\_TTL} TCV_1$$

$$f(M_2) = \frac{TTL_2}{Init\_TTL} TCV_2$$

In this context, the following cases are considered:

Case 1. $TTL_1 > TTL_2$ and $TCV_1 < TCV_2$

There are two possibilities:

1.1) $TTL_1 * TCV_1 > TTL_2 * TCV_2 \Rightarrow f(M_1) > f(M_2)$
In this case M_1 will be dropped. Assuming that $TTL_1 > TTL_2$, M_1 has a longer time to live and the task of transmitting it is left to other nodes.

1.2) $TTL_1 * TCV_1 < TTL_2 * TCV_2 \Rightarrow f(M_1) < f(M_2)$
In this case M_2 will be dropped. This message has already been sent to many nodes, which means that there are many copies of it in the network and it can be dropped.

Case 2. $TTL_1 = TTL_2$ and $TCV_1 < TCV_2 \Rightarrow f(M_1) < f(M_2)$

In this case M_2 will be dropped. Because $TCV_1 < TCV_2$, the results are the same as in case 1.2).

Case 3. $TTL_1 < TTL_2$ and $TCV_1 = TCV_2 \Rightarrow f(M_1) < f(M_2)$

Because $TTL_1 < TTL_2$, the message M_2 will be dropped for the same reasons as in case 1.1).

4.3 Dropping Algorithm

The following algorithm illustrates the simulation method that uses the presented utility function.

Algorithm 1 Make room for new message procedure.

```
procedure MAKEROOMFORNEWMESSAGE(buffer, msg)
    buffer.sort                    ▷ sorting is done according to utility function
    while buffer.size < msg.size do
        removeBufferTop;           ▷ the message with the higher utility is removed
    end while
end procedure
```

4.4 Simulation Results

In the proposed application, there are compared the performance metrics of the classical algorithms with the performances of the new implemented ones. These performance metrics are: delivery rate and average buffer time.

The simulation is done using Opportunistic Network Simulator [4] because it offers an easy to configure work environment regardless the requested scenario.

The chosen scenario was the one offered by the simulator, which contains three types of nodes: pedestrians, cars and trams. The nodes move on Helsinki map. The general characteristics of the system are:

- total simulation time - 12 h
- transmit speed - 250 Kbps
- transmit range - 10 m
- message size - between 500 Kb and 1 Mb
- message creation time - between 25 s and 35 s

There are 80 pedestrians with a speed between 1.8 km/h and 5.4 km/h, 40 cars with a speed between 9.7 km/h and 50 km/h and 6 trams with a speed between 25.2 km/h and 36 km/h.

The simulation is done by varying two of the system parameters: time to live(TTL) and buffer size. The TTL value is between 1 h and 9 h, with a 2 h step, and the buffer size is between 3 Mb and 9 Mb, with 1 Mb step.

By the variation of the buffer size, the obtained results are presented in Fig. 1. We can observe that the added buffer management policy brings an improvement in the message delivery rate, especially when using the PRoPHET routing algorithm (see Fig. 1b). In the case of the Epidemic routing algorithm, the values in both cases (the traditional FIFO one and the new implemented one) are approximately equal. However, there is a slight appreciation of the value when the node buffer has the size of 5Mb and a slight depreciation when the buffer has the size of 9 Mb (see Fig. 1a).

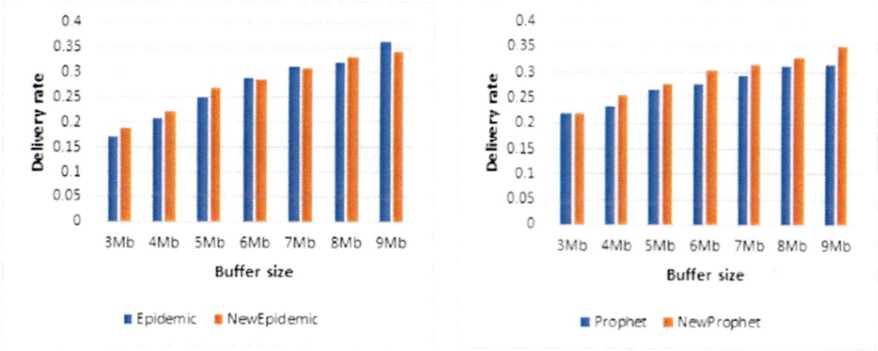

(a) Epidemic (b) PRoPHET

Fig. 1. Delivery rate efficiency with buffer size variation

By the variation of the TTL, the obtained results are presented in Fig. 2. It can be observed that when the message life time is greater or equal to 3 h, the new message drop policy manages to improve the message delivery rate for both routing algorithms: Epidemic in Fig. 2a and ProPHET in Fig. 2b.

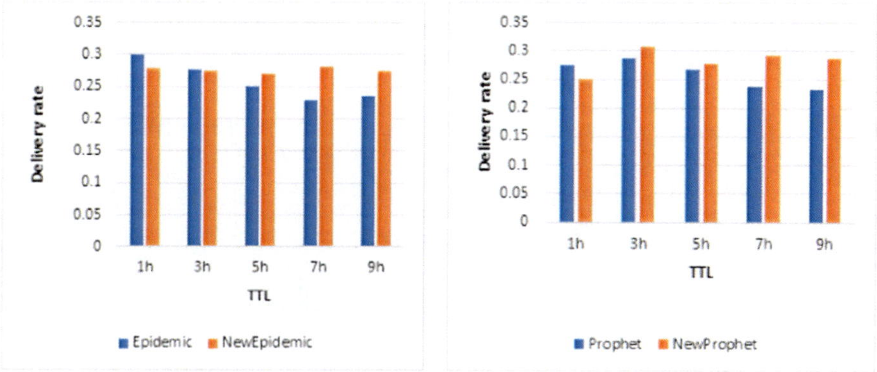

(a) Epidemic (b) PRoPHET

Fig. 2. Delivery rate efficiency with TTL variation

As you can see in Fig. 3 and Fig. 4, the buffer time average also has better values if the routing protocols have a drop message method implemented. These improved values are valid for buffer size variation and also for the TTL variation.

In these simulations, some tests were performed to check the latency of the messages. However, this performance metric had slightly lower values for routing algorithms with the new message drop policy. The lower values are due to the way of computing the utility function. It is possible to remove from the buffer a

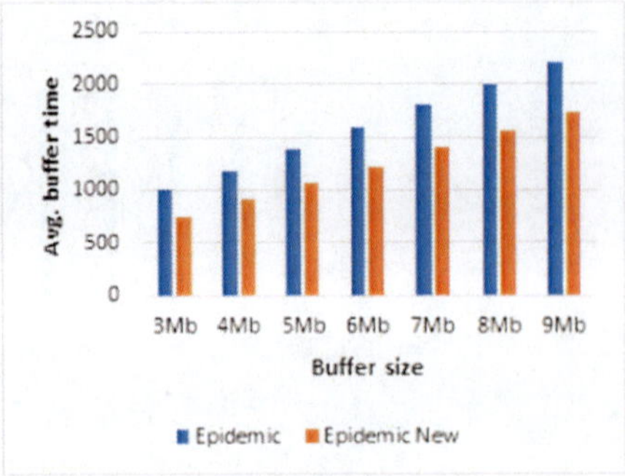

(a) Epidemic

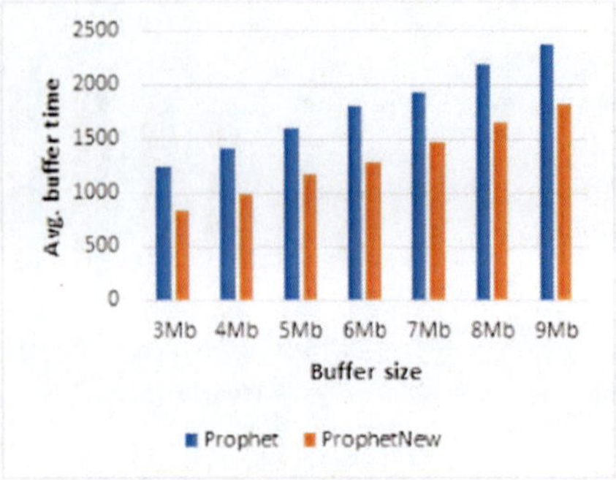

(b) PRoPHET

Fig. 3. Average buffer time with buffer size variation

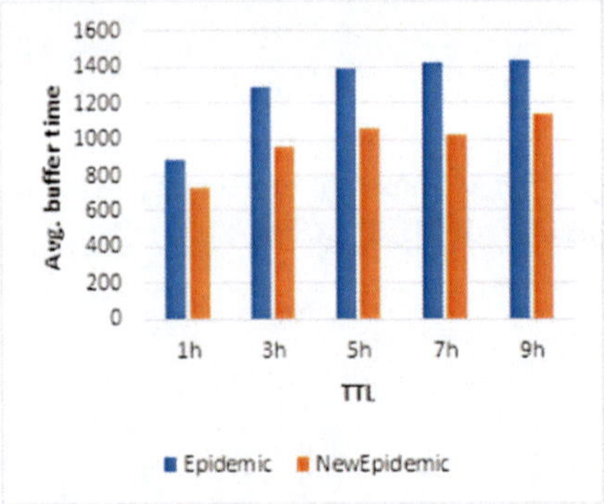

(a) Epidemic

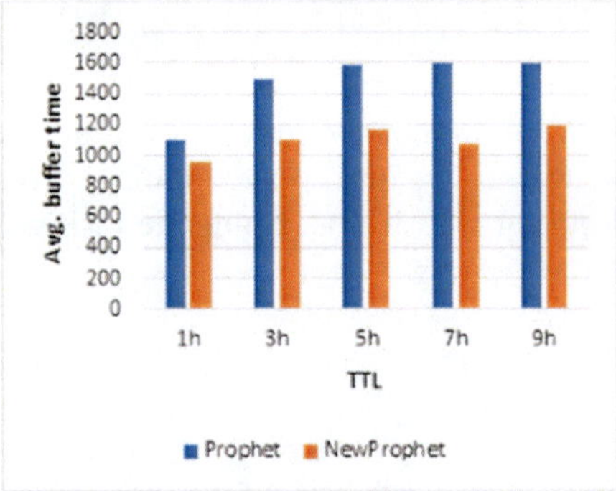

(b) PRoPHET

Fig. 4. Average buffer time with TTL variation

message with a small TTL and a big number of copies, but in the same time it can remove a few forwarded message with a big TTL number.

5 Conclusions and Future Work

The main goal of this paper was to develop a new drop policy to be paired with various traditional routing protocols and to bring an increase in the message delivery rate and in the mean time spent in the buffer by the messages. In the direction of the proposed goal, a new utility function was introduced. This function helps to create a priority queue that sorts the messages in the ascending order of the utility value. In this way, the messages with a big remaining TTL and a small forwarding number and the messages with a small remaining TTL and a high rate of forwarding are considered with high chances to have already been delivered by other nodes.

This approach was introduced into a concrete simulation framework and its results were tested over Epidemic and PRoPHET routing protocols. The improvement of these routing protocols could be seen in the figures in chapter four. The values in the figures were obtained by successive simulations.

One of the next steps of this research is to improve the presented utility function for dropping messages in order to increase the average delivery time of the messages. Another future work is to consider more real conditions of a DTN network, such as the unexpected interruption of the connection while sending a message. In this situation, the message fragmentation should be taken into account. Thus, the creation of a real life context for this application is another next step.

References

1. Cerf, V., Burleigh, S., Hooke, A., Torgerson, L., Durst, R., Scott, K., Fall, K., Weiss, H.: Delay-tolerant networking architecture. In: Network Working Group (2007). https://www.rfc-editor.org/rfc/pdfrfc/rfc4838.txt.pdf
2. Scott, K., Burleigh, S.: Bundle protocol specification. In: Network Working Group (2007). http://www.openrfc.org/rfc/5050.pdf
3. Fall, K.: A delay-tolerant network architecture for challenged internets. In: SIGCOMM, Proceedings of the 2003 Conference on Applications, Technologies, Architectures, and Protocols for Computer Communications, pp. 27–34 (2003). https://doi.org/10.1145/863955.863960
4. Keranen, A., Ott, J., Karkkainen, T.: The ONE simulator for DTN protocol evaluation. In: SIMUTools, Rome, Italy (2009). https://doi.org/10.1145/1537614.1537683
5. Conan, V., Leguay, J., Friedman, T.: Characterizing pairwise inter-contact patterns in delay tolerant networks. In: Proceedings of the 1st International Conference on Autonomic Computing and Communication Systems. ICST (2007). https://doi.org/10.5555/1365562.1365588
6. Warthman, F.: Delay-and Disruption-Tolerant Networks (DTNs). Warthman Associates, Version 3.2, 14 September 2015
7. Vahdat, A., Becker, D.: Epidemic routing for partially-connected ad hoc networks. Duke University, Technical report CS-2000-06 (2000)
8. Raghavendra, C., Spyropoulos, T., Psounis, K.: Spray and wait: an efficient routing scheme for intermittently connected mobile networks. In: Proceedings of the ACM SIGCOMM Workshop on Delay-Tolerant Networking (WDTN 2005), pp. 252–259 (2005). https://doi.org/10.1145/1080139.1080143
9. Lindgren, A., Doria, A., Schelen, O.: Probabilistic routing in intermittently connected networks. In: SIGMOBILE Mobile Computing and Communication (2003). https://doi.org/10.1145/961268.961272
10. Burgess, J., Gallagher, B., Jensen, D., Levine, B.N.: MaxProp: routing for vehicle-based disruption-tolerant networks. In: IEEE INFOCOM, pp. 1–11 (2006). https://doi.org/10.1109/INFOCOM.2006.228
11. Armero, C., Bayarri, M.J.: Queues. In: International Encyclopedia of the Social & Behavioral Sciences, pp. 12676–12680 (2001)
12. Ismailov, A.: Network monitoring in delay tolerant network, Masters thesis (2015)

13. Lenas, S.A., Dimitriou, S., Tsapeli, F., Tsaoussidis, V.: Queue-management architecture for delay tolerant networking. In: International Conference on Wired/Wireless Internet Communications, WWIC 2011, pp. 470–482 (2011). https://doi.org/10.1007/978-3-642-21560-5_39
14. Sehgal, R., Peyravy, H.: Delay tolerant networks modeling and analysis. In: Proceedings of the 30th International Conference on Computers and Their Applications, CATA 2015 (2015). https://doi.org/10.13140/RG.2.2.29407.92328
15. Jackson, J.R.: Networks of waiting lines. Oper. Res. **5**(4), 518–521 (1957). https://doi.org/10.1287/opre.5.4.518
16. https://en.wikipedia.org/wiki/M/G/1_queue#cite_note-1
17. Bertsekas, D.P., Gallager, R.: Data Networks. Prentice Hall, Upper Saddle River (1991)
18. Lindgren, A. and Phanse, K.S.: Evaluation of queuing policies and forwarding strategies for routing in intermittently connected networks. In: Proceedings IEEE COMSWARE, pp. 1–10, January 2006. https://doi.org/10.1109/COMSWA.2006.1665196

Proposal of a Multi-standard Model for Measuring Maturity Business Levels with Reference to Information Security Standards and Controls

Cristian Barria(✉), David Cordero, Lorena Galeazzi, and Alejandra Acuña

Centro de Investigación en Ciberseguridad CICS, Universidad Mayor, Santiago, Chile
{cristian.barria,lorena.galeazzi,alejandra.acuna}@umayor.cl,
david.cordero@mayor.cl

Abstract. The continuous security information risks force organizations to constantly update their security protocols. This implies, among other aspects, to base their monitoring mainly on their own maturity status in the SGSI (Information Security Managing System). When a Chief Information Security Officer elaborates a protection plan of IT assets, a wide and varied range of threats must be considered. These tasks are executed using conceptual models, which do not usually work in an integrated and systematic way. Thus, these models seek to increase maturity levels for protecting and safeguarding information security. Among the most common [1], we find COBIT 5, CSE-CMM, NIST-CRST to which we add the security standards like OWASP, ISO 27000-1, SANS. From here then, it is possible to see the lack of a multi-standard model that integrates systematically the individual actions with the expected results.

The present project proposes an integrated model that links and blends, on the one hand, the security standards and, on the other hand, the measurements of the organization's maturity levels. By doing this, it is possible to count with a set of relevant actions, classified by evaluation categories, which provide conditions for crossing regulations and standardized controls. This finally allows to explore how efficient these acquired measures are, and, when needed, the corrections that should be introduced ahead.

Keywords: Standards · Security · Organization's maturity levels · Controls

1 Introduction

Cyber attacks are currently on the rise, with great impact on emerging technologies, whether creating operation interruptions, causing sensitive data to be lost or compromised, damaging the quality of products or services, and even physical property [2]. To combat these cyber threats, there are cybersecurity controls, norms and standards, which seek to counteract or mitigate computer attacks on organizations; among others, the following should be highlighted:

I. Dzitac et al. (Eds.): ICCCC 2020, AISC 1243, pp. 121–132, 2021.
https://doi.org/10.1007/978-3-030-53651-0_10

ISO 27001 defines the concept of the Information Security Management System, and guarantees the execution of a set of processes which manage the information accessibility, developing a clear and structured work methodology based on four fundamental phases (Plan, Do, Verify and Act) according to the Deming Cycle; it applies to all types of organizations or parts of them and includes the organizational structure, policies, planning activities, responsibilities, practices, procedures, processes and resources, as well as specifying the requirements for establishing controls security related to the protection of information assets and that provide guarantees to interested parties.

The Open Web Application Security Project (OWASP), determines and combines the causes which make a software insecure, provides assistance to improve the security of different applications by spreading information about their vulnerabilities, and offering a Top 10 on the ten most critical risks. Each item describes the general probability and the factors used to classify the typical severity of the risk, and guides on how to check for problems, presenting different examples and information, and how to avoid them.

The SANS provides the guidelines to select and specify the security controls that organizations and informatics systems use when supporting companies with the processing, storage and/or transmission of information. These same security controls enable the generation of more effective information and safer risk management practices within companies, which boost an organization's effectiveness and trust that, at a long run, contributes to its same continuous improvement.

In the same way, the maturity states of organizations emerge as a complement to information security, i.e., due to their protection level of information security. These states are arranged in maturity charts, which group the areas where organizations have the most protection, and where they have the greatest deficiencies in the protection of their data. Within these charts we can find:

COBIT 5 helps IT to create value for the organization, by maintaining a balance between the production of benefits and optimization of the risk levels and the usage of its resources, building an effective governance and management framework that links a series of enabling principles which define in general terms how to help achieve the goals of a company of any size, be it commercial, non-profit or public sector.

SSE-CMM assesses and helps to improve an organization's security practices and methods by guiding it in evaluating its security practices, as well as being used as a standard mechanism for customers to evaluate a supplier's security; It also covers security practices taking into account their entire life cycle, organizational activities such as management, organization and engineering; interactions with other disciplines such as systems, software, hardware, human factors, testing, systems management, operation and maintenance, and interactions with other organizations, including acquisition, systems management, certification, accreditation, and evaluation.

NIST-CSRC is a cybersecurity risk management tool that allows evaluating the effectiveness of controls and their profitability; it adapts to different sectors and countries, making it easy to adopt in the audit processes; It can be used to generate a new cybersecurity program or as a tool to analyze the gap of existing cybersecurity programs and improve them. It is structured in such a way that it allows a comprehensive approach to cybersecurity governance, easily aligning it with business needs.

By conducting an analysis of both worlds, standards and controls vs. maturity charts, gaps arise that organizations must fill to mitigate cyber attacks [3].

The problem is given by the variety of existing security standards and maturity charts, since each of them specializes in a specific area, so when adopting them simultaneously they can demand a great effort of integration since most of them controls and models define their own scope, definitions and terminologies; In addition, its independent use prevents organizations from achieving all the benefits due to limitations in their application in certain areas. As these standards and maturity charts overlap, resources could be wasted by having different organizational departments managing different approaches independently, leading to some disadvantages due to the lack of adequate understanding, fewer resources available, and excess of required documentation. In the long run, this complicates the adoption of security standards and maturity charts, so it is necessary to build a multi-standard model that covers the crossing of standards and maturity levels. This model can become a support tool for organizations, with their information assets protected and in optimal conditions.

The model must then link the crossing of different security standards regarding their maturity levels, through which the activities required in the construction of a multi standard model are developed, providing support to the different organizations in the protection of their information assets so they can effectively approach a project of this level of importance, given the current context of organizations being based on international standards.

Regardless of the existing cybersecurity controls, norms and standards, these will require the approval and support of the management department, knowing in depth the priorities that the organization has in order to define what it wants to protect, preliminary determining the scope, times, resources and personnel where the top management of the organization is involved in the establishment, implementation, operation, monitoring, evaluation, maintenance and permanent improvement.

It is necessary to adopt a series of steps that allow compliance with the elements that are part of it, for example, in the ISO/IEC 27003: 2017 standard, which can be applied in any organization, regardless of type, size or nature. A series of phases with their respective stages are suggested, which help to incorporate different practical elements such as:

1. Define the risk management process, focusing on what is being protected in the organization
2. Define the security policy which helps to fulfill the security objectives, articulating them with the policies and within the foreseen scope, contemplating the applicable legal and regulatory requirements and taking into account the commitment of the management department to achieve them.
3. Justify and analyze information security requirements by identifying assets within the scope.
4. Assess and plan the treatment of risks by determining the cause of a potential loss in order to understand how, where and why this loss could occur, estimating vulnerabilities and confronting them against the level of risk acceptance.
5. Contemplate the documented information to be had, the implementation of the controls provided for in the risk treatment plan approved by senior management with

the resources assigned for this purpose, and the permanent monitoring of controls and new risk scenarios which arise and adjust them to the reality of the organization.

This research proposes the development of a multi-standard model that meets the needs mentioned above and consists of the following stages: Stage 1: Selection, categorization and modeling of a multi-standard structure based on information security rules and controls. Stage 2: Selection, categorization and modeling of a multilevel structure based on standards for the measurement of maturity levels based on its compatibility between functions and categories. Stage 3: Unification of the models in a macrostructure aimed at the measurement of maturity levels, crossed with the information security standards and controls that an organization requires. From this, propose an efficient multistandard model.

2 Information Security Controls

Information security controls are established with reference to the industry framework. They are created based on good practices, which are identified based on the experience of events or incidents that have been recorded. Each laboratory or cybersecurity research center evaluates the level of the threat and the measures adopted to face them. In this way, their suitability is corroborated, in relation to the standards established by the industry [4]. As a result, lists of recommended countermeasures emerge, which consider technical, administrative and operational aspects, integrated into the information system in question.

Basically the controls seek to identify and evaluate the associated risks as the case may be; that is why there is a diversity of controls to establish and measure the level of vulnerability that an organization may have [5]. The above allows establishing such controls, which are proposed by the different organizations in charge.

2.1 Crossing of Information Security Controls

With the results obtained through the compatibility matrix of the OWASP, ISO 27000-1 and SANS controls, new questions are generated which must be answered by the compatible controls, within the categories integrated in a multi standard model.

OWASP
The Open Web Application Security (OWASP) project is an open community dedicated to enabling organizations to develop, acquire, and maintain secure and reliable applications and APIS [6].

ISO 2700-1
ISO/IEC 27001 is an information security standard that specifies the necessary requirements to establish, implement, maintain and improve an information security management system (ISMS) according to the "Deming Cycle": PDCA (Plan, Do, Verify, Act), allowing the assurance, confidentiality and integrity of data and information, as well

as the systems which process it. The application of ISO-27001 means a differentiation from the rest, which improves the competitiveness and image of an organization [7].

SANS

SANS security controls are a recommended set of cyber defense actions, aimed at mitigating the most common and harmful attacks with the intention of maximizing their automation. They are aligned with the NIST cybersecurity framework and are a subset of the controls defined in publication NIST 800-53, a compendium of recommended controls for US agencies, and on which security audits of US state agencies are conducted. [8].

The defense mechanisms collected in these controls are based on the experiences of effective containment of real attacks. Actions are prioritized in order of mitigation, based on NSA score.

2.2 Compatibility Between Information Security Controls

The intersection between the norm and the id of each control, means the inclusion and compatibility of the norm in relation to the others, where:

* = It is included in the standard
+ = Compatible with comparison standards
− = Incompatible between standards

Controls are considered to be compliant, when a similarity is detected between the three standards to be compared, those controls in which 2 or less similarities are detected are considered as "Not compliant", as shown in Table 1.

Table 1. Compatibility between OWASP, ISO 2700-1 and SANS controls

Controls	Control ID	Standards			Compliance level
		OWASP	ISO 2700-1	SANS	
OWASP	A1:2017	*	+	+	Complies
	A2:2017	*	+	+	Complies
	A3:2017	*	+	+	Complies
	A4:2017	*	+	+	Complies
	A5:2017	*	+	+	Complies
	A6:2017	*	+	+	Complies
	A7:2017	*	−	+	Does not comply
	A8:2017	*	−	+	Does not comply
	A9:2017	*	+	+	Complies
	A10:2017	*	+	+	Complies
ISO 27001	ISO-C1	−	*	+	Does not comply
	ISO-C2	−	*	−	Does not comply

(*continued*)

Table 1. (*continued*)

Controls	Control ID	Standards			Compliance level
		OWASP	ISO 2700-1	SANS	
	ISO-C3	+	*	+	Complies
	ISO-C4	–	*	+	Does not comply
	ISO-C5	+	*	+	Complies
	ISO-C6	+	*	+	Complies
	ISO-C7	–	*		Does not comply
	ISO-C8	–	*	+	Does not comply
	ISO-C9	+	*	+	Complies
	ISO-C10	+	*	+	Complies
	ISO-C11	–	*	–	Does not comply
	ISO-C12	–	*	+	Does not comply
	ISO-C13	–	*	–	Does not comply
	ISO-C14	–	*	–	Does not comply
SANS	CSC1	–	+	*	Does not comply
	CSC2	–	+	*	Does not comply
	CSC3	+	+	*	Complies
	CSC4	+	+	*	Complies
	CSC5	+	+	*	Complies
	CSC6	+	+	*	Complies
	CSC7	+	+	*	Complies
	CSC8	+	+	*	Complies
	CSC9	+	+	*	Complies
	CSC10	–	–	*	Does not comply
	CSC11	+	+	*	Complies
	CSC12	–	+	*	Does not comply
	CSC13	+	+	*	Complies
	CSC14	–	+	*	Does not comply
	CSC15	+	+	*	Complies
	CSC16	–	+	*	Does not comply
	CSC17	–	–	*	Does not comply
	CSC18	–	+	*	Does not comply
	CSC19	–	+	*	Does not comply
	CSC20	–	–	*	Does not comply

3 Maturity Level

Maturity levels are a clearly defined evolutionary platform, aimed at achieving an organizational vision, related to what is defined as safe procedures. This way, each level of maturity provides a foundation layer for an organization's continuous improvement process. The maturity level standards chosen for this case study are described below [9].

Maturity Level Standards Crossing

Based on the crossing of the selected maturity level standards (COBIT5, CSM-CMM, NIST-CRST), an own model is established that takes into account the compatibility between them.

COBIT 5

Cobit 5 is a framework for understanding the state of governance and management of an organization's IT; it also evaluates the state of these technologies from support tools, where 5 maturity levels are established; These levels are: 1) meet stakeholder needs, 2) cover end-to-end organizations, 3) apply a single integrated framework, 4) enable a holistic approach, and 5) separate government from management. Finally, they are organized in one of the following categories: incomplete process, executed, managed, established, predictable and optimized [10].

SSE-CMM

The SSE-CMM (The Systems Security Engineering Capability Maturity Model) describes the essential characteristics of an organization's security engineering process, aimed at guaranteeing an optimal level of maturity in cybersecurity. It is developed with the premise that if you can guarantee the quality of the processes that the organization uses, you can guarantee the quality of the products and services generated by the processes. It has 5 maturity levels, which are: carried out informally, planned and tracked, well defined, quantitative control, and continuous improvement [11].

The SSE-CMM defines six levels of maturity. Each of these levels is considered to consist of a series of generic-based Practices that support the performance of the process areas: Level 0 - Not carried out, Level 1 - Informally carried out, Level 2 - planning and monitoring, Level 3 - well-defined description, Level 4 - Quantitatively controlled description, and Level 5 - continuous improvement.

NIST-CRSC

NIST-CSRC (National Institute of Standards and Technology) is a comprehensive exchange of information security tools and practices, providing resources for security standards and guidelines, and for identifying and linking key security web resources to support organizations. This framework is comprised of three main parts: the core framework, the framework implementation levels, and the framework profiles.

The basic framework employs five fundamental functions: Identify, Protect, Detect, Respond, Recover [12].

The Framework Implementation Tiers correspond to four maturity levels: Level 1 - Partial, Level 2 - Risks reported, Level 3 - Repeatable, Level 4 - Adaptive. Finally, the

Framework Profiles are used to describe the Current Profile and the Target Profile of certain cybersecurity activities.

Next (Table 2), the association of the standards for determination of maturity levels is shown, which makes possible the process of classification and determination of those functions, categories and subcategories which can be combined with other standards that carry out the same process.

Table 2. Compatibility between COBIT5, CSE-CMM and NIST-CRST maturity models. The acronyms correspond to the abbreviations in Spanish of each term.

Incorporation and compliance matrix of the standards best characteristics (Function)		Standard		
		COBIT 5	CSE-CMM	NIST-CRSC
Function	Plan and Organize (PO)	X	ACS, EI, ERS, EA, EV	ID, PR
	Acquire and Implement (AI)	X	ASC, CS	DE
	Deliver and Support (DS)	X	MPS, PES, ENS	RS
	Monitor and Evaluate (ME)	X	VVS	RC
	Manage Security Controls (ACS)	PO	X	ID, PR
	Assess Impact (EI)	PO	X	ID, PR
	Evaluate Security Risks (ERS)	PO	X	ID, PR
	Assess Threat (EA)	PO	X	ID, PR
	Assess Vulnerability (EV)	PO	X	ID, PR
	Construction Safety Argument (ASC)	AI	X	DE
	Coordinate Security (CS)	AI	X	DE
	Monitor Security Posture (MPS)	DS	X	RS
	Provide Security Entry (PES)	DS	X	RS
	Specify Security Needs (ENS)	DS	X	RS
	Evaluate Security Risks (ERS)	ME	X	RC
	Verify and Validate Security (VVS)	ME	X	RC
	Identify (ID)	PO	ACS, EI, ERS, EA	X
	Protect (PR)	PO	ACS, EI, ERS, EA	X
	Detect (DE)	AI	ASC	X
	Respond (RS)	DS	MPS, PES, ENS	X
	Recover (RC)	ME	ERS. VVS	X

4 Analysis to Measure Maturity Levels Regarding Information Security Standards and Controls

Next, the *Maturity Level Plan* (PNM) and the *Maturity Measurement Plan* (PMM) are developed from the previously obtained data. This way, the questions which answer the controls, categories and levels compatible between standards and controls are generated.

4.1 Proposed Security Model Cycle

From the information crossing between standards and controls, and the compatibility validation at maturity levels and controls, it is possible to generate 92 questions classified according to the security standards in evaluation categories, distributed in topics, namely: personnel-related security, access control, policy compliance, communication security and asset management, as shown in Table 3.

Table 3. Questions generated by knowledge areas

Classification	Topics	Related questions
C1	Personnel-related security	12
C2	Access control	24
C3	Policy compliance	32
C4	Communications security	12
C5	Asset management	12
	Total questions	**92**

In addition, these questions cover the classification of 63 (sixty-three) categories generated from the compatibility of standards and controls, being a conglomerate of results in an adaptive measurement towards the determination of the level of organizational maturity. The measurement of 5,796 (five thousand seven hundred and ninety-six) crossed data is necessary to determine the maturity level and the impact sectors affected in the organization. It should be noted that for each topic and associated question, the compatible controls are considered, and Table 4 shows the values to be evaluated for the PMN and PNM.

Table 4. Values to evaluate PMN and PNM

PMM	PNM	PMM * PNM matrix	C1	C2	C3	C4	C5
C1	**A1**	A1	204	408	544	204	204
12	17	A2	192	384	512	192	192
C2	**A2**	A3	60	120	160	60	60
24	16	A4	156	312	416	156	156
C3	**A3**	A5	144	288	384	144	144
32	5	**Total Crossing**	756	1.512	2.016	756	756
C4	**A4**						
12	13						
C5	**A5**						

(continued)

Table 4. (*continued*)

PMM	PNM	PMM * PNM matrix	C1	C2	C3	C4	C5
12	12						
Total PMM	**Total PNM**						
92	63						
Total data to evaluate							
5.796							

The proposed model ranges from the standards compatibility and maturity charts, to the aforementioned areas of knowledge, with the generation of questions in each of them, determining the results. These questions contain a specific number of controls related to it (standard), in this way it seeks to demonstrate the current maturity level in cybersecurity within the organization. Table 5 shows the number of controls involved.

Table 5. Crossing intervention rates

	Involved controls
PMM-C1 * PNM	537
PMM-C2 * PNM	1062
PMM-C3 * PNM	1421
PMM-C4 * PNM	527
PMM-C5 * PNM	533

These topics are oriented to the impact zones which cover both the information security pillars, defined by confidentiality, integrity, availability and authenticity, and those of organizations in the areas of systems, networks, data and users (Fig. 1).

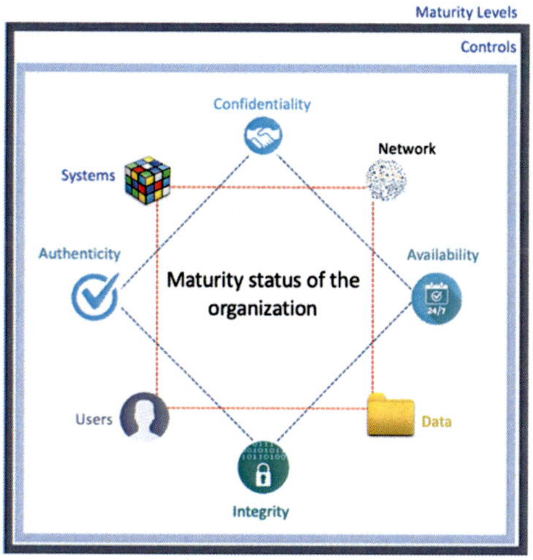

Fig. 1. Impact of controls and maturity levels on the pillars of information security and the areas of the organization.

5 Future Work

The constant study of standards, metrics and controls is essential to support the ISMS. Although this article studied the compatibility of certain standards and a matrix was generated from the compatible points (in the context of information security), it is necessary to incorporate other standards and controls, which can complement to a greater extent the approach for measuring cybersecurity in organizations, this procedure constituting a permanent cycle of continuous improvement. In addition to the integration and updating of standards and controls, it is intended to develop an API with these controls, in order to obtain the measurement results from any software that consumes said interface. Another proposal for improvement to the Multi standard Model proposed is the incorporation of associated risk levels aiming to strengthening this proposal.

6 Conclusions

The constant review, supervision and management of data security tends to a constant improvement in the processes involved. Just as cyber attacks evolve, the procedures with which we protect information must also evolve, adapt or be updated.

Throughout this article, a set of controls and metrics were presented based on 3 (three) maturity standards and 3 (three) sets of controls related to cybersecurity, which constitutes a multistandard model focused on security information. This allows the ISMS to know in a simple way what decisions it will consider based on vulnerable areas, and how to act based on the metrics obtained, always bearing in mind that each of these frameworks are "good practices" and that not all of its areas can affect an organization.

References

1. Rea, A., Sánchez, I., San Feliu, T., Calvo, J.: Maturity models in cybersecurity: a systematic review. In: 12th Iberian Conference on Information Systems and Technologies (CISTI), pp. 284–289 (2017)
2. Barría, C., Cordero, D.: Proposed classification of malware, based on obfuscation. In: IEEE 6th International Conference on Computers Communications and Control (ICCCC), Oradea, Romania (2016)
3. Chopra, A., Chaudhary, M.: Implementing an Information Security Management System. Apress, New York (2020)
4. De Haes, S., Van Grembergen, W.: Enterprise Governance of Information Technology, 2nd edn. Springer, Belgium (2015)
5. Thompson, E.C.: Building a HIPAA-Compliant Cybersecurity Program. Apress, New York (2017)
6. OWASP (Open Web Application Security Project). https://owasp.org/. Accessed 20 Jan 2020
7. ISO/IEC 27002:2013: Information technology — Security Techniques — Code of practice for information security controls, Switzerland (2013)
8. SANS (SysAdmin Audit, Networking and Security Institute). https://www.sans.org/critical-security-controls. Accessed 20 Jan 2020
9. Almuhammadi, S., Alsaleh, M.: Information security maturity model for NIST cyber security framework. In: Wyld, D.C., et al. (eds.) ITCS, SIP, CST, ARIA, NLP – 2017, pp. 51– 62. © CS & IT-CSCP (2017)
10. ISACA: Transforming Cibersecurity Using COBIT 5, ISACA, USA (2017)
11. SSE-CMM. http://www.sse-cmm.org/model.htm. Accessed 20 Jan 2020
12. CSRC-NIST. https://csrc.nist.gov/. Accessed 20 Jan 2020

IT Solutions for Big Data Processing and Analysis in the Finance and Banking Sectors

Catalin Ceaparu(✉)

Department of Economic Informatics and Cybernetics, The Bucharest University of Economic Studies, 6 Romana Square, 1st District, 010374 Bucharest, Romania
catalin.ceaparu@gmail.com

Abstract. This paper aims to give a general overview of the technologies used by two important trends in Business Intelligence nowadays, that continue to reshape the Data Architecture landscape worldwide. Bringing equally relevant value to businesses today, Fast Data and Big Data complete each other in order to enable both quick/short term as well as thorough/long term commercial strategies of companies, regardless of the industry they are part of. The body and conclusion of this paper will focus on the benefits of using the newest FinTech solutions for both aforementioned data processing models, while clearly stating the differences between the two. Both open source and proprietary type of solutions will be presented with the purpose to offer a thorough picture as to what the best architectural landscape of Big Data analytics should look like.

Keywords: Big Data analytics · Fast Data · Big data platforms

1 General Overview

Nowadays, many organizations collect, deposit and analyze data directly related to their customers and their commercial business model. The term "Big Data" is used to describe large volumes of data collection, the exponential speed they increase with, and the large variety for forms it can take.

In a world where the Digital Universe seems to be doubling its dimension every two years, the financial and banking data will be the corner stone of this growth. Therefore, it is crucial for these companies that activate in one of the fiercest competitive markets to have the ability to process all the information they have at their disposal, since it is pretty clear that the ability to analyze and leverage on data makes the difference between growing the business or going out of business.

Monetizing data, as a means of growth, is in its early days. The technical infrastructure on which the banking systems was built has started to shake under the pressure of the Big Data. That is because the amount of data turns out to be increasingly complex and more and more dynamic for the conventional processing instruments to handle.

With this in mind, the current paper will take the endeavor of providing a high-level study of a few relevant solutions for processing and analyzing large volumes of data. Moreover, this will focus on both Fast Data and Big Data together with the constraints they operate with and the benefits each of the two bring to businesses today.

I. Dzitac et al. (Eds.): ICCCC 2020, AISC 1243, pp. 133–144, 2021.
https://doi.org/10.1007/978-3-030-53651-0_11

2 Fast Data Concept Introduction

As data is collected at an ever-increasing rate and since most of these data are on the move, the benefits of processing large volumes of data to an organization can be lost if these data are not processed as soon as they become available.

To process data quickly out of large volumes requires, from infrastructure perspective, two things, namely: a system that is able to cope with developments as fast as they appear and a data repository that is able to process each item as it arrives [1].

One of the best ways to capture the value of the data received is to react at the time they arrive. Processing the data received in different batches, takes time and therefore may lead to the full or partial loss of the value these data bring.

The main goal of Fast Data is to quickly gather and analyze structured, but also unstructured data in order to sustain rapid actions (to address various problems or to increase the income of an organization).

Fast Data corresponds to the application of Big Data analytics on smaller datasets, real-time or almost real, to solve a specific problem. They play an important role in applications that require reduced latency.

Fast Data often arrives in data systems in streams. Currently the emphasis is on processing large data streams in high speed, considering that new flash drives are ready to exceed the current speed limit that is mostly limited by the performance of classic hard drive devices.

The combination of in-memory databases and new flash technologies will allow an increase in data flow processing capacity [2].

3 Big Data and Data Lake Concepts

Big Data consists of datasets whose size and structure exceed the processing capacities of traditional programs (databases, software, etc.) for the collection, storage and processing of data in a reasonable time. The data can be structured, semi-structured and unstructured, and this division makes it impossible to manage and process efficiently with traditional technology.

The criteria for determining the difference between Big Data IT and traditional IT technology are:

1. Volume – Very large volumes of data.
2. Speed – Very high data transfer rate.
3. Diversity – Partially structured data.

Four other features have been added to refine the definition of the concept: veracity, variability, value and visibility.

There are three ways to process large volumes of data [3]:

- Batch processing in pseudo-real time or soft processing when only stored data is processed.
- Real-time flow processing when the data processed is not stored and only the results of processing operations are stored.

- Hybrid processing using the hybrid model with three architectural principles: robustness (The system must be able to manage the hardware, software and human errors); Data immovability (raw data is stored forever and never changed) and recompositing (results can always be obtained by (re)-composing raw data stored) [4].

In the current architecture, the Big Data concept is usually associated with the Data Lake concept. Data Lake refers to a massive, scalable storage repository, that holds a vast amount of raw data in their native format, plus processing systems that can engage data without compromising their structure.

Data Lake is usually built to manage large and fast volumes of unstructured data (as opposed to data warehouses that contain structured data) from which future information is deducted/extracted. Data Lake uses dynamic analytic applications (not static, pre-constructed as in data warehouses). Data Lake data becomes accessible when created (again as opposed to data warehouses designed for slow data change) [5].

4 The Enterprise Architecture in the Big and Fast Data Context

Given the new context in which Fast Data are as valuable as historical data, organizations are required to adopt a new data management strategy. Traditional database architectures and systems are not able to cope with the challenges that Fast Data raises.

The Big Data architecture is centered around a Data Lake where the organization spills all its data. Data Lake is not necessarily unique because of its design and functionality, but rather its importance resides from the fact that it represents a cost-effective system in which an organization can store anything.

The development or purchase of high-performance applications that extract the benefits of Big Data and Fast Data represent a new challenge for organizations worldwide. The adoption of Fast Data and Big Data also implies important changes in the Enterprise architecture.

The Fast Data segment of the Enterprise architecture may include a quick in-memory database component. This data segment has a number of critical requirements that include the ability to ingest and interact with data streams, make decisions at every event in the

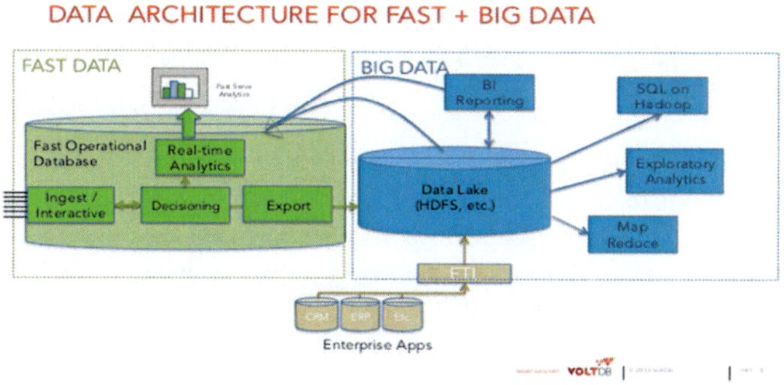

Fig. 1. Data architecture example for Fast Data and Big Data/ **Source**: Slide share website (https://www.slideshare.net/VoltDB/full-360webinar, Accessed December 2019)

stream (streams), and apply real-time analyses to ensure visibility in fast streams data received [1] (Fig. 1).

A successful Fast Data architecture must satisfy at least the following requirements:

- To ensure a well performing and reliable data acquisition and ingestion process
- To provide flexibility in storing and querying data
- To allow real-time data analysis using high-performance tools

5 Fast Data Technologies

As mentioned at the beginning of this work, in order to be able to process data quickly, we need a data flow management system (data streaming), with which data can be sent to applications that will consume or store them just as they are being generated.

Below are listed some open source and proprietary solutions, as well as the differences between them.

5.1 Fast Data – Data Flow Management Platforms (Data Streaming)

Data Flow Management platform – is a platform that publishes and subscribes to data streams (recordings) and it is similar to a message queue or to a messaging system. The platform processes data streams as soon as they are being generated and it stores them in a durable and error-tolerant way.

Apache Kafka

Kafka is an open source distributed streaming platform generally used for two large classes of applications [6]:

- Build real-time flow data pipeline that reliably sends data between systems or applications.
- Build real-time streaming apps that transform or react to data streams.

RabbitMQ

It is the most widespread and used open source messaging broker. Here is a list of some of the main features included [7]:

- High Availability – RabbitMQ has provided clustering and high availability queues for several major releases. The "Coor Quorum" version uses the "Shelf" consensus algorithm to provide data replication at a higher performance than classic HA queues.
- Performance – RabbitMQ can also process a million messages per second, but it requires more than 30 knots for that.
- Multiple subscribers – RabbitMQ can route messages to numerous queues, depending on the type of exchange (such as fanout or topic) and links between queues.

5.2 Fast Data – Data Streaming Processing Platforms

Another component needed in the fast data analysis, besides the data streaming solution, is the processing of the data received. This component can be a data processing platform or an in-memory database to process events (information) as they arrive.

Data flow processing is designed to analyze and act on data streams transmitted in real time, using continuous queries. An essential component of data flow processing is the Data Streaming Analytics, in other words, the ability to continuously apply mathematical or statistical analysis to the data flow. Flow processing solutions are designed to manage the high volume of data in real time, with a scalable, error-tolerable and a high availability architecture.

Below are listed some open source and proprietary solutions, as well as the differences between them.

Apache Storm

It is a distributed, real-time computing system, free and open source. Apache Storm is an open source framework that provides a massive and scalable collection of events. Apache Storm can be integrated with any data flow management system and any database system [8].

Apache Spark

Apache Spark is a unified processing engine for the data distributed. It has a programming model similar to MapReduce, but it extends by turning to abstract the shared data called Resilient Distributed Datasets or RDD.

RDD is a collection of read-only objects distributed via a set of machines, that can be rebuilt in case a partition is lost. The elements of an RDD need not exist as a physical storage, while a connector to RDD has enough information to calculate RDD based on the data safely stored. Moreover, this means that RDDs can always be rebuilt in case the nodes give in [9].

Spark is capable to perform in-memory computation and it also allows data memory storage which eliminates the maximum disk limitation (YARN) for iterative tasks.

Apache Spark can run using its standalone cluster mode, on EC2, on Hadoop YARN, on Mesos or on Kubernetes. It can access data from HDFS, Alluxio, Apache Cassandra, Apache HBase, Apache Hive and hundreds of other data sources [10].

IBM InfoSphere Streams

It is IBM's main product for flow processing. It provides a highly scalable event server, integration capabilities, and other typical features needed to implement flow processing use cases. IDE is based on Eclipse and it provides visual development and configuration.

5.3 In-Memory Data Bases for Fast Data Processing

VoltDB

VoltDB is a solution that aims to combine the best of both Big Data and Fast Data. It is a fast relationality-oriented base and smart flow processing platform.

VoltDB supports Big Data analysis capabilities through SQL and Java applications support, as well as ecosystem support for Hadoop-based offerings (Cloudera, MapR, Hortonworks), Apache Spark and database offerings from HPE Vertica, Teradata, IBM Netezza and others [11].

VoltDB combines ACID transactions, SQL capabilities, HA and DR clusters of traditional DBMS models with linear scaling, virtualization, and native NoSQL cloud.

Redis

Redis is an in-memory storage of data structure used as a database, cache and message broker. Redis stores all data in RAM, thus allowing for extremely fast reading and writing. It runs extremely efficiently inside the memory and manages high-speed data that requires simple standard servers to deliver millions of operations per second with a latency of under milliseconds. Redis does not have a schema, but when using one of its data structures (such as HASH or sorted sets), users can take advantage of memory operations to accelerate data processing [12].

6 Big Data – Technologies and Platforms

Big Data represents what we call 'data-at-rest' and there are large volumes of stored data to be processed. 'Data-at-rest' provides historical context, this context allows for complex data analysis (prescriptive, diagnostic, descriptive, predictive and outcome analytics), which in combination with the response received from the rapid data analysis further allows rapid decision making based on a broader historical setting, instead of the restrictive view that looking at what is happening at the present time gives.

A Big Data platform is an ecosystem of services and technologies, which allows the analysis of a big, complex and dynamic amount of data.

Below we briefly present some solutions that exist on the market and that can be taken into account when planning the new architecture:

6.1 Apache Hadoop

In most Big Data scenarios, Apache Hadoop has become the mark to beat when it comes to sharing and accessing the data and the computing input.

Apache Hadoop is a working frame that allows splitting the working processes on many hosting servers that are not necessarily high-performance computers. It has two main components: a MapReduce execution engine and a distributed files system called HDFS (Hadoop Distributed File System). The advantages behind Hadoop are: high flexibility, scalability, low costs and reliability in managing and processing Big Data in structured and unstructured Big Data sets.

Hadoop has evolved into YARN – Yet Another Resource Negotiator, a frame for tasks planning and cluster for resources managing [13].

6.2 Hortonworks Hadoop Distribution

HDP/Hortonworks Data Platform is an open source solution that offers immediate value like storage costs reduction by integrating YARN to its data center. In addition, this ensures the company's Data Center cost optimization by the YARN taking over less critical tasks such as the ETL process. YARN allows HDP platform to integrate all data processing engines both in the commercial version, as well as the open-source ones, to be able to constantly deliver services and resources in the entire platform. Ambari is the platforms Web interface and it is meant to ensure a simple, consistent and safe HDP management [14].

Apache Ambari is an example of the Hadoop Cluster Management console. Its' objects are the provisioning, managing and monitoring of such a cluster. Some of the companies that use the Hortonworks platform are Samsung, Spotify, Bloomberg and eBay.

7 Using Open Source or Proprietary Products? What's the Best Way to Choose?

Unfortunately, there is currently no "out of the box" solution on the market to combine the two concepts (Fast Data & Big Data). Such a solution can be built using the products presented above, while keeping in mind that choosing the most suitable components should consider a series of aspects, such as those detailed below.

Open source products can be tested for free, have the advantage of providing free access to the source code, free support through the entire community that helps develop them and, additionally, is available to answer questions about it. Moreover, these products have fewer defects which are fixed quicker, have better security and do not create the dependency relationship between the customer and the vendor [15].

Of course, the use of these products is not risk-free considering. We need to keep in mind that free support is not always as quickly available as needed and the fact that these are complex software packages that only highly skilled resources within organizations can handle.

In contrast to open source products, proprietary ones are much more stable, come with dedicated support and can be used by people who do not need special qualifications to handle them. Of course, they do not allow access to the source code and create dependence on the vendor given that it is hard to replace a product already implemented in the organization [15].

In order to be able to make the right choice, when choosing the product, characteristics such as those in the table below should be considered (Table 1).

These characteristics can also be used for the selection of a proprietary product, with the mention that for those we should consider the price component as well as the relationship with the vendor (Table 2).

After the selection of components has been carried out, you can move on with the implementation part. This can be done by creating an evolving prototype, further more improved by a series of initial version refinements until a minimum viable product is reached.

Table 1 Criteria and characteristics for open source products/Source: https://www.researchgate.net/publication/276008627_The_Selection_Criteria_of_Open_Source_Software_Adoption_in_Malaysia

Criteria	Characteristic	Descriptions (Related question)
Reliability	Maturity	Used to indicate the software maturity in the market. (Is the software new on the market?)
	Popularity	Used to indicate the software popularity and available references in the market. (Does this software have numerous users? Any books/website/forum/blog written about this software available in market?)
	Availability	Used to measure the system quality of support available for the software. (Does this software frequently release new version?)
Usability	Learnability	Used to indicate the level of the system learnability. (How easy is it to learn or understand the software without using user manual?)
	Operability	Used to measure the system operability. (Is the software easy to operate/handle?)
	Accessibility	Is this software easy to access without other third-party software or plug-in?
	User interface aesthetics	Is the user interface suitable with this software functionality?
Performance efficiency	Time behavior	Is this software easy to install/configure and operate within short time?
	Resource utilization	Does this software use minimal/limited resources or can be used with existing resources (e.g.: server, operating system)?
Functionality	Functional completeness	Does the software meet user's expectation and requirement?
	Functional correctness	Does the software provide correct output as user's expectation?
	Functional appropriateness	Does the software function appropriately?
Maintainability	Modularity	Used to indicate the quality of the source code. (Does the code structural and readable? How well is the software designed?)
	Modifiability	How easy the system can be customized to meet user's requirement?
	Reusability	How easy to reuse or extent the code for further extension or integration?
	Testability	Is the software error-free?
Security	Confidentiality	How secure is the data and the software? How confidence that software is free from vulnerabilities?
	Integrity	Does the software have any control mechanism to ensure system integrity?
	Authenticity	Does the software provide level of user's authentication?
Tangible	Support	Is there any community or commercial support provided?
	Documentation	Complete documentation provided? Both technical and user manual?
Reliability	Version	Used to measure if the community has developed clear thoughts and plans about features will be added in the future. (Does software version release as targeted or expected time with mainly new functionality?)
Responsiveness	Community	Used to indicate the responsiveness of the community. How active is the community for the software?
Assurance	Competence	Does the community possess of required skill and knowledge?
	Credibility	Does the development team and community have performed good track record? How many bugs were fixed in last 6 month?
Empathy	Communication	Does the community acknowledge your problems and help in solving it?
Competence	Skill	How many internal technical staff skilled with tools and language used by this software?

Table 2. Criteria and characteristics for proprietary products/Source: https://www.cloudave.com/25571/how-to-select-enterprise-collaboration-vendors/

Characteristic	Descriptions (related question)
Price	How much does the product/platform cost and how do they charge
People	How does the vendor treat you (other customers)?
Support and Maintenance	What options does the vendor offer for supporting your development and maintaining it?
Vertical Expertise	Does the vendor have expertise and clients in your specific vertical?

8 Use Case: Using the New Fast and Big Data Architecture to Identify Fraud in the Banking Financial System

One of the biggest challenges the banking financial system currently faces is to identify and fight in real-time against fraudulent transactions. Although important steps have been made to identify these transactions in due time, the losses incurred by them remain significant.

In 2018, for example, the UK banking system managed to prevent losses of £1.66 billion, but at the same time lose £1.20 billion [16].

In order to identify possible fraudulent transactions, analysis systems are needed to facilitate real-time reactions. Most of the times these systems must be able to react within a few seconds, a very narrow timeframe window available between the customer's transaction initiation and its approval by the financial institution.

As per the statistics presented above, not all fraudulent transactions can be identified at the time they occur, but fraud patterns can also be discovered by further investigation of completed transactions.

Therefore, we continue by presenting the new architecture, in which the two concepts of Fast Data and Big Data, that complement each other, can respond to the requirement of identifying and stopping frauds.

Real-time identification of fraud can be done using the Fast Data concept. The real-time analysis of transactions allows an immediate response that protects both the interests of the client and those of the financial institution.

We will detail below what are the technologies that can be used to achieve this.

Transactions initiated by customers, whether made online, in store, ATM payments or cash deposits or transfers between their own accounts, can be seen as applications that financial institutions must approve after checking them beforehand.

For this safety-proof process to be carried out so that the customer is satisfied by the fast services provided by the financial institution, it is mandatory for the technologies used for the applications analysis and approval to ensure real-time data checks and response.

The Fast Data concept is based on fast collection and processing of data at the time when they are produced. Therefore, the moment the application is initiated by the client, it is collected by the Apache Kafka streaming platform. This is capable of transmitting the application at the time it is initiated, and instantly sent it to the processing platform.

For the quick processing of these applications, the most suitable solution is the use of a fast-operational database (in memory). Processing must take place in real time and should not take more than a few seconds to allow sending quickly the answer to the customer. For this, one of the existing solutions on the market such as VoltBd, SAP or MemSql, can be used.

When a new transaction is initiated, it must be checked from several points of view before it is approved. One of the dimensions of the analysis is the customer-vendor relationship that verifies previous transaction history, using data provided by the Big Data component, as well as analysis of customer status such as geolocation analysis.

To perform such checks, the quick operational base must support the simultaneous application of a set of queries (over 50 complex queries) on the data stream.

Moreover, with the help of these fast-operational bases, live Datamarts can be built, thus allowing the ad-hoc query and visualizing of data streams by users from fraud prevention departments. Data visualization can be done using existing BI Reporting solutions in the Big Data component or using integrated solutions in the processing platform.

Once the data flow analyzed, the transaction can be approved or denied, and the result of the analysis is stored in the Big Data component, more precisely it becomes "date at rest" in Data Lake.

Of course, not all fraudulent transactions can be identified in real-time, a fact which is evidenced by the figures mentioned above. Once stored, transactional data may be subjected to further analysis to discover new patterns of fraud that have not been revealed at the time when the transactions took place.

For this type of analysis, one can use the Big Data component that allows the analysis historical datasets.

The Big Data component not only offers a historical analysis of transactions, but also a complex historical analysis if we are to consider the multitude of sources that fuel it. In order to analyze data at rest to identify new fraud patterns, relational (structured) data received from the internal systems are not enough. This data must also be linked and analyzes in the context of the information coming from external sources, information that can be stored and transmitted in an unstructured model. For both structured and unstructured data storing and analyzing, the most suitable enterprise solution is the one offered by Big Data Hortonworks Platform. It allows the analysis of large and complex data volumes.

The dataset can also be filled in with information from legacy systems, systems that in the new architecture vision are integrated into the concept of Big Data/Data Lake. The Hortonworks platform allows for quick integration with the already existing infrastructure and systems inherited, thus facilitating the required access to data for analysis.

The traditional method of detecting fraud schemes, whereby data analysts run SQL queries upon a data Warehouse containing a huge amount of information, is no longer enough due to the high response time required by such an analysis. The more time it takes the fraud to be detected, the bigger the damage associated to the fraud.

Big Data platform allows us to easily discover fraud patterns with Machine Learning and Predictive analysis, and with the help of graphical interfaces, users can conduct additional investigations for statistics or to consolidate evidences.

The use of two methods of analysis, of moving data and data at rest can significantly reduce the percentage of fraud that affects financial institutions globally. The two concepts are complementary in identifying fraud in the banking financial system, given that real-time data analysis is correlated with scenarios and patterns discovered from the analysis of historical sets (Fig. 2).

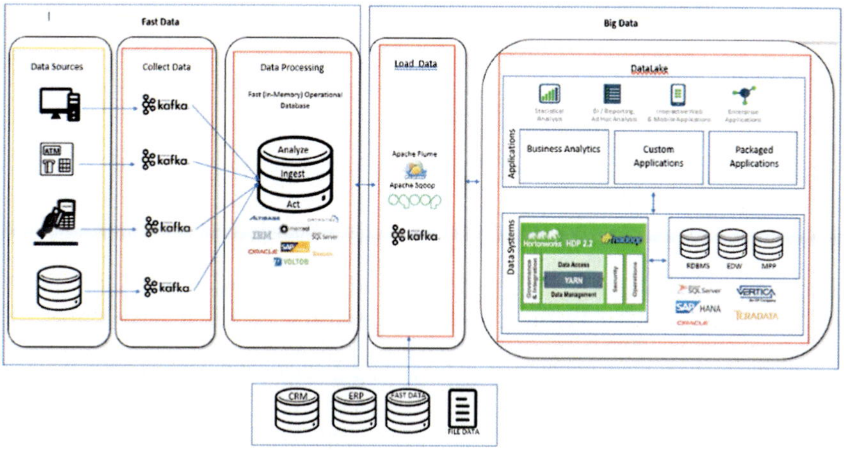

Fig. 2. Fast & Big Data architecture (for fraud analysis)/ Source: architectural proposal of the author for the presented use case

9 Conclusions

Given the high volume of data generated nowadays and the fact that its crucial for financial institutions to respond to the needs of the market by analyzing these large volumes of data, it becomes more and more obvious that both fast data analysis combined with historical data analysis is required in order for these companies to be able to benefit from the full data value. Although, at the present time, the architecture of these institutions does not have a clear focus on integrating these 2 concepts (fast data and big data), the status quo will definitely change in the near future if these institutions want to remain competitive in the market.

The diversity of the current technologies allows and, moreover, encourages the exploration of both concepts for the benefit of commercial performance and profit maximization. Depending on the strategy that each financing institution chooses to go for, there is always a choice to be made between open source and proprietary solutions.

Each of the solutions presented in this document tries to respond to the business needs and challenges of handling large amounts of data. It is recommended that the actual implementation of any of them is to be done after a careful analysis of all implementation and maintenance related costs, and only after multiple reviews with all stakeholders within the organization of the risks and opportunities that the implementation of such solutions entails.

In the end, as the tendencies seem to indicate, one could say that the future will most likely belong to open source solutions, those where multiple users contribute to keeping it up to date, stable and flexible to answer quickly to changing business needs.

References

1. Jarr, S.: Fast Data and the New Enterprise Data Architecture, 1st edn. O'Reilly Media, Sebastopol (2014)
2. Maas, G., Kontopoulos, S., Glover, S.: Designing Fast Data Application Architectures, 1st edn. O'Reilly Media, Sebastopol (2018)
3. Hornbeck, R.L.: Batch Versus Streaming: Differentiating Between Tactical and Strategic Big Data Analytics (2013). http://datatactics.blogspot.ru/2013/02/batch-versus-streamingdifferentiating
4. Marz, N., Warren, J.: Big Data: Principles and Best Practices of Scalable Real-Time Data Systems. Manning Publication Co., Shelter Island (2015)
5. Gorelik, A.: The Enterprise Big Data Lake, 1st edn. O'Reilly Media, Sebastopol (2019)
6. Apache Kafka Official Website. https://kafka.apache.org/. Accessed 20 Nov 2019
7. RabbitMQ Official Website. https://www.rabbitmq.com/. Accessed 22 Nov 2019
8. Apache Storm Official Website. https://storm.apache.org/. Accessed 23 Nov 2019
9. Zaharia, M., Chowdhury, M., Franklin, M.J. Shenker, S., Stoica, I.: Spark: Cluster Computing with Working Sets (2011). https://amplab.cs.berkeley.edu/wp-content/uploads/2011/06/Spark-Cluster-Computing-with-Working-Sets.pdf
10. Apache Spark, Official Website. https://spark.apache.org/. Accessed 24 Nov 2019
11. Volt DB, Official Website. https://www.voltdb.com/. Accessed 29 Nov 2019
12. Redis, Official Website. https://redis.io/. Accessed 29 Nov 2019
13. Apache Hadoop Official Website. https://hadoop.apache.org/. Accessed 25 Nov 2019
14. Hortonworks Data Platform Official Website. https://hortonworks.com/products/data-platforms/hdp/. Accessed 26 Nov 2019
15. Optimus Information Official Website. http://www.optimusinfo.com/downloads/white-paper/open-source-vs-proprietary-software-pros-and-cons.pdf. Accessed 05 Mar 2020
16. UK Finance Official Website. https://www.ukfinance.org.uk/press/press-releases/banking-industry-prevented-%C2%A3166-billion-fraud-2018. Accessed 13 Jan 2020

Automatons Immersed in Ocean Currents for Transformation of Biomass into Fuel

Lucio Cañete Arratia[1(✉)], Felisa Córdova[2], and Andrés Pérez de Arce[3]

[1] Universidad de Santiago de Chile, Av. Ecuador 3769, Santiago de Chile, Chile
lucio.canete@usach.cl

[2] Universidad Finis Terrae, Av. Pedro de Valdivia 1509, Santiago de Chile, Chile
fcordova@uft.cl

[3] Trade Plus S.A., Carlos Vildósola 279, Santiago de Chile, Chile
andresperezdearce@gmail.com

Abstract. When observing the Chilean sea from both biotic and mechanical perspective, oceanographers note that the Humboldt Current carries abundant biomass and that the movement of the water itself has the capacity to do work. Taking advantage of these two qualities of the ocean current, this article exposes the sketch of an automated device, the computational simulation when it was conceived and its mathematical model to make efficient the capture of biomass that will be processed, stored and dispatched as biodiesel. Said submerged automaton has a structural configuration that was outlined by cybernetic design resulting in a body that carries out the transformation process by itself, which starts on the side that faces the current with its content of biomass. This raw material is trapped thanks to an intelligent system that informs the reactor about the relative importance of the state variables that its body can control, stimulating those swimming organisms to move in the desired direction. The captured biomass begins its process until it becomes biodiesel by virtue of the mechanical energy provided by the same flow of seawater that affects the reactor. The rear part of the reactor releases both incident water and by-products into the sea without harmful environmental consequences. Some users of this new type of device are armies in time of conflict and merchant marines during algae bloom.

Keywords: Ocean current · Biomass · Mathematical model · Automated reactor · Biodiesel · Cybernetic design

1 Introduction

The growing demand for energy by various economic agents has forced them to look for new and non-traditional alternative sources. Some of these sources are the oceans which covering more than 70% of the Earth's surface, offer a potential of 93,000 terawatt-hours per year of electricity generated by thermal gradients, salinity gradients, tides, waves and currents [1]. One of the latter is the Pacific Ocean current that flows along part of the South American coast. This movement of seawater called Humboldt Current (also

I. Dzitac et al. (Eds.): ICCCC 2020, AISC 1243, pp. 145–157, 2021.
https://doi.org/10.1007/978-3-030-53651-0_12

called the Peru Current) not only has a regular movement, but also biomass, both usable as a form of energy [2, 3].

Since there is no technology to simultaneously use both forms of energy from said Humboldt Current, the present work outlines at conceptual level an automatic device capable of capturing, processing, storing and dispatching of biomass transformed into biodiesel; using the kinetic energy of the water flow to generate this kind of fuel. Such an artifact, simply referred to in this article as a Reactor, for its successful performance has various mechatronic and biochemical challenges among others [4].

This paper addresses the challenge of the state that the automatic device must have for a better management of energy. Specifically, the objective of this article is to conceive a mathematical model so that the Reactor knows how important are some controllable variables. That is, to know the weights of the state variables that are under its control in each changing context.

Furthermore, since not all the variables are controllable by the Reactor when it operates, they must be controlled in the gestation of the apparatus and for this reason the present work also accepts another challenge: generate an intelligent procedure to design a complex automaton.

The next section of this article describes Humboldt Current with its qualities as an energy supplier. The third section outlines the artifact itself, while the fourth section details the two challenges of this work. Section five develops a conceptual model of the state of the submerged Reactor facing the current. Section six outlines how to design this new artifact in an austere way through an iterative simulation of different assembly elements. Finally, in the seventh section the conclusions are delivered.

2 The Allure of This Current: Plankton, Nekton and Water Flow

The Humboldt Current, or more strictly speaking the Humboldt Current System because it is a complex of flows with a resulting northward [5], is a continuous movement of cold (~18 °C) and low-salinity (~34.7%) mass of sea water that flows north along the western coast of South America. This current extends from southern Chile (~45th parallel south) to northern Peru (~4th parallel south) and reaches 500–900 km offshore. Its maximum speed is about 0.2 m/s and its deep ranges between the 200 m and 300 m.

A show in Fig. 1 [6], it is one of the main ocean currents in the world, bringing oxygenated water north from the South Pacific for thousands of kilometers before it dissipates in the warmer waters around the equator. The Humboldt Current creates one of the largest and most productive marine ecosystems in the world because cold waters with low salinity and high levels of nutrients are brought to the surface through upwelling, providing sustenance to fish and marine mammals. The coasts of Chile and Peru are therefore one of the largest fisheries in the world, with approximately 18% to 20% of the world's fish catch [7].

The Humboldt Current has two major sources of energy, at least in theory: biological and mechanical. The biological source is composed of living beings, which for the present work are classified into two large groups: those that are able to swim and those that are not able. As shown in Table 1, the former is known as Nekton and the latter as Plankton.

Why this taxonomy? Because the automatic Reactor must attract the Nekton that is attractive to the process and repel the nekton that is not attractive. Indeed, some marine

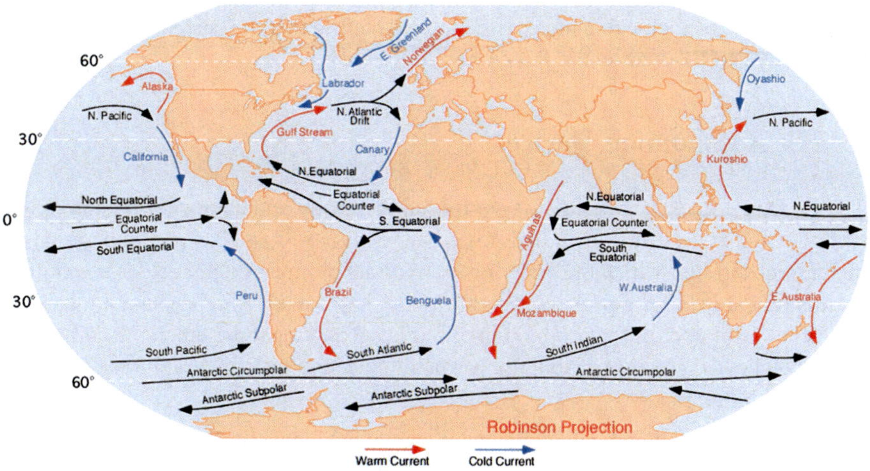

Fig. 1. The ocean currents in the world where Humboldt (here called Peru) is one cold of them. (Public Domain. U.S. government publication)

Table 1. Comparison between the potential sources of biological energy

Issue	Peacetime	Wartime
Common definition	Marine organisms that drift in the water	Marine animals than can swim against water currents
Reynolds Number	Greater than 1000	Less than 10
Size	Either microscopic or macroscopic	Macroscopic
Kingdom	Either plants or animals	Animals
Examples	Phytoplankton: diatoms, dinoflagellates, coccolithophores, and green algae. Zooplankton: crustacean, jellyfish, and small protozoa that feed on other plankton	Fishes, whales, sea-turtles, dolphins, crabs, lobsters and squids

organisms may have low bio-energy potential and/or be environmentally sensitive. These must be avoided through some stimulus that takes them away from the path to the Reactor. On the contrary, organisms that are easy to process and rich in biomass that do not have ecological value, must be attracted to the Reactor. This attraction or repulsion, as the case may be, occurs thanks to the swimming capacity of the Nekton.

The Plankton, on the other hand, is not able to move by its own means, indeed they are pushed by the current. So this kind of organism cannot be stimulated at a distance by the Reactor to alter their routes.

About the mechanical energy, the water of Humboldt Current as an average density of 1.024 g m^{-3}, greater than freshwater. Regarding a maximum speed of 0.2 m/s, the kinetic energy of each cubic meter is 21 J. Considering that the current flows incessantly this energy can be used by the Reactor at all times.

3 The Autonomous Device

The Reactor consists of an artifact composed of a body and an anchor. The body is the component that performs the transformation process itself and has a front part that faces the current through which the water enters with its biomass content (Fig. 2 and Fig. 3). At this first stage of the process, a light system has a main role because it must attract the Nekton suitable for the production of biodiesel and repel the others. In addition, this automatic device can move along the vertical plane and leans for capturing the greatest amount of biomass.

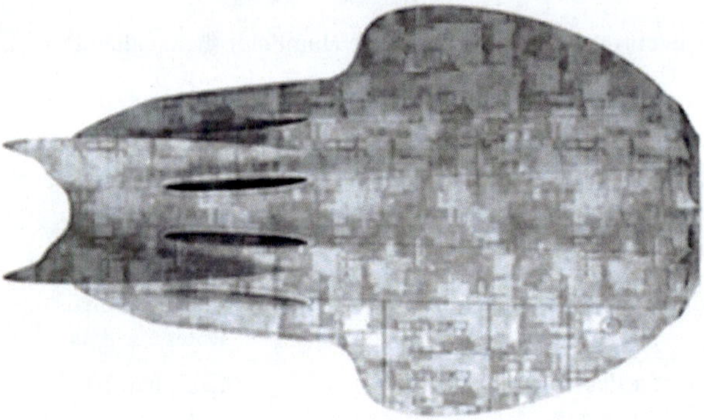

Fig. 2. Top view of Reactor where water and organisms enter from the right side of the figure.

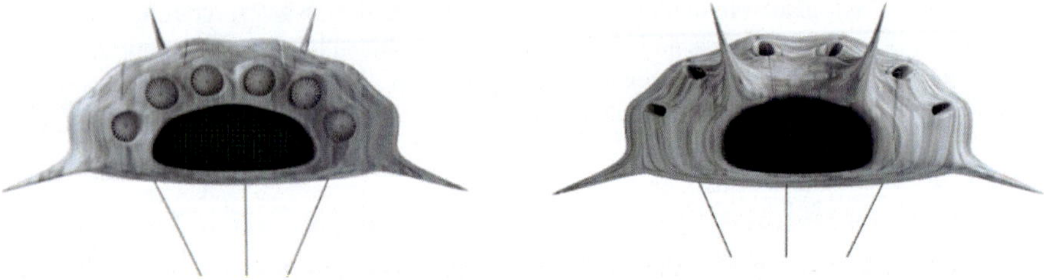

Fig. 3. Front view and back view of Reactor.

Once the flow enters through a grid located on the front, the raw material begins its process until it becomes a fuel by virtue of the mechanical energy provided by the same flow of seawater that reaches the Reactor. This energy is caught by six axial-flow turbines placed in the front part of the Reactor. Both incoming water and by-products are released into the sea from the rear of the Reactor. An overview of the internal process and its relation to the environment is shown in Fig. 4.

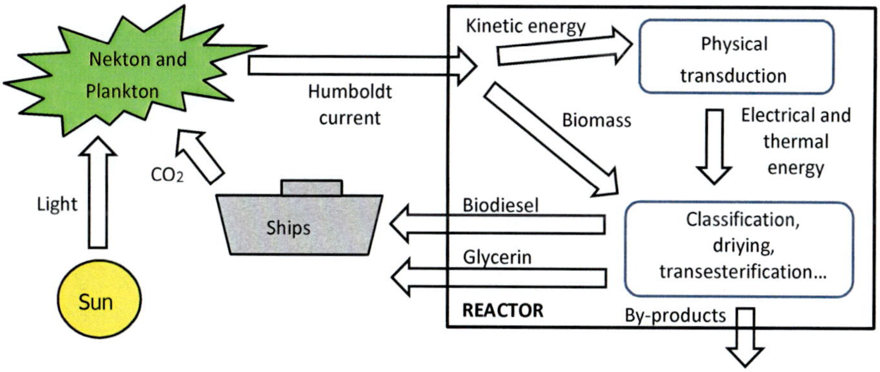

Fig. 4. Overview of the internal process and its relation to the environment

The main users of this type of Reactor are those armies in time of conflict and merchant marines during bloom algae. Even when the project is at the profile level, it is estimated that an initial version of the prototype will involve an investment of US$ 0.4 million.

4 Challenges

The Reactor as an intelligent device must perform a good management of energy, trying to generate the greatest amount of biodiesel taking full advantage of the oceanic sources from its environment, both the kinetic energy of the waters and the marine organisms that are the raw material.

The variables of the environment such as Plankton concentration and its biomass content are very difficult to measure. However, at every moment the Reactor can measure its own variables such as its inclination in two horizontal axes of space (pitch and roll) or attitude, depth and intensity of the light as shown in Fig. 5 and Fig. 6.

Fig. 5. View of Reactor with lateral lights on for altering the route of Nekton

In addition, the Reactor can estimate the amount of biomass entered through its "mouth", without the ability to differentiate which part of that total is due to its own merit (attracted Nekton) and which part is circumstantial (Plankton and accidental Nekton).

Then, it is possible to assume that in a certain period, the biomass entered into the Reactor depends on its posture, depth, and intensity of light. In turn, other variables depend on changing environmental conditions, which, as previously stated, are very difficult to measure and are impossible to control. For instance, in a research the zooplankton biomass presented a significant seasonal pattern with a maximum in spring (48.3 g m^{-2}), a minimum in autumn (25.5 g m^{-2}) while summer and winter biomass were alike with 30.4 g m^{-2} and 30.6 g m^{-2}, respectively [8]. About spatial variation, other research noted that diatom (phytoplankton) concentrations are maxima near the surface during strong upwelling periods, but small phytoplankton tend to have subsurface maxima during relaxation/downwelling periods, when the vertical stratification is the strongest [9].

Faced with this changing scenario, the Reactor must have an active role making adjustments to the variables that are under its control to capture the largest biomass whose space-temporal arrangement is not constant. Then, the intelligent system of the reactor to act successfully on these variables, must know how important they are in each context. That is, a challenge for this autonomous artifact is to determine the weight of the state variables under its control that allows the best energy performance.

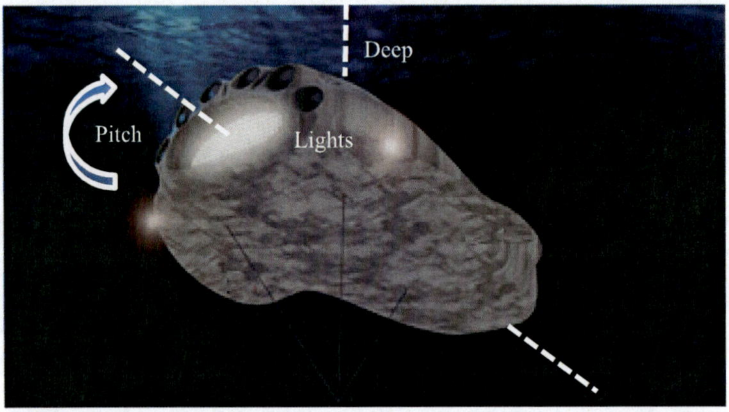

Fig. 6. Reactor and some state variables when it operates

5 Model for the Intelligent Control of Reactor When It Operates

The reactor as an autonomous entity must be endowed with artificial intelligence that allows it to know in real-time the weight or relative importance of the relevant variables of which it has control. One way to achieve this intelligence is through expert knowledge. How can the Reactor acquire that knowledge if it is a debutant in a changing environment?

Thanks to the proprioceptors (sensory receptors which receive stimuli from within the body, especially one that responds to position and movement), this device can measure the status of the variables under its control at every moment and can also measure with satisfactory both precision and accuracy the preference that the Nekton has for each arrangement of those controllable variables. Indeed, in its mission of attracting the greatest amount of Nekton to the "mouth" of the Reactor, the biomass arrived can be estimated according to the state of the artifact.

Then, changing its state variables the Reactor can be more or less attractive for the Nekton. That is, the Nekton's preference for moving towards the Reactor will depend on its status. If Pn are the n preferences of the Nekton for each status of Reactor characterized by a set m of variables V, the following functions may be written:

$$\begin{aligned} P1 &= \text{f}\ (V11,\ V12 \ldots V1m) \\ P2 &= \text{f}\ (V21,\ V22 \ldots V2m) \\ &\vdots \qquad\qquad \vdots \\ Pn &= \text{f}\ (Vn1,\ Vn2 \ldots Vnm) \end{aligned} \tag{1}$$

The function is the same in each of the equations and as previously established, the different values acquired by P will depend on the values taken by each variable V and its weight. So Pi will be the preference for the state i given by its Vij understood as the amounts of the variables that generate the preference (with $i = 1,2,3\ldots n$ and with $j = 1,2,3\ldots m$)

For simplicity, it will be assumed that the function is a linear summation where each variable Vij will have an always constant relative importance Wj (with $j = 1,2,3\ldots m$). Therefore, the set of n preferences will be a collection composed of linear additive equations as in (2).

$$\begin{aligned} P1 &= W1V11 + W2V12 \ldots WmV1m \\ P2 &= W1V11 + W2V22 \ldots WmV2m \\ &\vdots \qquad\qquad \vdots \\ Pn &= W1Vn1 + W2Vn2 \ldots WmVnm \end{aligned} \tag{2}$$

Since the amount Vij is measurable, it is enough to determine the Wj and thus also quantitatively determine each Pi from $i = 1$ to $i = n$. The procedure for determining such weights and therefore knowing quantitatively the level of importance in this case, is known as Linear Preference Mapping [10]. This fairly robust procedure is composed of three sequential stages described below.

The first stage consists on the one hand, in measuring each state variable in question regardless of the units in which they are expressed this Vij. As previously stated, this measurement is carried out by the Reactor through its proprioceptors in each state.

On the other hand, at the same stage simultaneously the Reactor measures the preference of organisms in each of its states, symbolized as Pi. One way for estimating the amount entered to the Reactor is the Generalized Volumetric Method [11].

Once the first stage is completed, the set of Eqs. (2) will have a fuzzy estimate on the left member (Pi) and clear numbers on the right member (Vij), except for the unknown weights (Wij). How to unveil those weights?

The second stage consists in discovering those weights or relative importance. Before, it is necessary to remember that each *Pi* is an estimate and therefore it is not possible to have a quantitative assurance of them since the Nekton is not the only biomass entered into the Reactor (in a premeditated manner), but also is the Plankton which by chance enters into this artifact. That is, there is no certainty of the cardinality of the measurements but of the ordinality. Therefore, the *Pi* can be compared with each other.

Assume for example that there are four preferences, each corresponding to one state of Reactor: *P1, P2, P3* and *P4.* The possible comparisons are these:

$$\begin{aligned} \mathrm{P1} &> \mathrm{P2} \\ \mathrm{P1} &< \mathrm{P3} \\ \mathrm{P1} &> \mathrm{P4} \\ \mathrm{P2} &< \mathrm{P3} \\ \mathrm{P2} &< \mathrm{P4} \\ \mathrm{P3} &> \mathrm{P4} \end{aligned} \tag{3}$$

However, the presence of errors in the results of the comparisons is accepted, which are assumed to be minimal. Returning to the previous example: if $P1 > P2$ is the same $P1 - P2 > 0$, even entering this error: $\mathrm{P1} - \mathrm{P2} + \text{some error} > 0$.

Therefore, the third stage of the procedure consists in minimizing the sum of all these errors. Thus in this example, six comparisons are made, therefore, there are six errors: *E1, E2, E3, E4, E5,* and *E6*. Therefore, the following Linear Programming model is assembled:

To minimize

$$R = E1 + E2 + E3 + E4 + E5 + E6 = \sum Ei \tag{4}$$

Under this set of six restrictions: (5) reliability of the comparison because although there is no security about the quantities of captured biomass, it is certain which is greater or less than another; (6) weights greater than or equal to zero, (7) biomass greater than or equal to zero, (8) errors are always greater than or equal to zero and (9) the sum of all weights must be equal to unity.

$$\begin{aligned} P1 - P2 + E1 &> 0 \\ P1 - P3 + E2 &< 0 \\ \vdots \quad \vdots \quad &\vdots \\ P3 - P4 + E6 &> 0 \end{aligned} \tag{5}$$

$$W1 \geq 0, W2 \geq 0, \ldots Wm \geq 0 \tag{6}$$

$$\begin{aligned} P1 &\geq 0 \\ P2 &\geq 0 \\ &\vdots \\ P4 &\geq 0 \end{aligned} \tag{7}$$

$$E1 \geq 0, E2 \geq 0, \ldots E6 \geq 0 \tag{8}$$

$$W1 + W2 + \ldots Wm = 1 \tag{9}$$

Finally, the third stage consists of inserting Eq. (2) in the first group of restrictions, configuring the definitive model. The minimization is resolved and as a result are obtained the relative weights of each variable (Wj), which are the information sought.

6 Cybernetic Design for the Reactor When It Is Conceived

The model just described allows the Reactor to hierarchize the variables that are under its control while it operates and thus improve its performance. But, what happens to other relevant variables that this automaton cannot alter? Certainly there are constants for the Reactor given during its conception such as the number of turbines and the dimensions of the keels that can affect this behavior. Therefore, the structural design process must also tend to the optimum [12]. Thus, thanks to computational advances, it is possible to channel the design according to certain standards through a process that in this work is called Cybernetic Design. However, this process must face complexity in a harmonious way according to the Parsimony Principle in order to simplify the Reactor configuration as much as possible [13].

This design participates in the simulation of the Reactor during early phase, before it is built. In other words, the Cybernetic Design is a virtual medium which produces the prototype as a result of an iterative process that includes structural adjustments.

This design considers various parameters where the Technical Specifications are essential, which in this case are the quantitative expressions of the document that defines the Reactor requirements. Since such specifications need a certain materiality to be fulfilled, Cybernetic Design requires Assemblable Elements obtained from the available universe. Energy is also required to make such material assemblies. Then the Cybernetic Design procedure displays the following quantifiers:

- Arrangements (A): Whole made up of the Assemblable Elements trying to conceive the ideal Reactor each one expressed ordinally ($i = 1,2,3\ldots n$), such that the first Arrangement is called $A1$, the second $A2$ and the nth An.
- Non-compliance (I): Measure of dissatisfaction of the Technical Specifications determined in percentage with respect to an ideal conception of the Reactor, such that the dissatisfaction of the first Arrangement is $I1$, that of the second is $I2$ and the nth is In.
- Consumption (C): Energy consumed in each Arrangement, expressed as Ci for each Ai.
- Wealth (T): Available energy for the entire process.

The procedure in question consists of constructing with the Assemblable Elements a whole which satisfies the Technical Specifications at the minimum energy cost. This procedure consists of the following main steps:

a) Consider the Assemblable Elements obtained from the industrial ecosystem.
b) Organize the Assemblable Elements putting together a whole (Ai) with the purpose that the result meets $Ii = 0$ and with the restriction $Ci < T$.

c) If this Arrangement Ai complies with $Ii = 0$, save it.
d) If the coupling Ai does not comply with $Ii = 0$: carry out another Arrangement.
e) Repeat steps b), c) and d)
f) Of all the saved Arrangements, the minor will be the Prototype.

A simplified and graphic representation of the Cybernetic Design is presented in the Fig. 7. It shows an iterative procedure that tends to maximize net profitability.

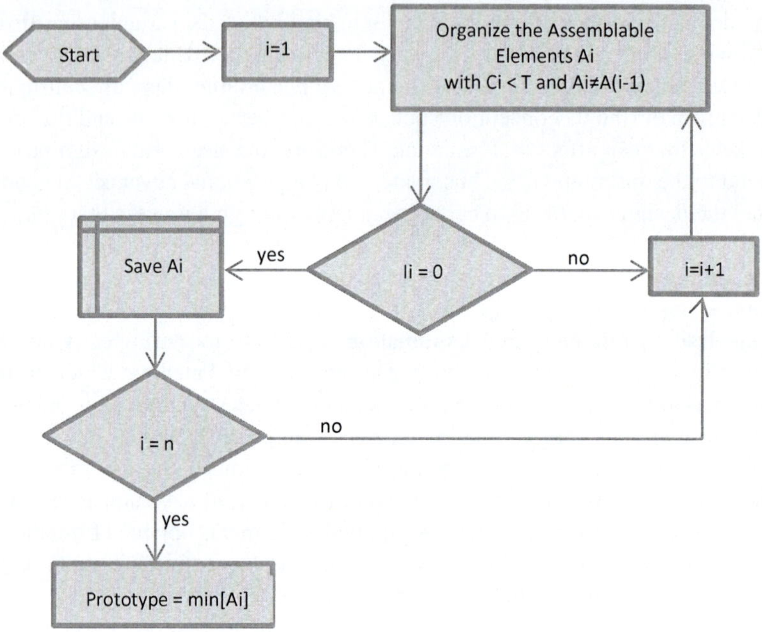

Fig. 7. Simplified algorithm of the Cybernetic Design

7 Applications

One of the main criticisms of this type of automaton is that it gets its inputs from the bases of the marine food chain and therefore, its operation negatively imbalances the ecosystem. However, since this Reactor can know the relative importance of each variable, it can reach the best state for two different scenarios: peace and war, as shown in Table 2.

In peacetime it can operate occasionally when the presence of organic inputs is abundant, especially when this raw material is in excess. In fact, sometimes the abundance of certain organisms due to various factors such as climate change, can large population became in marine plagues. In this scenario the Reactor makes its best contribution by intelligently capturing that type of biomass.

In wartime there is a different economic situation and therefore artificial intelligence provides other answers [14]. In this case the hierarchy of variables is not the same as during peace and the reactor will be able to make other decisions. Even in dire straits the Reactor can shift mode, drop anchors; and use its turbines as propellers to move against an enemy ship acting as a torpedo.

Table 2. Comparison between two scenarios of performance of the Reactor

Issue	Peacetime	Wartime
Vertical location	Emerged, submerged	Submerged
Main clients	Mercantile marine, pharmaceutical and cosmetic industry	Navy
Operation	Regularity	Conflict
Valued products	Biodiesel, glycerin	Biodiesel
Preferred inputs	Bloom algae	Anything

This Reactor is useful only in marine currents? No, this device can be used in any stream that drains organic material. Thus, for example, it can be arranged to capture part of the 14 μg/L of phytoplankton that contain the rivers of a region of Spain [15]. Furthermore, the reactor can act as a river purifier using as inputs to the organic waste discharged into the riverbed.

8 Conclusions

The present work describes the pioneering idea of simultaneously taking advantage of biomass and water flow to generate biodiesel through a submerged device that faces an ocean current, which was conceived from early stages through an artificial intelligence procedure. This artifact has the following characteristics:

- It occupies the kinetic energy of abundant oceanic water as a public good, without generating negative externalities.
- It can influence the organisms that contain biomass by means of altering its state variables.
- It is autarkic and autonomous when operates.
- It is the result of parsimonious design.

Since it is an autonomous entity, the Reactor in charge of such a task must know how important are the variables over which he has control, according to each particular situation. Faced with this challenge, this work developed at a conceptual level a mathematical model that delivers the weight of the state variables of the artifact.

The model has an adequate syntax because it maintains an internal coherence regardless of the variables considered. The model also has semantics because it successfully represents the phenomenon. Finally, it has praxis because it is easy to apply.

The main contribution of the model is that it is a mathematical representation that allows quantifying the importance of the variables of a phenomenon, an essential activity for proper decision-making.

All this mathematical model of automaton behavior was elaborated on a structural configuration that resulted from a simulation algorithm. Said procedure dealt with complexity from the early stages of the development of the Reactor, complying with the technical specifications, using for this purpose the least amount of energy as a universal resource. In other words, the device not only manages energy while it operates, but it was also conceived through an intelligent energy management process. Thus, there is second-order cybernetics, since not only is an automatic entity designed; but also its design was automatic.

As it can be seen in Fig. 1, there are dozens of ocean currents in the world and therefore this project is also applicable to those that are rich in biomass and preferably which have higher speeds. However, since the capture of biomass affects the base of the ecological chain, the operation of the reactor is preferred in times of risk to National Security and/or when there is algae bloom.

Acknowledgments. This work received the valuable contribution of the Reactor drawings made by the architect and designer Álvaro Sebastián Aguirre Boza, a full-time professor at the University of Santiago de Chile.

References

1. International Energy Agency: Implementing Agreement on Ocean Energy Systems. IEA-OES, Paris (2015)
2. Bahaj, A.S., Myers, L.E.: Fundamentals applicable to the utilization of marine current turbines for energy production. Renew. Energy **28**(14), 2205–2211 (2003)
3. Charlier, R.H., Justus, J.R.: Chapter 10 (407–432) Marine Biomass Energy in Elsevier Oceanography Series, Elsevier (1993)
4. Manaf, I., Embong, N., Khazaai, S., Rahim, M., Yusoff, M., Lee, K.: A review for key challenges of the development of biodiesel industry. Energy Convers. Manag. **165**(1), 508–517 (2019)
5. Montecino, V., Lange, C.: The humboldt current system: ecosystem components and processes, fisheries, and sediment studies. Prog. Oceanogr. **83**(1–4), 65–79 (2009)
6. https://commons.wikimedia.org/wiki/File:Corrientes-oceanicas.png
7. https://geography.name/humboldt-current/
8. Aronés, K, Grados, D, Ayón, P., Bertrand, A.: Spatio-temporal trends in zooplankton biomass in the northern Humboldt current system off Peru from 1961–2012, Deep Sea Research Part II: Topical Studies in Oceanography, pp. 169-170 (2019)
9. Gómez, F.A., Spitz, Y.H., Batchelder, H.P., Correa-Ramirez, M.A.: Intraseasonal patterns in coastal plankton biomass off central Chile derived from satellite observations and a biochemical model. J. Mar. Syst. **174**(1), 106–118 (2017)
10. Xia, H.-C., Li, D.-F., Zhou, J.-Y., Wang, J.-M.: Fuzzy LINMAP method for multiattribute decision making under fuzzy environments. J. Comput. Syst. Sci. **72**(4), 741–759 (2006)
11. Benoist, N.M.A., Betta, B.J., Morris, K.J., Ruhl, H.A.: A generalized volumetric method to estimate the biomass of photographically surveyed benthic megafauna. Prog. Oceanogr. **178**(1), 1–12 (2018)

12. Glanville, R.: A (cybernetic) musing: design and cybernetics. Cybernetics and Human Knowing **16**(3–4), 175–186 (2009)
13. Sober, E.: Ockham'razors. Cambridge University Press, Cambridge (2015)
14. Cummings, M.L.: Artificial Intelligence and the Future of Warfare. Chatman House, London (2017)
15. García-Prieto, J.C., Burguillo, F.J., Roig, M.G., Proal-Najera, J.B.: Prediction of phytoplankton biomass in small rivers of Central Spain by data mining method of Partial Least-Square Regression. In: 4th International Conference on Water Science 2019, vol. 48, no. 1, pp 2–8. MDPI, Basel (2019)

Challenges Porting Blockchain Library to OpenCL

Grigore Lupescu(✉) and Nicolae Tapus

Faculty of Automatic Control and Computer Science, University Politehnica Bucharest, Bucharest, Romania
grigore.lupescu@gmail.com

Abstract. This article discusses the complexities of porting a performance blockchain library, encompassing core cryptographic operations, to the OpenCL framework. We present the solution we developed as a general guideline and we highlight the limitations of the OpenCL framework. Given the potential use case of multiple platforms and devices, the effective portability of the library for end users is presented. Finally, a comparison with a CUDA variant of the library is discussed, both in terms of code complexity, runtime and performance.

Keywords: Port · GPU · OpenCL · Blockchain · Hash

1 Introduction

A blockchain encompasses a list of records which are linked together using cryptography. Usually each record holds the hash of the previous record together with a timestamp. A typical usecase for blockchains is via a peer-to-peer architecture where new blocks are constantly being validated and the list of records is constantly growing.

In a peer-to-peer descentralized system, the concept of permissionless blockchain means that a node which contributes to the system may be run by any entity, and the transactions are handled by a consenus mechanism based on votes. As previously stated, one important part of the blockchain is the hashing operation. A hash is a mathematical algorithm which maps data from a variable size to a fixed size. Examples of common hash functions are MD5, SHA etc.

The Solana project represents a new blockchain architecture based on Proof of History (PoH) and is thoroughly described in its coresponding whitepaper [1]. The goal of the project is to obtain a distributed system that can scale transactions proportionally with the network bandwith. For example the whitepaper [1], describes a theoretical upper bound of 710 k transactions per second (tps) for a gigabit network. As opossed to initial blockchain architectures the transactions do not rely on a Proof of Work (PoW) scheme, and thus specialized architectures are not required (like was the case with Bitcoin using ASICs for example). Initial blockchain technologies and the PoW scheme have been analyzed in depth by literature [6–8].

I. Dzitac et al. (Eds.): ICCCC 2020, AISC 1243, pp. 158–166, 2021.
https://doi.org/10.1007/978-3-030-53651-0_13

Current design specifies 2 types of entities to sustain the Solana distributed system: validators and archivers. At the base of the Solana validator cluster, are a set of computers that verify the output of submited untrusted programs. The cluster maintains a set of records, called the ledger, which is the base for reproductibility. The role of the archivers, which are specialized thin clients, is to download a part of the ledger and store it. The library presented in this paper relates to the validators, to the signing and verification of transactions.

1.1 Solana Performance Libraries

The performance of the Solana validator cluster is measured as average transactions per second (tps) and how long it takes for a transaction to be confirmed by the super majority of the cluster. To track the performance, each cluster maintains several counters which are futher uploaded to the cloud.

Because Solana allows for transactions to scale with network bandwith, the requirements for a validator are many core CPUs (e.g. Intel Xeon or AMD Threadripper) coupled with fast large SSDs (e.g. 4TB size). Proof of History is based on SHA hashing [5] and architectures such as AMD Threadripper (CPU x86_64) support SHA hardware instructions. An alternative to CPU processing is GPU processing which can better keep up with the higher throughput of a network, given the large number of cores and fast video memory.

Our variant is based on the solana performance libraries which define the CUDA implementation and the CPU parallel variants. The Solana performance libraries source code [2], as well as the newly ported implementation [3] are open source and publicly available via github.

The goal of the our variant was to extend the current set of supported target devices. Before this, only devices that could run on, were CPUs and Nvidia CUDA GPUs.

There are several architectures that can perform hashing at a high throughput. Examples include: multicore CPUs, GPUs, FPGAs, ASICs. The most widely available processor oriented to raw number crunching with respect to power consumption is the GPU. GPUs can be found either in discrete add-in PCIe boards (PCI express bus) or directly integrated into a CPU SoC (System on Chip). As opposed to a CPU core, the GPU is highly oriented towards the SIMD (Single Instruction Multiple Data, Flynn taxonomy) processing model having most of its die space occupied by vector processing units as opposed to caches. The SIMD architecture is usually well suited towards block hashing.

The state of solana performance library consisted of a CPU C variant and a GPU CUDA variant [2]. This meant execution could happen on either a CPU (typically x86 architecture) or an Nvidia CUDA GPU. This limits the scope of participating nodes, given there are several other vendors of GPU units (e.g. AMD, Intel), as well as potentially other accelerators (e.g. FPGAs).

2 Candidate Compute Frameworks

Though multiple APIs exist, only a limited subset are more widely used and implemented by vendors. Currently CUDA (Nvidia only GPUs) is widely used in the HPC and server segment. At the present time, OpenCL is the de facto API standard, defined by Khronos and implemented by most major GPU IP vendors such as AMD, Intel, Vivante, Qualcomm etc. Other APIs are DirectCompute, SyCL, OpenACC, Mantle. It is slowly becoming superseded by other competing APIs, mainly do to its ineherent complexity and the fragmentation of its software ecosystem.

In our view, OpenCL is the current alternative to the CUDA framework [10], [14]. It is developed by a non profit technology consortium, Khronos Group, and aims to define a framework for writing programs that execute across several platforms from several vendors. With regards to this, OpenCL was a natural candidate for extending support of the library. By having an OpenCL implementation as well, most other GPUs automatically become supported (e.g. AMD, INTEL, QUALCOMM), as well as potentially other accelerators.

Both the hardware as well as software stack complexity make programming GPUs difficult and limit application portability. Developing an application means handling both the host side flow (i.e. CPU, memory transfers) as well as the device side flow (i.e. GPU, processing).

With OpenCL, a user must handle both the host side (e.g. CPU) as well as the device side (e.g. GPU). The host side is required since the GPU lacks the capability to access most of the OS services including memory allocation, thread preemption or file accesses. This increases the solution complexity for the software developer which must fine-tune the whole communication process.

The OpenCL framework defines a set of functions for the host side management, as well as a C99 based language for the device side. In OpenCL functions which run on the device are called kernels. OpenCL also defines a C99 based language for writing the device kernels. The kernels are managed and compiled on the host side, and issued for execution, on the device side, through the OpenCL stack (kernel execution is enqueued by the host side). The OpenCL kernel language defines a set of built-in scalar, vector types, preprocessor directives, operators, attributes, qualifiers as well as a set of built-in functions like math, synchronization primitives, work-group related, relational etc. Each new version of OpenCL in general extends on the previous one and adds new functionality through modified or added functions (both for the host and device side) [10,11].

Because of the general specification of OpenCL to support a wide range of architectures and vendors, there are general pitfalls related to portability, performance and as well added complexity. Limitations of OpenCL both in terms of API, vendor implementation as well as application portability have been throughfully analyzed [9,12,13]. Results usually point to the fact that it is difficult to develop high performance code that is portable accross platforms and operating systems.

3 Library Adapted Structure

The initial code in the library comprised of the ED25519 public-key signature system [4] using SHA-512 and Curve25519 algorithm, implemented both for the CPU HOST side through the C programming language as well as for the GPU DEVICE through CUDA. The majority of the code was shared having compiler directives indicating where the function was to be compiled both for HOST/CPU and for DEVICE/GPU. The major 2 operations that had to be ported to OpenCL where the sign operation and verify operation. These are also used to evaluate the performance of the library in terms of number of signs/verifies per second.

3.1 Device and Host Code Separation

OpenCL frameworks entails a separation of HOST and DEVICE code. Further because the library is delivered statically or dynamically, the whole DEVICE code (kernels) must be embedded inside the library and already present in memory to be compiled. This will result that any future updates to either the HOST or DEVICE code will require a full recompilation and distribution of the library. We have thus embedded the OpenCL device code in C header files, stored effectively in constant strings, which get compiled at init OpenCL runtime. The compilation only happens at init, with resulting executable kernels being available throughout the lifetime of the program. An alternative for this would be script automation, for the device code (OpenCL kernels) to be assembled through a script by including both specific device code as well as potentially generic (which can be used by the host as well), prior to the host program compilation.

Code duplication should be avoided and this can get complicated, due to the fact that the DEVICE kernel code needs to be embedded for compilation inside the HOST code (stored in C header files as strings). Conversely, each call to a DEVICE kernel needs to be specifically managed as per the OpenCL standard. In contrast to C code or CUDA, kernel function calls do not happen directly and need to be defined explicitly with the OpenCL API (having the kernel compiled, setting the arguments, doing an NDRANGE execution and waiting for the execution to end).

The code duplication problem and added complexity is a limitation of the OpenCL design. In general constants and generic functions could be made available on both the host side as well on the device side, and do not require special adjustments, using simple includes. We conclude the following issues, because of device and host code separation:

- code duplication
- added code management complexity
- cumbersome update of device code

3.2 Memory Regions

In the OpenCL API the following memory regions are defined: global memory, constant memory, local memory and private memory. Conversely a modern GPU architecture would have the following memory regions available: register (equivalent to private), cache L1/L2 (equivalent to local), constant cache, texture cache, video RAM and RAM. Each region in the GPU has its physical limitations, depending on the GPU. The compiler and runtime together with hardware would adjust how the program actually executes (e.g. if register spills will happen).

The algorithms in question do contain large tables of precomputed values that would not fit in most GPU caches. In this case we defined them in the constant memory space, which would translate to them being placed in global memory (video RAM or RAM, depending on GPU). There is of course the implication of performance, since reads are slow from global memory, but this would allow for the best portability.

Another issue is handling pointers from one address space to another (e.g. private vs global). In OpenCL, functions need to clearly have each pointer address space specified accordingly. This does complicate the size of the code as well as the overall solution complexity. Because the number of functions to modify and adapt for each address space can grow very fast, we resorted to allocating and converting the data from one address space to another, where possible. For example if a private address space was required but we only had the global address, we would have a private region defined, copy the global values there, process it and copy the result back. This may however have a negative impact performance.

The following complications arise, because of the OpenCL API limitations on memory regions:

- added code complexity and code duplication
- potentially poor performance

4 Library Runtime Execution

The solana performance library is used by the blockchain architecture in processing and validating transactions and should provide the required functions in a transparent manner, handling in the background all the system details (hardware and software), with minimal input from the user or system. The manual search and setup of platforms of devices is an issue in this particular case. Due to the fact that we are targetting the implementation to be contained fully inside the library, it results that the init part must be executed only once throught the lifetime of the program.

To handle this all the functions which require GPU acceleration have at the begining a call to an OpenCL init function. This will check if the OpenCL environemnt has been setup and if not will start the platform and device query, the context and queue creation as well as the kernel compilation. The overhead of this is minimal given it is only once executed and the entities it creates are used throughut the whole execution.

The current implementation just queries for a specific accelerator type (usually GPU), through all the available platforms and devices, and proceeds with the setup. The platform and device defaults to the last device from the last platform but is also capable of selecting the platform/device id. Going towards multi device setup the following situations would need to be considered:

- Single platform, multiple similar devices
- Single platform, multiple different devices
- Multiple platforms, multiple different devices
- Multiple platforms, multiple different devices repeating

Our major concern in a multiple platform, multiple device is how to ensure the load is adequete so as to distribute the work. We need to ensure a device is not queried back from several different platforms, create thus a list of unique devices and keep track of each device limitations be it performance, memory etc. This would complicate the library which would basically need to define a scheduler/load balancer. To avoid this complication, we have only implemented the single platform single device variant.

Another concern with regards to performance is if the executing GPU has any shared hardware with other processing entities including the HOST/CPU. An example of such a situation is if the GPU is integrated on the same die as the CPU, inside the SoC (System on Chip). Performance constraints and techniques to improve performance have been explored in literature [12], but usually present a significant complexity increase in code.

We encountered differences in compilation between the tested platfoms (Intel, AMD and Nvidia). These differences vary from duration differences (e.g. very long compilation time, order of minutes compared to seconds on other platform) to functionality differences sets of warnings to errors in compilation, errors of the compiler (e.g. "Shader compiler had memory allocation problem", "Codegen phase failed compilation.") or even the compiler failure altogether with SIGSEGV. Likewise there have been differences even between compilation of devices of different architectures, given the same compiler. Such problems have been highlighted in literature [13]. This limits portability and conversely increase the effort in development, due to the large set of platforms and devices that must be checked.

Regarding runtime and portability we conclude:

- added code complexity
- potentially unused resources
- unexpected delays
- discrepancies in compilation accross platforms and devices

4.1 Performance Analysis

In this section we analyze the performance of the ported library using single device targets. While the CUDA variant is able to distribute work amongst multiple devices, this was not yet implemented in the port, given the complexity discussed in the previous section. When comparing with the CUDA variant, we limit its execution to a single device.

Results have been obtained by running the verify_25519 variant (binary program for solana perf library corectness and performance validation) for OpenCL and have been compared to the previous existing CUDA variant. For the OpenCL variant we have implemented a platform and device selection mechanism and during execution we have monitored several GPU metrics and compared them with the CUDA program execution. Performance results are consistent with current expectations given architecture types. We found that performance of the ported solution is scaling well with device capability and its current load, with no significant differences over CUDA single device.

Figure 1 shows the performance (measured in thousands of signs/verifies per second) of the current OpenCL port, relative to its initial CUDA variant, running on the single device Nvidia Tesla K40M. We can see that the performance is similar, moving from CUDA to OpenCL. The measured time reflects only the full execution time (including memory transfers and allocations), but does not include init time of both variants. The init time is not relevant, given the sign and verify operations will be run continuosly throught the lifecycle of the program. In particular the init process would take longer on OpenCL because of the added platform setup, query as well as device compilation.

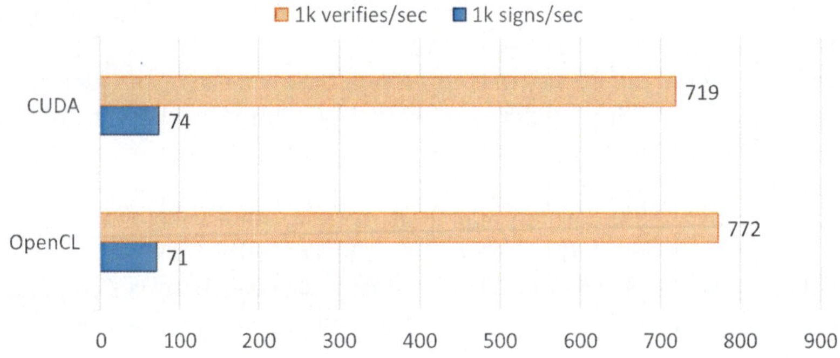

Fig. 1. Performance comparison of solana perf lib binary using OpenCL and CUDA, running on target single device GPU Nvidia Tesla K40M (5 Tflops and 288GB/s VRAM)

Figure 2 shows a performance comparison given the OpenCL port, on various architecture types, measured in number of signs per second (in the order of thousands 1 k signs/verifies per second). We can see that GPU architectures perform better compared to multicore CPU architectures and that performance is correlated with the parallel processing power of the device (number of cores and memory bandwidth).

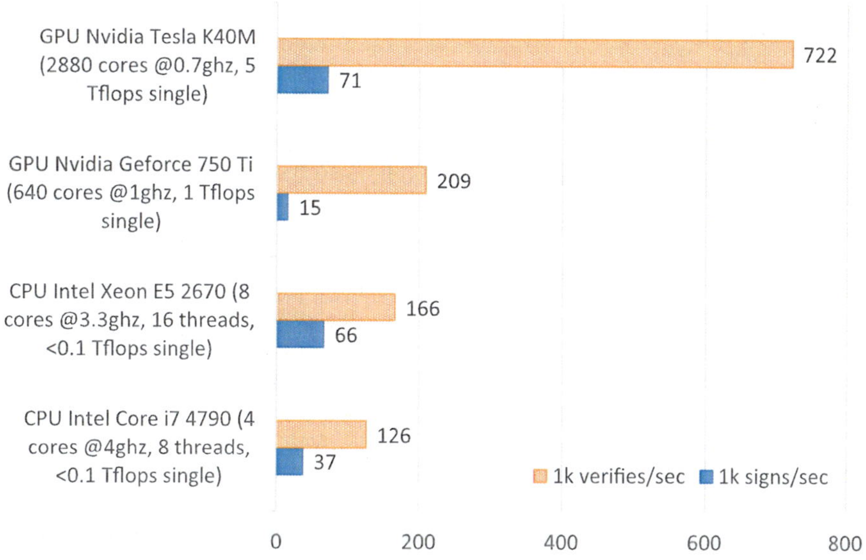

Fig. 2. Performance comparison of various architectures (CPU and GPU), of different parallel capabilities, running the solana perf library port binary using OpenCL

At the moment of writing this article the initial changes have been reviewed and merged to the main branch of the project [2]. Further changes are required to make use of the newly introduced OpenCL variant, by directing the execution in the solana validator to use the OpenCL variant, instead of the CUDA variant.

5 Conclusions

The Solana project aimes to obtain a distributed system that can scale transactions proportionally with the network bandwith. The current limitation to this, is keeping up with the high throughput network bandwith, for signing and verifying transactions. Architectures such as GPUs are ideal candidates and using an API such as OpenCL for portability is well suited, at least at a theoretical level.

In this article we have presented the issues with porting the Solana blockchain performance library to OpenCL. This library encompasses core cryptographic operations necesary for the validation and sign of transactions. The changes presented are related to both how the code is structured and optimized (e.g. memory qualifiers), how the runtime has to be fully managed as well as what runtime issues have been encountered.

Our solution aims to offer a very thin layer of logic and represents a fully functional and portable OpenCL variant of the blockchain library in question, with adequete performance. In the future we aim to further optimize and adjust the implementation.

References

1. Yakovenko, A.: Solana: a new architecture for a high performance blockchain v0.8.14. https://github.com/solana-labs/whitepaper/blob/master/solana-whitepaper-en.pdf
2. Solana, opensource project performance library. https://github.com/solana-labs/solana-perf-libs
3. Solana performance libraries OpenCL implementation, commit 539a67a6bbb62 51b005762ac704c0704d85732. https://github.com/solana-labs/solana-perf-libs
4. Bernstein, D.J., Duif, N., Lange, T., Schwabe, P., Yang, B.-Y.: High-speed high-security signatures. https://ed25519.cr.yp.to/ed25519-20110926.pdf
5. NIST, SHA-3 Standard: Permutation-Based Hash and Extendable-Output Functions, August 2015. https://nvlpubs.nist.gov/nistpubs/FIPS/NIST.FIPS.202.pdf
6. Mahony, A.O., Popovici, E.: A systematic review of blockchain hardware acceleration architectures. In: 30th Irish Signals and Systems Conference (ISSC). Maynooth, Ireland, **2019**, pp. 1–6 (2019). https://doi.org/10.1109/ISSC.2019.8904936
7. Ekbote, B., Hire, V., Mahajan, P., Sisodia, J.: Blockchain based remittances and mining using CUDA. In: International Conference On Smart Technologies For Smart Nation (SmartTechCon), Bangalore, **2017**, pp. 908–911 (2017). https://doi.org/10.1109/SmartTechCon.2017.8358503
8. Ghimire, S., Selvaraj, H.: A survey on Bitcoin cryptocurrency and its mining. In: 2018 26th International Conference on Systems Engineering (ICSEng), Sydney, Australia, pp. 1-6 (2018). https://doi.org/10.1109/ICSENG.2018.8638208.
9. Lupescu, G., Gheorghe, L., Tapus, N.: Commodity hardware performance in AES processing, pp. 82 - 86. IEEE ISDPC (2014)
10. Khronos OpenCL Working Group, OpenCL 2.0 spec, July 2015. https://www.khronos.org/registry/cl/specs/opencl-2.0.pdf
11. Tsuchiyama, R., Nakamura, T., Iizuka, T., Asahara, A., Son, J., Miki, S.: The OpenCL Programming Book. Addison-Wesley, Boston 26 April 2012
12. Doerksen, M.: Designing APU oriented scientific computing applications in OpenCL. High Performance Computing and Communications (HPCC), pp. 587-592 (2011)
13. Sorensen, T., Donaldson, A.F.: The Hitchhiker's guide to cross-platform OpenCL application development. In: IWOCL (2016)
14. Nvidia, CUDA C Programming Guide, August 2018. https://docs.nvidia.com/cuda/archive/9.1/pdf/CUDA_C_Programming_Guide.pdf

Using Two-Level Context-Based Predictors for Assembly Assistance in Smart Factories

Arpad Gellert(✉) and Constantin-Bala Zamfirescu

Computer Science and Electrical Engineering Department, Lucian Blaga University of Sibiu, Emil Cioran 4, Sibiu, Romania
{arpad.gellert,constantin.zamfirescu}@ulbsibiu.ro

Abstract. The paper presents some preliminary results in engineering a context-aware assistive system for manual assembly tasks. It employs context-based predictors to suggest the next steps during the manufacturing process and is based on data collected from experiments with trainees in assembling a tablet. We were interested in finding correlations between the characteristics of the workers and the way they prefer to assemble the tablet. A certain predictor is then trained with correct assembly styles extracted from the collected data and assessed against the whole dataset. Thus, we found the predictor that best matches the assembly preferences.

Keywords: Context-based predictors · Assembly assistance systems · Adaptive systems

1 Introduction

Many factories avoid the full automation due to either the flexibility of the human operators or lower production costs. Nowadays, machine-assisted human-centered manufacturing is the common approach. On the other hand, in the absence of full automation, modern factories might adapt to the workers profile for a cost-effective and resource-efficient production. The training stage of the workers can be more efficient if assembly assistance systems are used instead of human trainers. A certain level of information assistance can help to structure, guide and control manufacturing processes [1]. Smart assembly assistance systems can improve the overall performance, and simultaneously reduce the skill requirements of the workers [14]. The automation might be designed and implemented in an adaptive manner, since the adaptability of the automation can mitigate some of the costs of human-machine interaction, such as unbalanced mental workload [11]. The dynamic configuration possibility of automation levels is another requirement [21]. A comprehensive review of the research challenges in the product assembly domain is presented in [23].

Assembly assistance systems designed for monotonous manufacturing tasks should not over-challenge or under-challenge the worker, it might adapt to the needs of the human operator in real-time [4] and to the constraints of the task [22]. Moreover, it must consider the skills and a possible functional decrease of the human worker's capabilities

I. Dzitac et al. (Eds.): ICCCC 2020, AISC 1243, pp. 167–176, 2021.
https://doi.org/10.1007/978-3-030-53651-0_14

[17]. Thus, interactive and context-aware instructions during the assembly processes become more and more important [5].

In this work, we propose an adaptive assembly assistance system able to dynamically adapt the production process to the worker's actual condition, his/her general characteristics, preferences and behaviors in assembling products. This human-oriented assembly assistance system is using a context-based predictor to recommend the next assembly step based on the current state of both: the semi-product, and the worker. From the experiments performed with 68 trainees, we extracted the relevant correlations between certain human characteristics and assembling behaviors. The obtained dataset was subsequently used to train and evaluate different predictors. A detailed description of the experiment is out of the scope of this paper focusing on the techniques to engineer the prediction-based assembly assistance. The goal is to identify the most efficient prediction scheme in order to be physically integrated into a real assembly manufacturing process. Thus, the final adaptive human-centered training station will be able to receive real-time information about the worker and use it to suggest the manufacturing steps. Our training station will finally allow the following functionalities: capability of recognizing through sensors the product components and human features and actions, ability of learning patterns and correlate human operator contexts with the assembly states of a certain product, possibility of assisting the trainee in correct product assembly (either by recommending the next step or by detecting wrong steps), capability of connecting the relevant data systems for an easy training set-up.

The reminder of the paper is organized as follows. The next section presents some state-of-the-art manufacturing concepts and a related work in context-based prediction techniques. Section 3 describes the two-level context-based prediction schemes used to provide suggestion for the next assembly step. Section 4 discusses the experimental results. Finally, the last section concludes the paper and presents further work directions.

2 Related Work

In [20], the authors proposed a human-machine centered assembly station and presented a case study in a mini-factory laboratory, switching from manual production of pneumatic cylinders to a hybrid assembly system combined with a lightweight robot. In [10], the authors proposed an automated virtual training system in the automotive manufacturing domain. The goal was to increase the production process transparency for the human operator and to allow fast adaptation to new manufacturing requirements. Their cost-effective scalable hardware-setup relies on game-based user interaction. In [13], the gamification is used to achieve a mental state in which a worker is fully immersed in activity, with energized focus and the belief in the success of that activity. In [15], the authors presented a semantic service discovery and ad-hoc orchestration system which is adaptable to contextual information changes. The processes are generated taking into account the current structure of the production plant and ability of the field devices in applying semantic discovery and service selection. Thus, they applied context-based service orchestration to control the manufacturing process. None of these works are using context information to suggest the next assembly state. In contrast, our adaptive training station is using context-based prediction to recommend the next manufacturing step, based on the previous steps and the operator's features.

Simple and hardware-efficient predictors have been successfully used for branch prediction in the microprocessors' domain. A good description of these prediction structures was provided in [16] and [2]. The authors have shown that the two-level predictors are simplifications of prediction by partial matching, being in fact Markov predictors. Advanced branch prediction schemes were presented in [3]. Some of the branch predictors have been adapted and successfully used in ubiquitous systems, like person movement prediction in smart office buildings [18]. In our opinion, these simple and efficient prediction schemes are appropriate for our needs and, therefore, we adapt them in this work to be usable for assembly assistance. Their detailed description is provided in the next section.

Markov chains as stochastic models were successfully used in bioinformatics [12], web mining [9], image denoising [8], smart energy management [6], etc. Hidden Markov Models are doubly embedded stochastic processes consisting in a hidden stochastic process that relies on a set of observable stochastic processes. Hidden Markov Models were applied with very good results in speech recognition [19], smart buildings [7], computational biology [24], etc. These powerful stochastic models could be further used as prediction methods to provide the next assembly step in manufacturing processes.

3 Prediction-Based Assembly Assistance

In order to be able to recommend the next step of a manual assembly process, our system must recognize the current context consisting in the previously assembled components, but it might also correlate with the current features of the human operator. Since in this work we focus on a customizable modular tablet, we must codify all its possible states. The tablet consists in eight components (see Fig. 1): a mainboard, a screen and six modules which can be speakers, flashlights and power banks.

Fig. 1. Customizable modular tablet.

Thus, seven steps are necessary in a correct assembly process of such a table. We have chosen a binary codification (usable as decimal, too), considering "1" for the components already assembled on the mainboard and "0" for the ones not yet assembled. The first bit of the binary code represents the screen, the next three bits are the modules from the first row and the last three bits are the modules from the second row of the mainboard. Thus, based on the code of a certain state, we can anytime determine which components were already assembled. By comparing the codes of two consecutive states, we can determine which component changed between them. Figure 2 presents a model of the tablet in a certain assembly state: three modules are assembled on the mainboard, whereas other three modules and the screen are not yet assembled. Thus, the binary code of the exemplified assembly state is 0101001, its corresponding decimal code being 41.

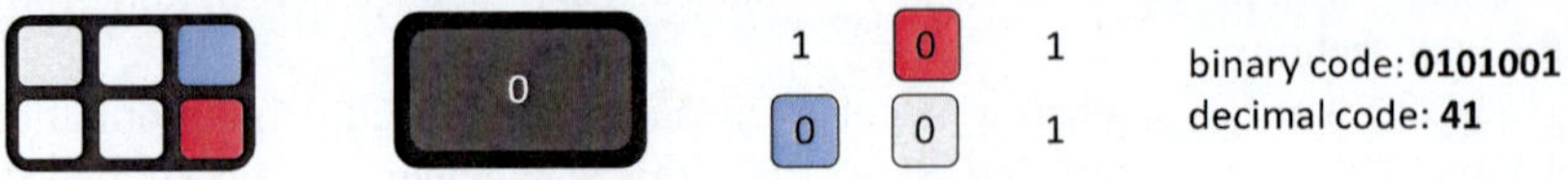

Fig. 2. Example of codification.

The assembly process codified through the above described method can now be modeled using a two-level context-based predictor. The first level consists in a left-shift register containing the last *R* states, *R* being the order of the predictor. The second level of the predictor is a pattern history table with two columns: the pattern and the state. Each entry of the table contains a context of *R* consecutive assembly states in the pattern column and the corresponding next state in the state column. The structure of the two-level context-based predictor with simple states is presented in Fig. 3.

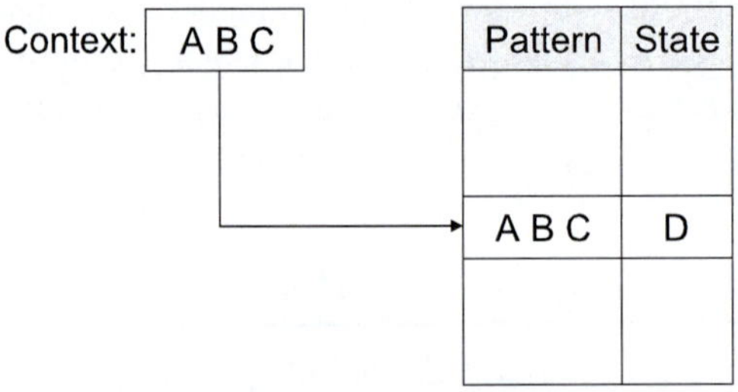

Fig. 3. The two-level context-based predictor of order 3 (R = 3).

In the example depicted in Fig. 3, *C* is the current state, *A* and *B* are the previous two states, thus *A*, *B* and *C* forming the current context, whereas *D* is the predicted next state. In the learning stage, the predictor is populated with the occurring contexts and their corresponding next states. After the learning stage, the stored data can be used for prediction during the assembly process. As an example, for the sequence *ABCDABABC*, the prediction with a third order two-level context-based predictor would be *D*, since after the context *ABC* we have last seen *D* in the past. In our application, the states are the 7 bits long decimal codes reflecting the assembly stage of the tablet, as we described above. Thus, a certain bit on 0 means that the corresponding component is not yet mounted, or a wrong component is mounted there, and 1 means correctly mounted component.

The two-level context-based predictor presented in Fig. 3 can be enhanced with two-state automata. The automata provide more stability with respect to the variations in the observation sequence. A two-state automaton has a weak state and a strong state. One bit is necessary to store the additional information: a logical 0 if the state is weak or a logical 1 if it is a strong one. When the same observation occurs again and the automaton is in weak state, it will change to strong state. When the same observation occurs again and the automaton is in strong state, it will remain in the strong state. When different observation occurs and the automaton is in weak state, it will store the new observation in a weak state. When different observation occurs and the automaton is in strong state, it will keep the old observation in a weak state. The structure of the two-level context-based predictor with two-state automata is presented in Fig. 4.

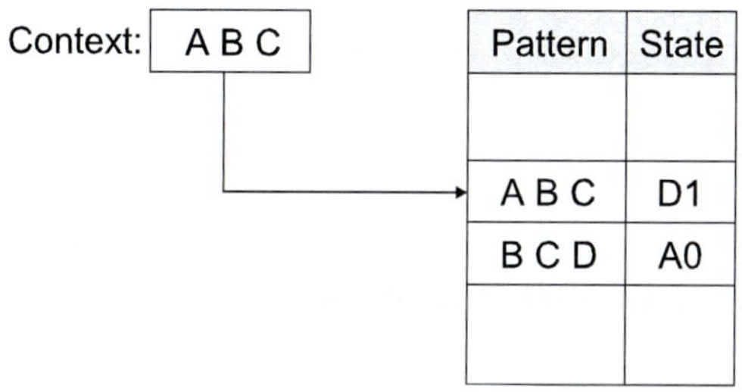

Fig. 4. The two-level context-based predictor of order 3 (R = 3) with two-state automata.

If we consider the observation sequence *ABCDBABCDABC*, the *BCD* pattern is followed first by *B* (the stored weak state is *B0*) and is also followed last time by *A* (*B0* is replaced with the weak state *A0*). The *ABC* pattern is followed first by *D* (the stored weak state is *D0*) and is followed last time by *D* again (the state is changed to the strong *D1*). In Fig. 4, *ABC* is again the current context, and the prediction is *D*. In our application, the states are 8 bits long decimal codes in which the first bit reflects the weakness (0) or the strongness (1) and the rest of the bits are describing the assembly state of the tablet (0 meaning unmounted or wrongly mounted component and 1 meaning correctly mounted component).

Another improvement of the two-level context-based predictor from Fig. 3 consists in allowing to store multiple possible states instead of only one. Thus, the pattern history table will contain for a certain pattern a list with all the states that followed that pattern during the learning stage. If a new state occurs for a pattern, it is appended to the list. When a prediction is necessary, the last seen state will be predicted for a given pattern. Therefore, this enhancement cannot increase the prediction accuracy, but we expect a significant improvement of the wrong step detection indicator, since we can easily check if the current assembly step of the worker is or not in the list of his actual assembly pattern. Figure 5 depicts the two-level context-based predictor with multiple states. If we consider the *ABCDABCABC* observation sequence, after the pattern *ABC* we have seen the states *D* and *A*, so these two states are stored in a list within the corresponding entry from the pattern history table. The prediction would be *A*. In the software implementation of the predictor we have not limited the size of the lists, nor the size of the pattern history table itself (implemented as a dictionary).

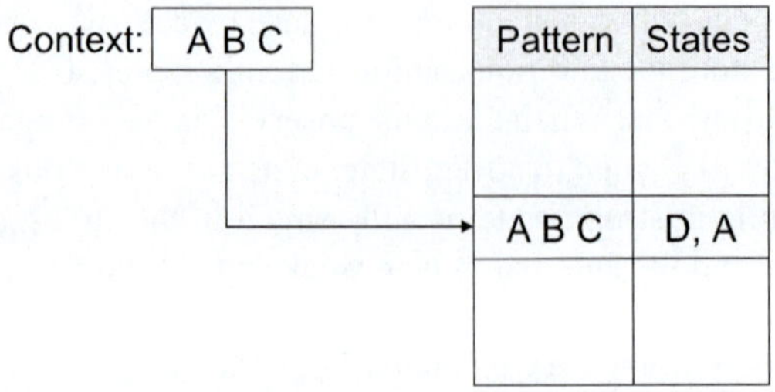

Fig. 5. The two-level context-based predictor of order 3 (R = 3) with multiple states.

4 Experimental Results

For the evaluations of the previously presented predictors, we used the data obtained from the experiments performed with 68 trainees (BSc students), in which they had to freely assemble the tablet (without guidance). The data were collected by observing their behavior in assembling the tablet. Later this process was decomposed in subsequent state transitions of the tablet and codified as previously explained. The proposed predictors, implemented in C#, are trained with the effective assembling steps extracted from the collected data. Note that the effective steps are a subset from the real ones which includes many intermediate wrong trials to assemble the tablet. The trained predictors are assessed against the whole dataset. We measured the prediction accuracy, the coverage, as well as the error detection indicator. The prediction accuracy is the percentage of correct predictions from the total number of predictions. The coverage is the percentage of correct predictions from the total number of assembly steps. The error detection indicator is the percentage of correctly detected assembly errors from the total number of errors (occurred when the real state and the prediction were different). Our final goal is to physically integrate the best predictor into the smart assembly station.

First, we evaluated and compared the two-level context-based predictor with simple states (depicted in Fig. 3) and the one with two-state automata (presented in Fig. 4). Figure 6 presents the prediction accuracy obtained with these two predictors considering different context sizes. We can observe that these two compared predictors have about the same accuracy. Only on the context sizes 1 and 2 the predictor with two-state automata is slightly better. On the higher context sizes both predictors provide the same accuracy. We can also observe that the accuracy is increasing up to a context size of 6. The reason of the accuracy decrease starting with the context size of 7 is that such long contexts are rarely matched during the assembly process, which is shown by the extremely low coverage (see Fig. 7).

We can see that both predictors have almost the same coverage. As we expected, as higher is the context size, as lower is the predictor's coverage. The chances are good to find short patterns, but low or very low to find long patterns. A limitation is introduced by the fact that only seven steps are necessary to correctly assemble the tablet. Analyzing Fig. 6 and 7, we can conclude that the two-state automata introduced an insignificant improvement. Thus, the simpler prediction scheme proved to be more efficient.

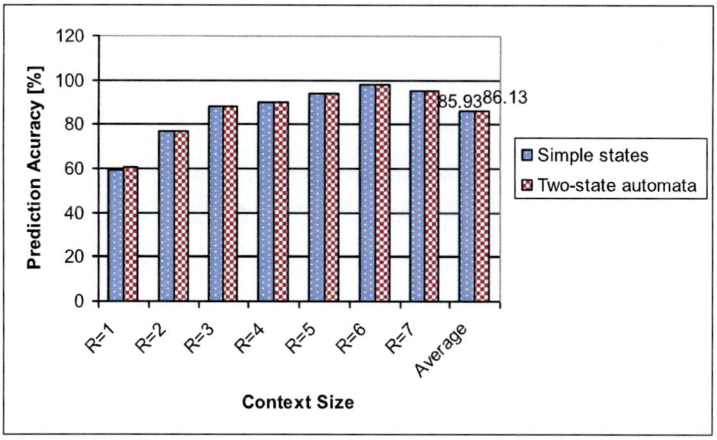

Fig. 6. Prediction accuracy of the two-level predictors by considering different context sizes.

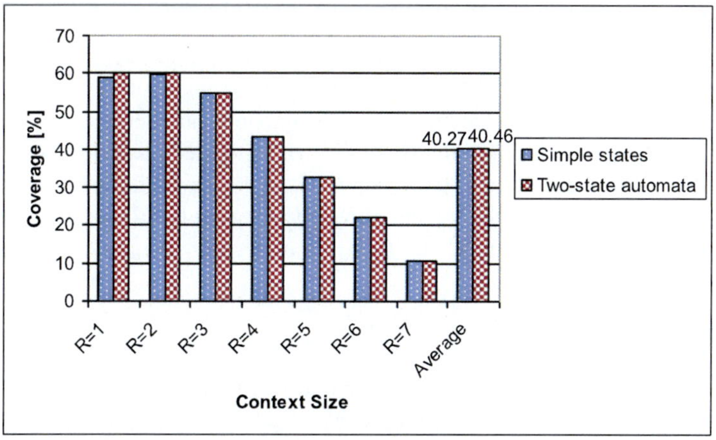

Fig. 7. The coverage of the two-level predictors by considering different context sizes.

Next, we present the error detection capability of the analyzed predictors. This is an important indicator, since we are going to use such predictors to detect also possible wrong assembly steps beside the useful indications they can provide during the manufacturing process.

As Fig. 8 shows, the error detection capability is quite good for both analyzed predictors. For long context sizes (6 and 7) it reaches even 100%. However, we would like to see such good error detection capabilities for shorter contexts, since they can assure better coverage and higher prediction accuracy. The two-level predictor with multiple states (depicted in Fig. 5) has the same accuracy and coverage as the previously analyzed predictor with simple states, since it predicts the last seen state for a certain pattern. However, storing multiple states for each pattern improves significantly the error detection capabilities, since the pattern history table is able to contain thus all the assembly styles.

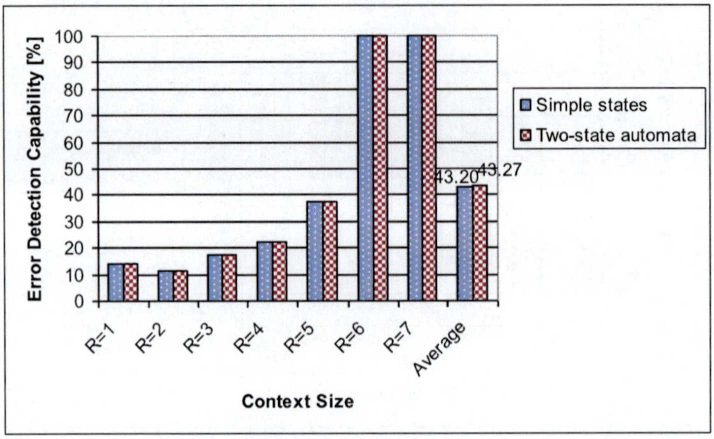

Fig. 8. Error detection capability of the two-level predictors with different context sizes.

If the learning stage of the predictor is applied on a sufficient number of assembly processes, it can learn all the assembly possibilities. With this improvement, we obtained an error detection capability of 95.65% considering a context size of 1 and 100% for longer contexts. Since we are interested in a good coverage (total number of correct predictions) and in accurate error detection, the two-level context-based predictor of order 1 (R = 1) with multiple states seems to be the optimal. Its lower prediction accuracy is not a drawback, since on mispredictions the provided next step is still acceptable – as it is a correct learned step – and can be a useful assembly recommendation for the worker.

5 Conclusions and Further Work

Context-aware assistive systems and collaborative robots can support people to manage the increased variability and complexity of products while reducing the number of errors. These systems can potentially provide a diversity of dissimilar opportunities for manufacturing, such as training employees with less experience or even eliminate the need for training, quality assurance, reducing the cognitive complexity associated with assembly tasks, and integrating elderly and disabled people into the workplace. Although there are various opportunities to employ advanced technologies such as augmented and virtual reality, artificial vision and biosensors, a user-centered approach is critical to the success of these systems. Beside the specific context of the assembly task and the manufacturing environment, assistive systems should adapt the instructions in real-time based on the workers' psychomotor capabilities. This becomes even more obvious in mixed-initiative human-robot collaboration where the proper allocation of tasks between humans and robots requires great flexibility to achieve optimal joint human-robot performance. It includes models and evaluation criteria for the optimal task-allocation between human and robot in a dynamic environment (i.e. the stochastic nature of the manufacturing process and models of human status and performances).

Pattern recognition in the manufacturing process is an essential enabler to provide the assembly assistance functionalities. Although prediction methods have been used in many domains, to our knowledge there is no research to predict the behavior of human operators in manufacturing assembly tasks. In this work, we have analyzed the possibility of using two-level context-based predictors as assembly assistance of the manufacturing operators. The evaluation results shown that the most efficient analyzed prediction scheme was the two-level context-based predictor of order 1 with multiple states, which can assure a very good error detection capability for short context sizes, providing also an enough high coverage. The proposed predictor can be either used to assist the trainees during their learning stage or to help the workers by detecting wrong assembly steps. The training stations enhanced with such predictive capabilities allow to replace a fixed and thus static assembly sequence with a dynamic one which is adapted to each human operator, providing flexibility and efficiency. Our approach fits well the widely applied machine-assisted human-centered manufacturing paradigm.

As a further work direction, we will develop and evaluate some more complex prediction schemes, like Markov chains, Hidden Markov Models or Bayesian Networks. We will choose the most efficient prediction method to be integrated into our smart assembly training station. Until now we did not find any significant correlation between the assembly style and the worker's profile. This issue requires further investigations with an extended set of experiments over different user types (i.e. personas).

Acknowledgments. This work is supported through the DiFiCIL project (contract no. 69/08.09.2016, ID P_37_771, web: http://dificil.grants.ulbsibiu.ro), co-funded by ERDF through the Competitiveness Operational Programme 2014-2020. We thank our BSc students Stefan Alexandru Precup and Radu Raducan for their help in preprocessing the data obtained from the experiment.

References

1. Aehnelt, M., Bader, S.: Information assistance for smart assembly stations. In: International Conference on Agents and Artificial Intelligence, vol. 2, pp. 143–150 (2015)
2. Chen, I.-C.K., Coffey, J.T., Mudge, T.N.: Analysis of branch prediction via data compression. ACM SIGPLAN Notices **31**(9), 128–137 (1996)
3. Florea, A., Gellert, A., Vintan, L., Veltan, M.: The impact of java applications at microarchitectural level from branch prediction perspective. Int. J. Comput. Commun. Control **4**(1), 27–40 (2009)
4. Funk, M., Dingler, T., Cooper, J., Schmidt, A.: Stop helping me – I'm bored! Why assembly assistance needs to be adaptive. In: 2015 ACM International Joint Conference on Pervasive and Ubiquitous Computing and 2015 ACM International Symposium on Wearable Computers, pp. 1269–1273. Osaka, Japan (2015)
5. Funk, M., Bächler, A., Bächler, L., Kosch, T., Heidenreich, T., Schmidt, A.: Working with augmented reality? A long-term analysis of in-situ instructions at the assembly workplace. In: 10th ACM International Conference on Pervasive Technologies Related to Assistive Environments, pp. 222–229. Island of Rhodes, Greece (2017)
6. Gellert, A., Florea, A., Fiore, U., Palmieri, F., Zanetti, P.: A study on forecasting electricity production and consumption in smart cities and factories. Int. J. Inf. Manage. **49**, 546–556 (2019)
7. Gellert, A., Vintan, L.: Person movement prediction using hidden markov models. Stud. Inform. Control **15**(1), 17–30 (2006)

8. Gellert, A., Brad, R.: Studying the influence of search rule and context shape in filtering impulse noise images with Markov chains. SIViP **12**(2), 315–322 (2017). https://doi.org/10.1007/s11760-017-1160-1
9. Gellert, A.: Web access mining through dynamic decision trees with Markovian features. J. Web Eng. **16**(5–6), 524–536 (2017)
10. Gorecky, D., Khamis, M., Mura, K.: Introduction and establishment of virtual training in the factory of the future. Int. J. Comput. Integr. Manuf. **30**(1), 182–190 (2017)
11. Hancock, P.A., Jagacinski, R.J., Parasuraman, R., Wickens, C.D., Wilson, G.F., Kaber, D.B.: Human-automation interaction research: past, present, and future. Ergon. Design Q. Hum. Factors Appl. **21**(2), 9–14 (2013)
12. Jääskinen, V., Parkkinen, V., Cheng, L., Corander, J.: Bayesian clustering of DNA sequences using Markov chains and a stochastic partition model. Statist. Appl. Genet. Mol. Biol. **13**(1), 105–121 (2014)
13. Korn, O., Schmidt, A., Hörz, T.: Assistive systems in production environments: exploring motion recognition and gamification. In: 5th International Conference on Pervasive Technologies Related to Assistive Environments, Heraklion, Crete, Greece, pp. 1–5 (2012)
14. Korn, O., Schmidt, A., Hörz, T.: Augmented manufacturing: a study with impaired persons on assistive systems using in-situ projection. In: 6th International Conference on Pervasive Technologies Related to Assistive Environments, Rhodes, Greece (2013)
15. Loskyll, M., Heck, I., Schlick, J., Schwarz, M.: Context-based orchestration for control of resource-efficient manufacturing Processes. Future Internet **4**(3), 737–761 (2012)
16. Mudge, T.N., Chen, I.-C.K., Coffey, J.T.: Limits to branch prediction. technical report, electrical engineering and computer science Department, University of Michigan, Ann Arbor, Michigan, USA (1996)
17. Peruzzini, M., Pellicciari, M.: A framework to design a human-centred adaptive manufacturing system for aging workers. Adv. Eng. Inform. **33**, 330–349 (2017)
18. Petzold, J., Bagci, F., Trumler, W., Ungerer, T., Vintan, L.: Global state context prediction techniques applied to a smart office building. In: Communication Networks and Distributed Systems Modeling and Simulation Conference, San Diego, CA, USA (2004)
19. Rabiner, L.R.: A tutorial on hidden markov models and selected applications in speech recognition. Proc. IEEE **77**(2), 257–286 (1989)
20. Rojas, R., Rauch, E., Dallasega, P., Matt, D.T.: Safe human-machine centered design of an assembly station in a learning factory environment. In: International Conference on Industrial Engineering and Operations Management Bandung, Indonesia, pp. 403–411 (2018)
21. Romero, D., Noran, O., Stahre, J., Bernus, P., Fast-Berglund, Å.: Towards a Human-Centred reference architecture for next generation balanced automation systems: human-automation symbiosis. In: IFIP International Conference on Advances in Production Management Systems, Tokyo, Japan, pp. 556–566 (2015)
22. Stork, S., Schubö, A.: Human cognition in manual assembly: Theories and applications. Adv. Eng. Inform. **24**(3), 320–328 (2010)
23. Tan, Q., Tong, Y., Wu, S., Li, D.: Anthropocentric approach for smart assembly: integration and collaboration. J. Robot. **2019**, 1–8 (2019)
24. Yoon, B.-J.: Hidden Markov Models and their Applications in Biological Sequence Analysis. Curr. Genomics **10**(6), 402–415 (2009)

Computational Intelligence and Soft Computing

Visual Analysis of Multidimensional Scaling Using GeoGebra

Martynas Sabaliauskas(✉) and Gintautas Dzemyda

Institute of Data Science and Digital Technologies, Vilnius University, Akademijos str. 4, 08412 Vilnius, Lithuania
{martynas.sabaliauskas,gintautas.dzemyda}@mii.vu.lt

Abstract. The paper deals with the multidimensional scaling (MDS) that depends on the class of nonlinear projection methods for a visual representation of multidimensional data. The performance of a new MDS-type method for multidimensional data dimensionality reduction and visualization (Geometric MDS) has been investigated visually using GeoGebra. Dynamic geometry program GeoGebra is a non-commercial and interactive software for the visual representation of algebra and geometry. We made specific GeoGebra scripts for the visual representation of the convergence process of Geometric MDS. This allows us to analyze the optimization of the stress function visually, describing the visualization quality and find the basins of attraction to the local minima. The results allow an easier comprehension of the MDS stress optimization by the anti-gradient search. Moreover, the results deepen the understanding of Geometric MDS, in general.

Keywords: Visual analysis · Multidimensional scaling · GeoGebra · Dimensionality reduction · Geometric MDS

1 Introduction: Multidimensional Scaling

People comprehend information presented visually easier than a large set of numerical data. In this paper, we deal with multidimensional data: many data points dependent on many features. There is a variety of methods for such data visualization (see review in [1]). Projection methods, linear and nonlinear, are among the most popular methods. The goal of the projection (visualization) methods is to represent the multidimensional input data points in a lower-dimensional space preserving specific properties of the structure of the data set. Examples of such visualization are given, e.g., in [1].

Multidimensional scaling (MDS) depends on the class of nonlinear projection methods for a visual representation of multidimensional data [1,2]. Suppose, we have a set $X = \{X_i = (x_{i1}, \dots, x_{in}),\ i = 1, \dots, m\}$ of n-dimensional data points, $X_i \in \mathbb{R}^n$.

I. Dzitac et al. (Eds.): ICCCC 2020, AISC 1243, pp. 179–187, 2021.
https://doi.org/10.1007/978-3-030-53651-0_15

Dimensionality reduction and visualization requires to find the coordinates of new points $Y_i = (y_{i1}, \ldots, y_{id})$, $i = 1, \ldots, m$, in a lower-dimensional space $(d < n)$ by holding proximities d_{ij} (e. g. distances) between multidimensional points X_i and X_j, $i, j = 1, \ldots, m$, as much as possible. MDS finds the coordinates of new points Y_i representing X_i in a lower-dimensional space $\mathbb{R}^d$ by minimizing so-called stress function depending on coordinates of the projected points $Y_1, \ldots, Y_m$. Usually, this function is multimodal. In this research, we consider the raw stress that is equal to sum of all squared differences between the proximity d_{ij} of pair of points X_i and X_j in the original space and the Euclidean distance d^*_{ij} between the corresponding pair of the points Y_i and Y_j in the projected space:

$$S(Y_1, \ldots, Y_m) = \sum_{i=1}^{m} \sum_{j=i+1}^{m} (d_{ij} - d^*_{ij})^2. \tag{1}$$

2 Geometric MDS

The new – geometric – interpretation of MDS method (denote it by Geometric MDS) is proposed and discussed in [3]. In Geometric MDS, the step size and direction forward the minimum of the stress function (1) are found analytically for a separate point Y_j in a projected space without reference to the analytical expression of the stress function (1), numerical evaluation of its derivatives, and the linear search. It is proved in [3] that the step direction coincides with the steepest descent (anti-gradient) direction, and the analytically found step size guarantees almost the optimal step in this direction.

The simple realization of Geometric MDS is based on fixing the initial positions of points $Y_i = (y_{i1}, \ldots, y_{id})$, $i = 1, \ldots, m$ and further changing the positions of separate points Y_j (once or several times) in consecutive order from $j = 1$ to $j = m$ iteratively. Initial positions of points $Y_i = (y_{i1}, \ldots, y_{id})$, $i = 1, \ldots, m$ may be chosen at random, using principal component analysis, etc. The stop criteria is the number of runs from $j = 1$ to $j = m$. Algorithm stops also, if the difference between the stress function $S(\cdot)$ values between two consecutive runs decreases less than some small constant ϵ. Consider the iterative process, where the positions of all points $Y_i = (y_{i1}, \ldots, y_{id})$, $i = 1, \ldots, m$ are changed consecutively once. The complexity of such part of the simple realization of Geometric MDS is $O(m^2 d)$. From the optimization theory view-point, Geometric MDS is some version of the coordinate-wise descent using d-coordinate blocks. The block is formed by the point $Y_j = (y_{j1}, \ldots, y_{jd})$ that has d coordinates. It is shown experimentally in [3] that the efficiency (obtained minimal values of the stress) of such realization of Geometric MDS is the same as compared with the well-known realization of MDS – the SMACOF algorithm.

The core of Geometric MDS method lies in computing of new point Y_j^* for current point Y_j as average of points A_{ij}, $i = 1, \ldots, m$, $i \neq j$, where A_{ij} are points lying on the line between point Y_j and Y_i on the distance d_{ij} from Y_i. Here, d_{ij} is the proximity between two multidimensional points X_i and X_j.

Let we have some initial configuration of points $Y_1, \ldots, Y_m$, and tend to optimize the position of the particular point Y_j only when the position of remaining points $Y_1, \ldots, Y_{j-1}, Y_{j+1}, \ldots, Y_m$ is fixed. In this case $S(\cdot)$ in (1) will be minimized if we minimize the so-called local stress function $S^*(\cdot)$ dependent on Y_j, only:

$$S^*(Y_j) = \sum_{\substack{i=1 \\ i \neq j}}^{m} (d_{ij} - d_{ij}^*)^2. \tag{2}$$

The following formulae are derived in [3]:

$$Y_j^* = Y_j - \frac{1}{2(m-1)} \nabla S^* |_{Y_j}, \quad S^*(Y_j) = \sum_{\substack{i=1 \\ i \neq j}}^{m} \left(d_{ij} - \sqrt{\sum_{k=1}^{d} (y_{ik} - y_{jk})^2} \right)^2, \tag{3}$$

$$Y_j^* = \frac{1}{m-1} \sum_{\substack{i=1 \\ i \neq j}}^{m} A_{ij} =$$

$$= Y_j + \left(\frac{1}{m-1} \sum_{\substack{i=1 \\ i \neq j}}^{m} \left(\frac{d_{ij} (y_{jk} - y_{ik})}{\sqrt{\sum_{l=1}^{d} (y_{il} - y_{jl})^2}} + y_{ik} - y_{jk} \right), \ k = 1, \ldots, d \right). \tag{4}$$

Note, that $S^*(Y_j^*) > S^*(Y_j)$ and

$$S(Y_1, \ldots, Y_{j-1}, Y_j^*, Y_{j+1}, \ldots, Y_m) > S(Y_1, \ldots, Y_{j-1}, Y_j, Y_{j+1}, \ldots, Y_m), \tag{5}$$

if Y_j is not a point of local minimum of the local stress $S^*(\cdot)$.

See Fig. 1 for graphical illustration of one step of Geometric MDS. Here, we consider case $m = 4$ and $d = 2$. For better understanding of the ideas and relation with formulas (3) and (4), we use Y_j and Y_j^* instead of Y_4 and Y_4^* in Fig. 1.

3 Visual Analysis Using GeoGebra

Dynamic geometry program GeoGebra is a non-commercial and interactive software for the visual representation of algebra and geometry. It relates algebra with geometry. GeoGebra is a free dynamic geometry software program, with tens of millions of users worldwide, and a great impact both in mathematics education and in applications [4]. There can be made lots of constructions, including points,

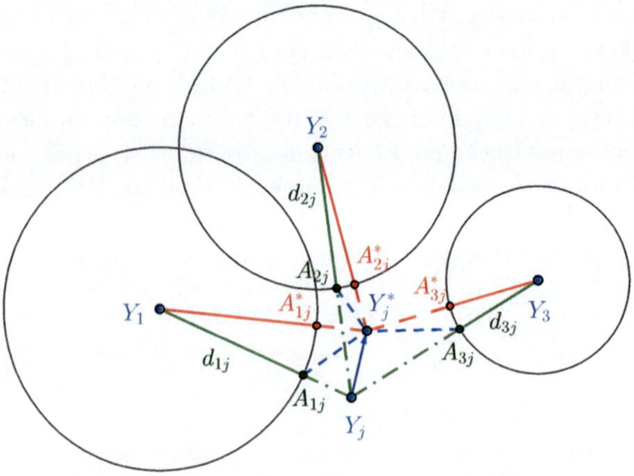

Fig. 1. An example of a single step of Geometric MDS method

lines, geometric shapes, implicit or explicit functions, algorithms, etc. The result is rendered as a real-time drawing that directly depends on the parameters of these constructions. There is a lot of recent applications of GeoGebra. The potential of GeoGebra Automated Reasoning Tools for studying mechanical linkages is shown in [6]. GeoGebra is applied for expressions having non-negative quantities (like distances) in Euclidean geometry theorems to be usable in a complex algebraic geometry prover [7].

Geogebra offers the effective facilities for easy development of visual applications as compared with commercial software, e.g. MatLab, Maple, Wolfram Mathematica. The nearest competitor to GeoGebra is Desmos [5]. It provides free online possibilities for interactive visualization of mathematics. However, its functions are poor and these functions are insufficient for complex tasks. There are two open source libraries in Python: Matplotlib and python(x,y). However, they need for an intense programming efforts.

We aim to use the interactive abilities of GeoGebra to discover new properties of the multidimensional scaling and Geometric MDS visually and dynamically.

For the analysis, the set of six n-dimensional points has been chosen ($m = 6$). Proximities (e.g. distances) between n-dimensional points are completely defined by radius of circles in Fig. 2. Like an example in Fig. 1, let us fix positions of first $m-1$ points $Y_1, \ldots, Y_{m-1}$ on a plane. We will vary position of the last point Y_m. In Fig. 2, such a point is denoted by Y_j for better understanding of the ideas and relation with formulas (3) and (4). The additional information, necessary to compute the local stress (2), is presented in Table 1.

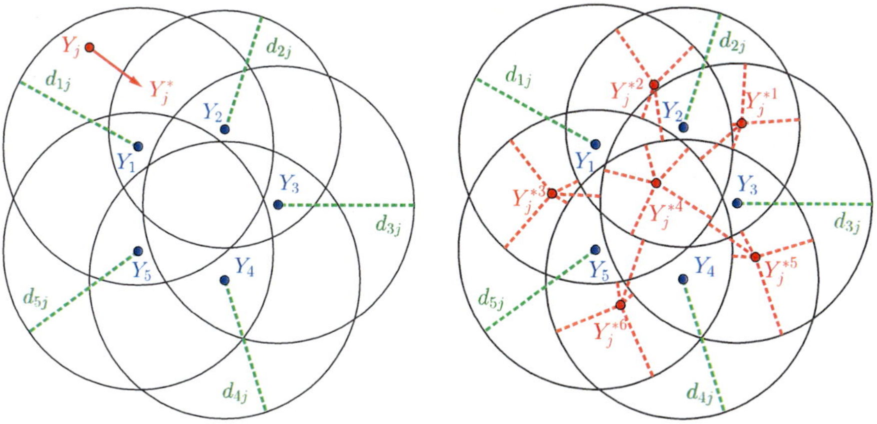

Fig. 2. Data for analysis of multi-modality of local stress of MDS

Table 1. Additional information for computing the local stress

i	Y_i	d_{i6}
1	$(4.0000, 0.0000)$	7
2	$(1.2361, 3.8042)$	6
3	$(-3.2361, 2.9389)$	7
4	$(-3.2361, -2.3511)$	7
5	$(1.2361, -3.8042)$	7

Table 2. Minima of the local stress

i	Y_j^{*i}	$S^*(Y_j^{*i})$
1	$(4.2235, 4.0189)$	27.9526
2	$(-0.2591, 5.9286)$	30.7175
3	$(-5.4505, 0.4519)$	34.2956
4	$(-0.1331, 1.0036)$	36.9138
5	$(4.9622, -2.6907)$	39.2012
6	$(-1.9484, -5.1015)$	42.3623

At first, we illustrate the multi-modality of the local stress function $S^*(Y_j)$ given in (2). In this case, we do not minimize $S^*(Y_j)$ by Geometric MDS. The local stress for different values of Y_j is evaluated, only. The local stress function depends on 2-dimensional variable Y_j. The position of Y_j (in the case of example in Fig. 2, $j = 6$) was varied using GeoGebra means and 3D visualization of the local stress is created (see Fig. 3). 2D contour of the local stress function is given in Fig. 4. This contour is obtained by Maple – it is the only case where Maple overcome GeoGebra. We see six minima. Their places are numbered in 2D and 3D plots, where the numbers correspond to the data on minima in Table 2. Y_j^{*i} is the point of i-th local minimum of the local stress (2), and $S^*(Y_j^{*i})$ is the corresponding value of the local stress.

Our application of GeoGebra allows us to fix a particular point Y_j by a computer mouse, to move this point to any place of the investigated plot and to see immediately a position of Y_j^* after one or several steps of Geometric MDS. It was applied in further experiments with the local stress seeking to disclose its peculiarities.

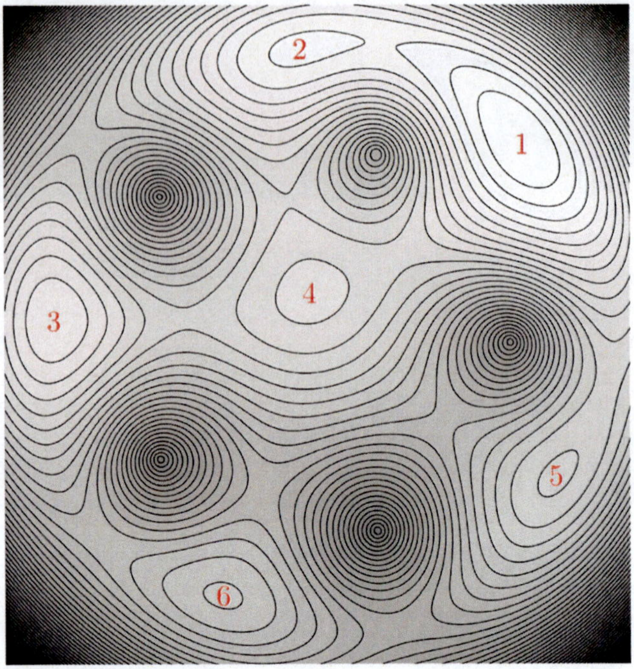

Fig. 3. 2D contour of the local stress

At the next stage, we started one-step optimization from various positions of Y_j. As a result, a set of corresponding positions of Y_j^* was obtained. The results are presented in Fig. 5. Dots indicate starting positions of search (points Y_j). Arrows show the resulting point of one-step optimization: points Y_j^* obtained from Y_j. Starting positions of Y_j were chosen randomly. The positions of local minima and maxima are well seen on the 2D plot in Fig. 3.

The third stage of investigation differs from the second one that 12 iterations of one-step optimization, starting from the same initial points Y_j as in the experiments above, were performed. Such a number of iterations is chosen because it results in the finding of sufficiently close point to the local minimum of the stress. The lines in Fig. 6 show the optimization trajectory from the starting point to the local minima. Arrow indicates the direction and the resulting point after the first step of optimization. Note that Fig. 6 is an extension of Fig. 5.

Figures 5 and 6 allow us to discover regularities and consistent patterns in the process of minimization of the local stress function $S^*(\cdot)$ of MDS. As mentioned above, the decreasing value of $S^*(\cdot)$ decreases the raw stress $S(\cdot)$, too.

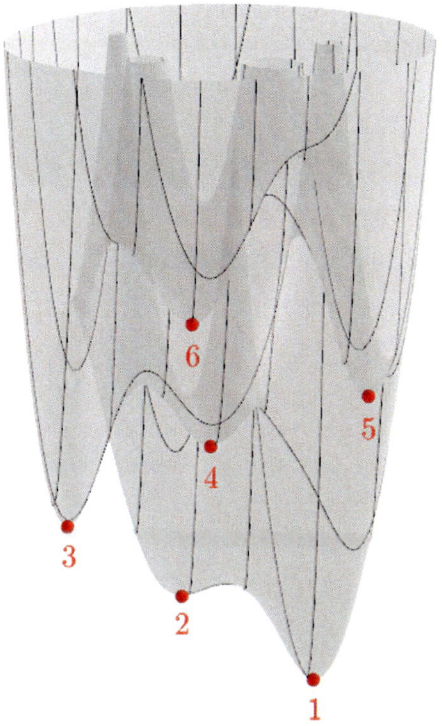

Fig. 4. 3D visualization of the local stress

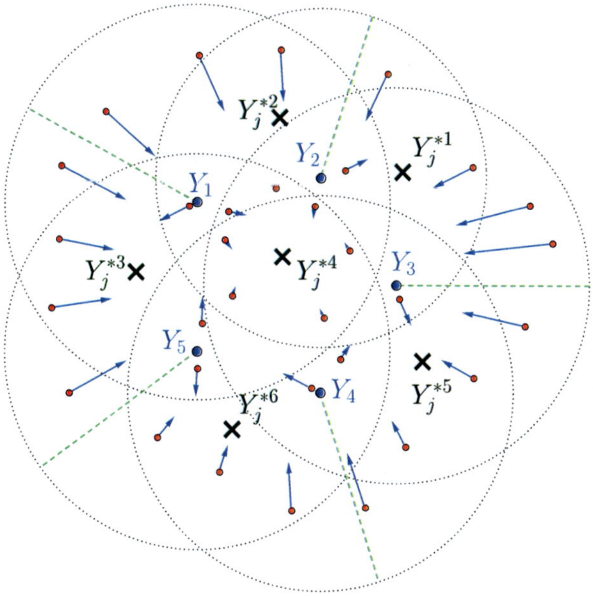

Fig. 5. One-step minimization of the local stress using GeoGebra

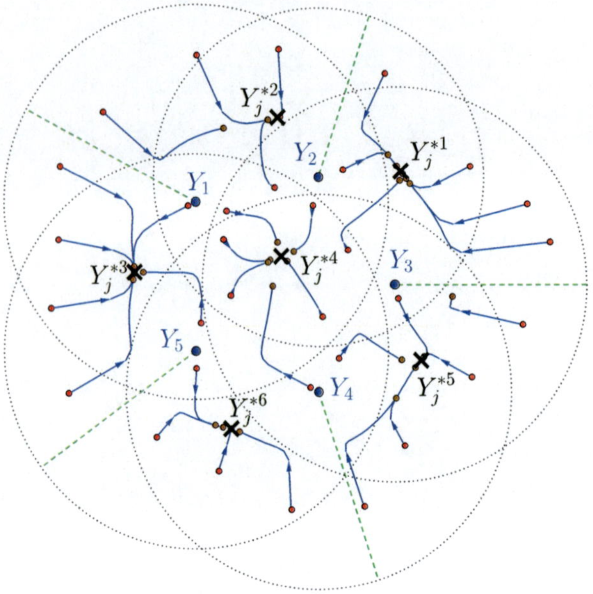

Fig. 6. Multi-step minimization of the local stress using GeoGebra

4 Conclusions

The performance of Geometric MDS has been investigated visually using dynamic geometry program GeoGebra. We made specific GeoGebra scripts for the visual representation of the optimization convergence process from the chosen point. This visualization allows us to find the basins of attraction to the local minima. The results allow an easier comprehension of the MDS stress optimization by the anti-gradient search that is a particular case of steepest descent. Moreover, the results deepen the understanding of Geometric MDS, in general.

Video illustration of multi-step minimization of the local stress using GeoGebra is given in [8]. Like in Fig. 6, the lines show the optimization trajectory from the arbitrarily chosen starting points to the local minima during 12 steps.

Further research should lead to getting the gain from the consistent patterns obtained using GeoGebra potential in the process of minimization of the local stress function and seeking to find a global minimum of the stress or some area close to this minimum. Analysis of the consistent patterns would help in choosing better positions of starting points for optimization as well as in improving the descent trajectory and Geometric MDS method in general.

Acknowledgements. This research has received funding from the Research Council of Lithuania (LMTLT), agreement No S-MIP-20-19.

References

1. Dzemyda, G., Kurasova, O., Žilinskas, J.: Multidimensional data visualization: methods and applications. Series: Springer Optimization and its Applications, vol. 75. Springer (2013)
2. Borg, I., Groenen, P.J.F., Mair, P.: Applied Multidimensional Scaling and Unfolding, 2nd edn. Springer, Cham (2018)
3. Dzemyda, G, Sabaliauskas, M.: A novel geometric approach to the problem of multidimensional scaling. In: Sergeyev Y.D., Kvasov D.E. (Eds.) Numerical Computations: Theory and Algorithms, NUMTA 2019. Lecture Notes in Computer Science, Vol. 11974, pp. 354-361. Springer (2020). https://doi.org/10.1007/978-3-030-40616-5_30
4. The GeoGebra Team (2019). https://www.geogebra.org/
5. Explore math with Desmos. https://www.desmos.com/
6. Kovacs, Z., Recio, T., Velez, M.P.: Reasoning about linkages with dynamic geometry. J. Symbolic Comput. **97**, 16–30 (2020)
7. Kovacs, Z., Recio, T., Solyom-Gecse, C.: Rewriting input expressions in complex algebraic geometry provers. Ann. Math. Artif. Intell. **85**(2–4), 73–87 (2019)
8. Sabaliauskas, M., Dzemyda, G.: Visual analysis of multidimensional scaling using GeoGebra: video example of local descent. Midas (2020). https://doi.org/10.18279/MIDAS.gmds-gg.159420

Edge Computing in Real-Time Electricity Consumption Optimization Algorithm for Smart Grids

Răzvan Cristian Marales(✉), Adela Bâra, and Simona-Vasilica Oprea

Department of Economic Informatics and Cybernetics, The Bucharest University of Economic Studies, Bucharest, Romania
{razvan.marales,simona.oprea}@csie.ase.ro, bara.adela@ie.ase.ro

Abstract. Nowadays the electricity consumption optimization represents a big improvement point for the electricity supplier, but also for the consumers. Both sides can benefit from the progress of sensors and ICT technologies and gain benefits if an automatically process is put in place. Hence, in this paper, we propose an algorithm which will monitor the electricity consumption and provide optimizations for each consumer, all in real time. For accurate monitoring outputs and better computation, the algorithm will run into a smart grid environment, where smart meters, actuator and appliances can be found and easily integrated. The proposed solution will be deployed in an edge computing environment. This architectural decision will make the final implementation more performant and less costly.

Keywords: Edge computing · Real-time electricity consumption optimization · Smart grids · Internet of Things · Controlling architecture · Smart devices

1 Introduction

The emerging smart grid technologies and modern appliances that allow connectivity encourage the consumers to become more active, monitor and control their electricity consumption. Prosumers are encouraged by governmental incentives and smart metering system to generate electricity and inject the surplus to the grid. The smart meters will bidirectionally counter the electricity flows improving the settlement process. The new ICT software platforms for blockchain allow peer-to-peer exchanges fostering competition. On the other hand, prosumers and pure consumers benefit from smart plugs that monitor and control de operation of appliances. The control consists in on/off and schedule modes. Moreover, prosumers benefit from better prediction tools that estimate the generation potential and schedule the consumption to take advantage of the local generation and consume less from the grid. Thus, the electricity is no more transmitted over long distances from large power plants to the consumption places, stressing the grid infrastructure; it is locally generated and consumed by the owners of the generating facilities (PV panels) or their neighbors.

Demand side management is a broad concept that comprises measures and strategies that stimulate the prosumers and consumers to use the controllable appliances, distributed

I. Dzitac et al. (Eds.): ICCCC 2020, AISC 1243, pp. 188–197, 2021.
https://doi.org/10.1007/978-3-030-53651-0_16

energy resources such as renewable small-scale generators, storage devices, etc. and increase the flexibility to meet an objective, such as electricity usage or peak consumption reduction.

By controlling the flexible appliances directly or via remote control at the grid operator or supplier level, the consumers benefit from significant bill reduction. They also contribute to a better renewable energy sources (RES) integration consuming local green electricity or adapting their consumption intervals to RES intermittent availability. Considering that most of the modern appliances are low energy intensive or have batteries, they are more flexible, adapt to the RES availability and store the electricity for a period of time.

Such devices generate large volume of data that could be locally processed and stored requiring less centralized computing resources.

The paper is divided into four sections. The first section of the paper is the introductive section, where theoretical foundation is created. Various topics are approached in this section, such as demand side management for electricity consumption, smart grid overview and Internet of Things (IoT). The second section reviews the most recent similar researches pointing out the IoT and edge computing solutions.

The third part presents the data processing, edge computing, design of the solution, the flow interaction, control mechanism, the proposed architecture and implementation.

Last section presents the conclusions. The proposed architecture is set up together with the electricity consumption algorithm and data processing flows. Base of the implementation, the results of the algorithm are presented and analyzed. At the end, the final conclusions of the paper are presented and correlated with future works.

2 Literature Survey

The computational resources and current challenges in a smart grid environment requires new ICT solutions for data processing as close as possible to its generation sources such as an edge computing solution [1, 2]. A survey of edge computing is provided in [3] considering the large volume of sensors and appliances that continuously generate data and communicate to each other to a sustainable smart environment. Yu W. et al. show that the edge computing improves the efficiency of IoT, classifying the edge computing based on its architecture, analyzing the performance in terms of latency, bandwidth occupation, consumption, etc. Lin J. et al. provide a couple of applications of edge computing based-IoT implementation at the smart grid and smart city level [4].

Zahaf H.E. et al. approach the real-time data processing with edge computing aiming to reduce the energy consumption. They address the dynamic voltage and frequency scaling [5], parallelization, real-time scheduling and resource allocation techniques. The data, collected from the appliances, requires to be locally pre-processed before being sent to the main servers for further processing and storage [6]. Three edge computing technologies are analyzed in [7], such as: mobile edge computing, fog computing and cloudlets concentrating on comparing the standardization requirements, principles, applications and architectures. The solution of edge-computing and energy efficiency of buildings and smart cities are studied in several scientific researches such as [8–11]. A IoT infrastructure is proposed in [12] to handle electricity consumption in relation with demand response and optimization opportunities facilitated by the advancement of sensors and actuators.

3 Proposed Solution

To extend the above related works, in this paper we propose an end to end edge computing solution which aims to optimize the electricity consumption in real time. The proposed architecture and control mechanisms for the smart appliances are presented below.

3.1 Edge Computing Architecture

Edge computing is a powerful paradigm, which can drastically decrease the computational resources and overall costs of an ICT system by bringing the computing processes closer to the end consumer, providing a much rapid processing time in comparison with a classical client-server or cloud solution. From the point of view of designing the architecture, the edge computing come with some important challenges. These challenges are due to the fact that an important part of the infrastructure components is not located near the core system, thus increasing the complexity of the entire system.

Our proposed architecture is trying to bring a simple and decoupled solution for an edge computing system. We design the architecture considering the scope of each component together with the physical placement. As outcome, the architecture is divided into three main areas:

- Household area contains all components which are physically placed inside the household and are interacting indirectly or directly with the outside components of the architecture;
- Community area contains all components which are physically placed near the householders' area and have the purpose of collecting data from several householders. The number of integrated householders depends on their proximity and density. This area represents the edge computing core environment, aiming to be as close as possible to the end consumer and in the same time to centralize and aggregate the consumption data of multiple consumers;
- Enterprise Cloud area represents all components which are located into a public cloud and have the main propose of collecting data from the community areas. These components are providing a global overview over all householders and communities' data. This overview will serve the electricity supplier by offering a better overview of real time consumption and historical information.

Each component of the architecture together with the areas can be visualized in Fig. 1. The components have unique purposes and present different kind of entities or implementations. Some of the components represent physical devices which are procured from an external source, some are external tools that are installed and operated by the system administrator and others are internal applications which are developed specifically to meet a business or technical purpose. The components are detailed below:

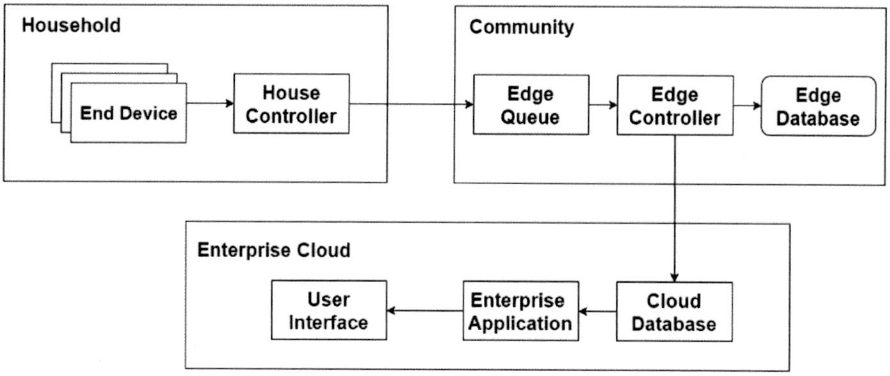

Fig. 1. The proposed architecture for the edge computing in a smart grid context

- End device – All physical smart devices placed inside household, which are able to measure the electricity consumption of appliances using different methods and expose these measurements through various interfaces (Wi-Fi, Bluetooth, ZigBee or LoRA). Each end device is linked to a home appliance and is able to interact with it;
- House controller – The device which is placed inside the household and it is the central point of data collection for each house. It is able to connect to all end devices in the proximity and measure the electricity consumption that is sent in streams to the components outside the household. Besides this, the controller can receive commands from the components outside the household and forward them to the end devices (commands such as power on/power off, dim or pause);
- Edge queue – It is the first component of the edge computing infrastructure. The edge queue is placed in the community area, very close to the households. The main purpose of it is to form an exchange channel between the households and other components outside the house. This kind of communication concept represent a real benefit for both sides (data sender and receiver), removing the bottlenecks in the communication and providing a simple solution for horizontal scaling;
- Edge controller – Main component of the community area, it is the core of the communication flow. The controller consumes all the data regarding electricity consumption coming from the edge queue and take decisions based on this data regarding switching off or on the appliances using the end devices or rescheduling them to another point in time. The commands are sent back to the end devices through edge queue and house controller. The edge controller is also able to aggregate the data and expose it to other components that are interested in it, in our case the components from the enterprise cloud area;
- Edge database – It is placed in the community area, can be implemented as single instance database or multiple instance database, depending on the data quantity of the community. This component comes as a tool for the edge controller, providing a persistence solution for all the information;
- Cloud database – The interaction between the community area and the enterprise cloud area is done via the database connectors of the cloud database. All the aggregated

data created by the edge controller is stored here in order to be used by the cloud applications;

- Enterprise application – The component is working with all the data stored in the cloud database, trying to provide useful information regarding consumption to the electricity supplier. The application displays alerts, statistics and metrics from historical periods of time to real time insights.
- User interface – The User interface or UI component is integrating with the Enterprise application to provide visualizations over the data. These visualizations can be pie charts, histograms, graphs, etc. The electricity supplier is using the component to have a better knowledge over the consumption of the the householders and communities.

3.2 Interaction Flows

One of the main goals of the architecture is to transfer the information between components in real time or near real time. The interaction flows need to be designed by focusing on performance, because this factor can drastically decrease or increase the transfer rate. An irrelevant call done in a synchronous communication during any interaction can represent a major issue that will increase the latency of data delivery.

One of the interaction flows is the registration process of the end devices presented in Fig. 2. The process is trigger once for each end device, when the device is booted for the first time. This procedure is mandatory because the components of the system need to be aware of the end device existence before any other interaction occurs. The process has the following steps:

1. The end device emits a signal to house controller, announcing that it is ready for registration;
2. House controller sends the registration as a message to the community controller via the community queue;
3. Community controller validates the registration, checking the combination between end device and house controller identifiers, the request date and the end device location;
4. Community controller finishes the validation and persists the registration request together with the end device details;
5. Community controller acknowledges the end device registration, taking it in consideration for future optimizations;
6. Community controller sends the registration feedback together with the end device identification details to the house controller via the community queue;
7. House controller receives the response and caches all the end device details internally. After this step, the end device is consider fully registered in the system.

Beside the registration process, another important interaction flow is the data flow from the end devices to the cloud database. This kind of information is critical for the performance of the system and determines the future action of the community controller. The propagation starts with the house controller, which is scheduled to scan all the end devices from the house and ask for consumption data at a precise time interval set by the system administrator. After the consumption is read by the house controller, the flow

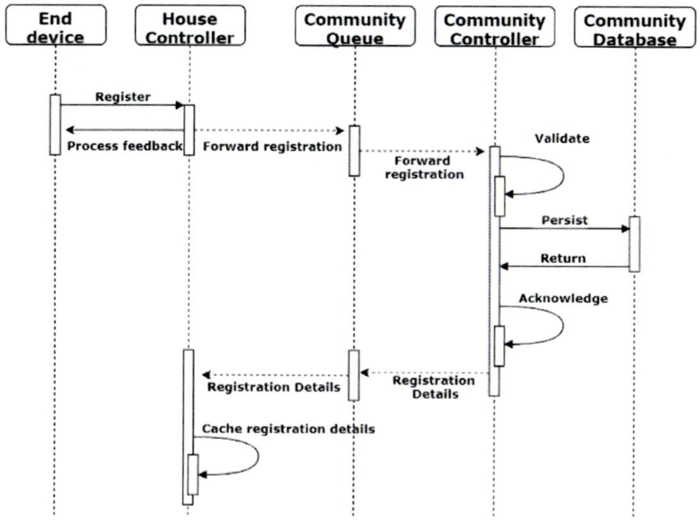

Fig. 2. Registration process diagram

splits into two asynchronous actions. First action stores the data locally inside of the house controller. Thus, the data persistency is assured in case the other external components are not available. In this scenario, the house controller may cache the electricity consumption data for several days, until the external components are functional. The other action forwards the data to the community controller using the community queue. Based on this data, the controller will analyze the overall statistics and take optimization decisions.

3.3 Control Mechanism

An important feature of the proposed architecture is the control mechanism over the end devices. This will support the optimization algorithms for the electricity consumption and allows the consumers to switch on/off their electric appliances when needed.

The starting point of the control mechanism is the community controller, which in our case runs periodically the optimization algorithms to decide if any interaction is needed. In Fig. 3 a graphical representation of a simple use case is presented.

The control mechanism steps are the following:

1. Community controller periodically calculates the total instant electricity consumption based on real time collected data;
2. If the real time consumption is approaching a threshold set by the optimization algorithms, then a list is created containing the end devices that will be switched off. These end devices are considered as flexible loads from the registration process and their main characteristic is that they can be interrupted at any time without disturbing the consumer's comfort;
3. For each end device in the list, a switch off command is sent via community queue to the house controllers;

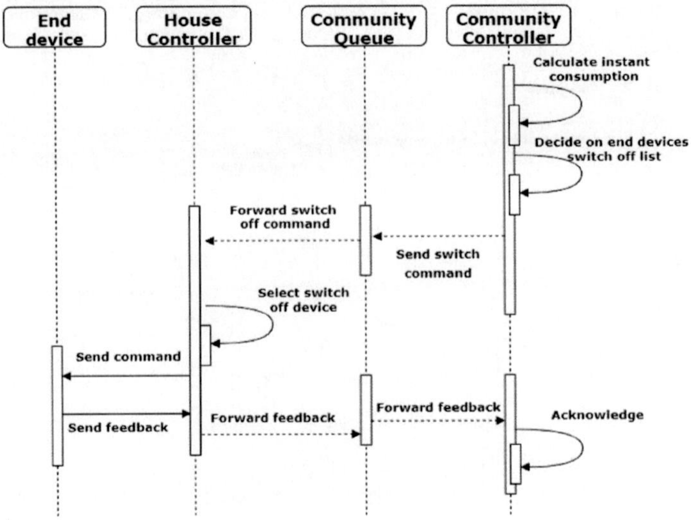

Fig. 3. The diagram of a sample case using the control mechanism

4. The house controllers receive the command and forward it to the end devices;
5. End devices switches off the corresponding home appliance and responds with a feedback message to the house controller;
6. The house controller takes the feedback message and forward it to the community controller via the community queue;
7. The community queue acknowledges the feedback message and saves the state of the end device.

3.4 Implementation

For the implementation of the proposed solution which combines the architecture together with the interaction flows and control mechanism we use an isolated environment, each component of the architecture being replace with a real-life implementation. The architecture implementation and its components are listed below:

- End devices – TP-Link HS110 smart plugs, which are able to read the electricity consumption data in real time and expose the information via a HTTP web service;
- House controller – Raspberry Pi model 3 smart controller, running a Java application;
- Community queue – Apache Kafka, which is an open source tool for distributing queuing;
- Community controller – Apache Spark cluster running an Apache Spark streaming application which is consuming all the messages from the queue in time frames (seconds interval);
- Edge database – Elasticsearch, distributed document type database, with the power of fast indexing for a rapid search functionality;
- Cloud database – Elasticsearch, selected for the same characteristics as the edge database;

- Enterprise application – Small group of micro services implemented in Java, which are serving all the required features;
- User interface (Enterprise) – ReactJs application which in background is connected with the Enterprise application via a REST interface.

The proposed optimization algorithm is running as Spark stream job, interrogating consumption metrics in real time. The total electricity consumption is calculated on the community level by aggregating the consumption of all end devices. Base on a threshold predefined by the electricity supplier, the algorithm decides when to switch off some home appliances that represent the flexible loads. The switching off process is done based on a flexibility order determined by a priority mark set at the registration time.

Meanwhile, if the energy consumption total on community level is going back under the defined threshold, the algorithm starts to switch on the home appliances, one by one, taking again in consideration the priority mark.

Figure 4 presents an overview of the developed user interface for controlling the end devices.

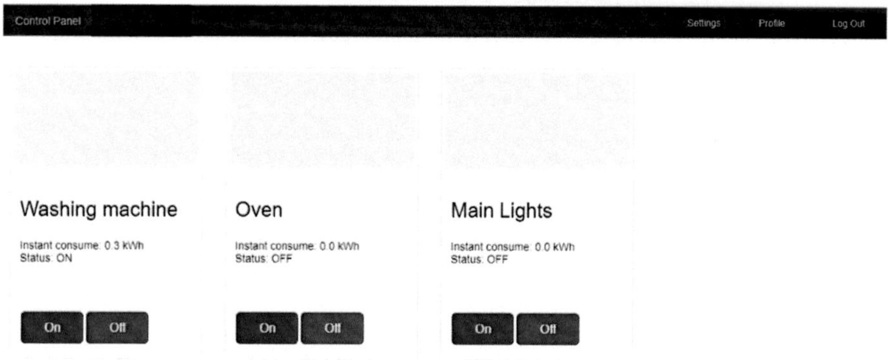

Fig. 4. User interface control panel

The proof of concept implementation was demonstrated in an isolated environment for 3 weeks. The outcome was good, all the interaction flows, control mechanism and optimization algorithm had robust results.

We identified after comparing the historical datasets of electric energy consumption, first one not having the optimization algorithm in place and seconds one using it activity, that the consumption was reduce with almost 10% in the case where the optimization algorithm was used. We are very pleased with the outcome or the algorithm in our demonstration, but we should take in consideration always that the results may be different base on community setup, consumer consumption strategy and openness and multiple other variables.

4 Conclusion

The proposed architecture for edge computing in the smart grid context contains three main areas for data and flows processing: household, community and enterprise cloud.

These areas group together several components used for communication and control of the end devices in the interaction with the edge core and cloud components. The interaction flows transfer the data between these components in real time, increasing the computing performance, the data persistency and enabling the control mechanism. Based on a threshold set by the optimization algorithm, the end devices are switch off according to their flexibility and priority mark. The proposed components of the architecture enable fast control and real-time monitoring of the electricity consumption, reducing the latency and the computational resources for data processing and algorithms. The implementation of these components uses open source software such as Apache Kafka, Spark, Java and Elasticsearch and affordable hardware components such as Raspberry Pi and well-known smart plugs. Therefore, the proposed architecture is open, flexible and can be extended easily with new modules or components such as distributed generation sources and storage devices.

Acknowledgement. This work was supported by a grant of the Romanian National Authority for Scientific Research and Innovation, CCCDI – UEFISCDI, project title "Multi-layer aggregator solutions to facilitate optimum demand response and grid flexibility", contract number 71/2018, code: COFUND-ERANET-SMARTGRIDPLUS-SMART-MLA-1, within PNCDI III.

References

1. Salman, O., Elhajj, I., Kayssi, A., Chehab, A.: Edge computing enabling the Internet of Things. In: IEEE World Forum on Internet of Things, WF-IoT 2015 - Proceedings (2015)
2. Minoli, D., Sohraby, K., Occhiogrosso, B.: IoT considerations, requirements, and architectures for smart buildings-energy optimization and next-generation building management systems. IEEE Internet Things J. **4**(1), 268–269 (2017)
3. Yu, W., et al.: A survey on the edge computing for the Internet of Things. IEEE Access **6**, 6900–6919 (2017)
4. Lin, J., Yu, W., Zhang, N., Yang, X., Zhang, H., Zhao, W.: A survey on internet of things: architecture, enabling technologies, security and privacy, and applications. IEEE Internet Things J. **4**(5), 1125–1142 (2017)
5. Zahaf, H.E., Benyamina, A.E.H., Olejnik, R., Lipari, G.: Energy-efficient scheduling for moldable real-time tasks on heterogeneous computing platforms. J. Syst. Archit. **74**, 46–60 (2017)
6. Sittón-Candanedo, I., Alonso, R.S., Corchado, J.M., Rodríguez-González, S., Casado-Vara, R.: A review of edge computing reference architectures and a new global edge proposal. Futur. Gener. Comput. Syst. **99**, 278–294 (2019)
7. Ai, Y., Peng, M., Zhang, K.: Edge computing technologies for Internet of Things: a primer. Digit. Commun. Networks **4**(2), 77–86 (2018)
8. Morán, A.J., Profaizer, P., Zapater, M.H., Valdavida, M.A., Bribián, I.Z.: Information and Communications Technologies (ICTs) for energy efficiency in buildings: review and analysis of results from EU pilot projects. Energy Build. **127**, 128–137 (2016)
9. García, Ó., Alonso, R.S., Prieto, J., Corchado, J.M.: Energy efficiency in public buildings through context-aware social computing. Sensors (Switzerland) **17**(4), 826 (2017)
10. Ejaz, W., Naeem, M., Shahid, A., Anpalagan, A., Jo, M.: Efficient energy management for the internet of things in smart cities. IEEE Commun. Mag. **55**(1), 84–91 (2017)

11. Ferrández-Pastor, F.J., Mora, H., Jimeno-Morenilla, A., Volckaert, B.: Deployment of IoT edge and fog computing technologies to develop smart building services. Sustainability **10**(11), 3832 (2018)
12. Barbierato, L., et al.: A distributed IoT infrastructure to test and deploy real-time demand response in smart grids. IEEE Internet Things J. **6**(1), 1136–1146 (2019)

The Study of Trajectories of the Development of State Capacity Using Ordinal-Invariant Pattern Clustering and Hierarchical Cluster Analysis

Alexey Myachin[1,2](✉) and Andrei Akhremenko[1]

[1] National Research University Higher School of Economics, 20 Myasnitskaya Street, Moscow 101000, Russia
amyachin@hse.ru

[2] Institute of Control Science of Russian Academy of Science, 65 Profsoyuznay Street, Moscow 117997, Russia

Abstract. This work is devoted to the methodology for identifying structurally close objects of the type "country_year" based on a system of indicators characterizing the state capacity 1996–2015. A comparison of clustering methods (including hierarchical clustering) with methods of analyzing patterns based on a pairwise comparison of indicators, ordinal-fixed and ordinal-invariant pattern clustering, is proposed. The possibility of sharing the methods of clustering and pattern analysis to obtain interpretable results from the point of view of political science is demonstrated. Groups of countries with similar development paths by reference years on the basis of a dynamic analysis of patterns are identified. The dynamic change in state capacity (from the point of view of the selected indicator system) of 166 countries of the world is determined.

Keywords: State capacity · Clustering · Pattern analysis · Ordinal-invariant pattern clustering

1 Introduction

Currently, there is an increasing interest in the study of state capacity. For example, the request "State Capacity" in Google Scholar gives 3,930,000 results (143,000 since the beginning of 2019). Despite the large number of works devoted to this topic, many problems arise with the operationalization of the concept. The following definition is proposed in [7]: "the ability of the state to choose and effectively implement its own decisions, including those related to domestic and foreign policy". In many works, for example [2] this (or a very close) definition is used.

Due to the complexity of defining the concept of "state capacity" (although there are many approaches [2, 4]), difficulties arise in defining a system of indicators that allows a quantitative assessment of the countries studied. For this purpose, a basic system of indicators was compiled, a correlation analysis was performed, and indicators having highly

I. Dzitac et al. (Eds.): ICCCC 2020, AISC 1243, pp. 198–205, 2021.
https://doi.org/10.1007/978-3-030-53651-0_17

correlated values were excluded. The result is the identification of 5 basic indicators: the quality of public institutions; the share of tax revenue in GDP; the share of military spending in GDP; the inverse of the total number of victims of conflict and homicide; the share of military personnel in the total population. This system is designed to identify groups of countries which are similar (using certain proximity measures) and study the dynamic development the paths of each of them. To compare state capacity in different countries, it is possible to use several basic approaches. The most popular include:

- the use of data classification methods. The application, for example, of neural networks to the initial set of countries allows us to build a model for correlating the studied objects into different classes. Despite the many advantages of such methods, their implementation requires the formation of a training sample of objects (which is very difficult for the task under consideration). In general, in order to use data classification methods, we need to know in advance the number of classes sought and their typical representatives;
- the compilation of a single aggregated rating using linear convolution (and analogues). In the general case, the correlation of a single numerical rating, which makes it possible to compile a generalized rating of countries depending on the value of this indicator, is a very popular method. Despite the simplicity of using this approach, its significant drawback (after the difficulty in determining the initial system of indicators) is the selection of weights for each criterion. Even assuming that weights can be determined (or made equivalent), when using this approach to search for objects with similar structures, the problem of low values of some parameters being compensated by the high values of others arises;
- the use of the theory of individual and collective choice. Since multidimensional numerical data are studied in the problem, it is possible to rank countries for each individual indicator, and then use various aggregation models (for example, the Borda's Rule, Hara's, Coomb's, Copeland's, Nanson's). If it is not necessary to compensate for the low values of some parameters with high values of others, it is possible to use threshold aggregation. However, in the presence of the same numerical values for different indicators, the studied objects will be assigned to a single group;
- the use of cluster analysis methods. The application of various clustering methods to the initial data set allows groups of objects with similar (to some extent proximity) parameter values to be revealed. From [9]: "A cluster usually means part of the data (in a typical case, a subset of objects or a subset of variables, or a subset of objects characterized by a subset of variables), which is distinguished from the rest by the presence of some uniformity of elements";
- the use of pattern analysis methods. Such methods can serve as a possible complement to data clustering methods for comparing and identifying objects related to a single group based on various metrics. A feature of this approach is the unification in a single group of objects which differ significantly in the absolute values of indicators but having similar data structures. From [10]: pattern is "a combination of certain qualitatively similar features".

This paper proposes a comparison of the results of ordinal-invariant pattern clustering and hierarchical clustering methods (using various proximity measures) for the formation of the dynamic development paths of countries. The main results of the decomposition are presented in [3].

2 Methodology

2.1 General Statement of the Problem

One of the main goals of this study is to determine the trajectory of the development of state capacity in 150 countries on the basis of the selected system of indicators and comparing the results obtained by 2 different methods: clustering and pattern analysis. Let us describe the problem in general form.

For the initial data, we study a set of objects of "country_year" type $W{:}|W| = 482$. The reference points are the data for 1996, 2005 and 2015. In total, 150 countries are studied in 1996, and 166 in 2005 and 2015. For convenience, the objects are set in vector form $w_i = (w_{i1}, w_{i2}, w_{i3}, w_{i4}, w_i\ w_{i5})$, where:

- w_{i1} is the quality of state institutions of the i-th object;
- w_{i2} is the share of tax revenues in the GDP of the i-th object;
- w_{i3} is the share of military spending in the GDP of the i-th object;
- w_{i4} is the inverse of the total level of victims of conflicts and killings of the i-th object;
- w_{i5} is the share of military personnel in the total population of the i-th object.

The objective of the analysis is to identify countries that are similar in terms of state capacity (in this case, different predetermined metrics are used to determine the similarity measure), and to identify the paths of development of state capacity in the studied countries (i.e. the change of development strategy in relation to the chosen system of indicators).

To so this, an analysis of the data is carried out and emissions are eliminated. Next, a correlation analysis is performed (for pattern analysis it is recommended to use data with relatively low correlation coefficients) and the subsequent normalization. In this work, we used linear scaling according to the formula:

$$w_{ij}^{*} = \frac{w_{ij} - w_{\text{min}\_j}}{w_{\text{max}\_j} - w_{\text{min}\_j}} \tag{1}$$

where w_{ij}^{*} is the normalized value of the j-th parameter of the i-th object obtained using linear scaling.

Further, using certain proximity measures described below (which are different for the methods of pattern analysis and clustering used in this work), groups of countries with a similar structure of state capacity are determined. Since there is no universally recognized system of indicators characterizing state capacity (or a general quantitative method for comparing levels in different countries of the world), the work compares the results obtained based on methods of pattern analysis and cluster analysis to identify relatively stable groups.

The last step is to study the path of the development of the level of state capacity. This work does not build a general rating of countries according to the level of state capacity or compare the obtained groups as "better or worse", but identifies groups with a similar structure of state capacity and identifies changes in the structure 1996-2015.

2.2 Pattern Analysis

Many works are devoted to the methods of pattern analysis. Some works demonstrate the possibility of using this method in analyzing electoral behavior (see [1]), however, this method is not used very often in political science.

In general, a pattern is understood as "a combination of certain qualitatively similar features" (see [10]). Thus, these methods are based on combining precisely qualitatively (and not quantitatively) close objects. Some of the requirements for the methods considered in this paper are also presented: the stability of the results to the choice of a sequence of input data; no need to set the number of final patterns to the implementing algorithm; the relatively high speed of the algorithm; no need to specify the composition (or typical representatives) of the final patterns.

As described above, 462 objects of the type "country_year" are examined, with 5 parameters characterizing state capacity. The methods of pattern analysis involve the use of a parallel coordinate system [8] to visualize the studied objects. This system contains (in the present study) 5 axes (according to the number of parameters), each of which describes a single parameter. Since each country-year object is assigned a 5-dimensional vector $w_i = (w_{i1}, w_{i2}, w_{i3}, w_{i4}, w_{i5})$, we can construct a curve through the points w_{i1}, w_{i2}, w_{i3}, w_{i4}, w_{i5}. In other words, a function g^w: R → R such that $g^w(i) = w_{i1}$ for all $i = 1,2,\ldots 5$.

After visualization, the question arises of choosing a partition metric for obtaining the results. There is an important feature in the analysis of patterns: the result should not depend on the absolute values of the selected parameters. For example, if hypothetical objects $w_1 = (1, 2, 3, 4, 5)$ and $w_2 = (10, 20, 30, 40, 50)$ are investigated, then (depending on the statement of the problem) most of the clustering methods will relate these objects in different groups (clusters). However, if you look at the values of the parameters, it is obvious that the objects w_1 and w_2 have a similar structure (the values of the parameters of the second object are the values of the parameters of the first object multiplied by 10). Methods of pattern analysis should combine such objects into a single group.

In this paper, we use an ordinal-fixed and an ordinal-invariant pattern clustering, described in detail in [10]. These methods are based on a pairwise comparison of all the studied parameters and the formation of object encodings for combining into groups. As a measure of proximity, the Hamming distance between the encodings of objects is used. However, there is an important addition: adjusting the results using centroids (the methodology is described in [11]).

After the objects are divided into groups, a dynamic analysis is performed. To this end, the development trajectories of the studied object are formed according to the selected indicators and objects with stable/unstable trajectories are identified.

2.3 Clustering

To compare the results obtained on the basis of ordinal-invariant pattern clustering, we use the methods of cluster analysis that have been successfully applied in a number of areas (including political science). Cluster analysis methods are the subject of many works [5, 12], in connection with which we will not describe in detail in this paper. However, we briefly describe the methodology used here.

To begin with, a specific metric, and, accordingly, clustering methods are chosen to solve the tasks posed in the work. The question of choosing the most suitable method has been asked for a long time. One of the classic examples is a hypothetical example of splitting a set of points shown in Fig. 1 [6]. This example shows that using different clustering methods on the same data set can lead to very different results. In this case, the question of the appropriateness of using certain methods and what method to use in each particular case is reasonable. Moreover, when comparing the results, it is important to understand which one can be used to solve the problem.

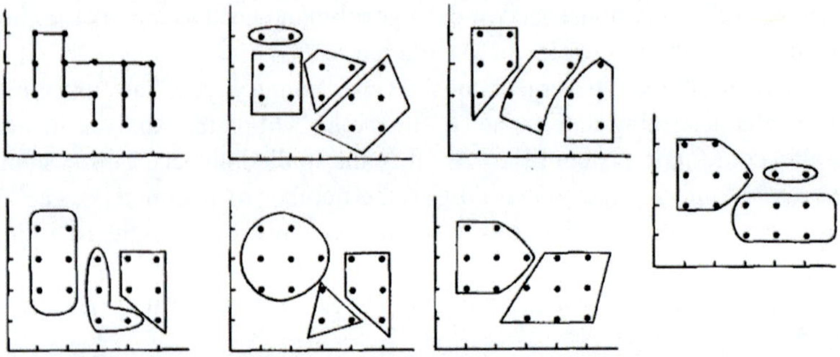

Fig. 1. An example of using different clustering methods on one data set. Figure modified from [6].

Answering this question is very difficult. As a general recommendation, you can use a result that has a well-grounded interpretation.

In the present work, the use of hierarchical clustering was chosen. For this purpose, based on the available data, a dendrogram was built (see Fig. 2).

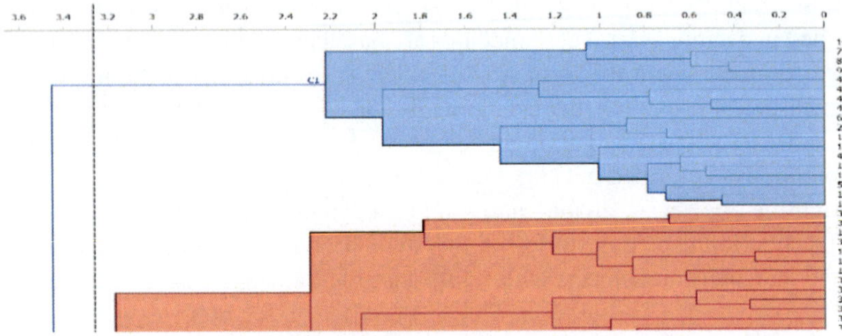

Fig. 2. An example of a part of constructing a dendrogram for studied objects.

As a measure of proximity using hierarchical clustering in this paper, the Euclidean distance is used (however, the research was conducted using other metrics). As a result, the initial set of objects was divided into 9 main clusters, which were subsequently compared with groups obtained using ordinal-invariant pattern clustering (with an adjustment of the results based on centroids of the obtained ordinal-invariant pattern clusters). This comparison is made to obtain sustainable results.

3 Results

3.1 A Comparison of the Results of Ordinal-Invariant Pattern Clustering and Hierarchical Cluster Analysis

Here are some groups of objects obtained, based on the methods of pattern analysis and clustering. Despite the difference in the approaches used, when correcting the results of ordinal-invariant pattern clustering on the centroids of each group, very different methods gave very similar results (Fig. 3).

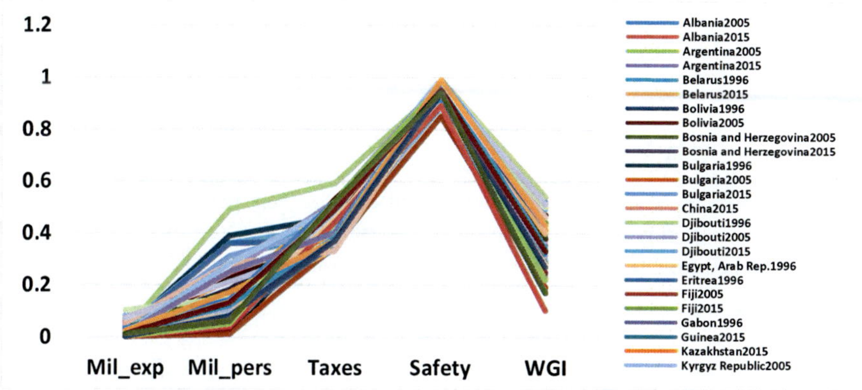

Fig. 3. Group 1. An example is the union of countries in terms of state capacity.

The methods used assigned these objects to a single group. Visually, one can note the distinguishing features: the relatively low values of the Mil_exp and Mil_pers indicators, the relatively average Taxes and WGI values, as well as the relatively high Safety values. This group characterizes 34 countries (63 objects).

Let us demonstrate one more example. Figure 4 shows the union of objects based on 2 different methods. The extremely high values of "WGI" and "Safety" indicators, the high and relatively high values of "Taxes", and the low and relatively average values of "Mil_exp" and "Mil_pers" are characteristic. This group includes 17 countries (total 50 objects), all of which (with the exception of Ireland) do not change their membership in this group for the study period: Denmark, Austria, Luxembourg, France, Norway, Belgium, Sweden, Australia, New Zealand, Canada, Germany, Ireland, Switzerland, the Netherlands, Iceland, Finland and the United Kingdom.

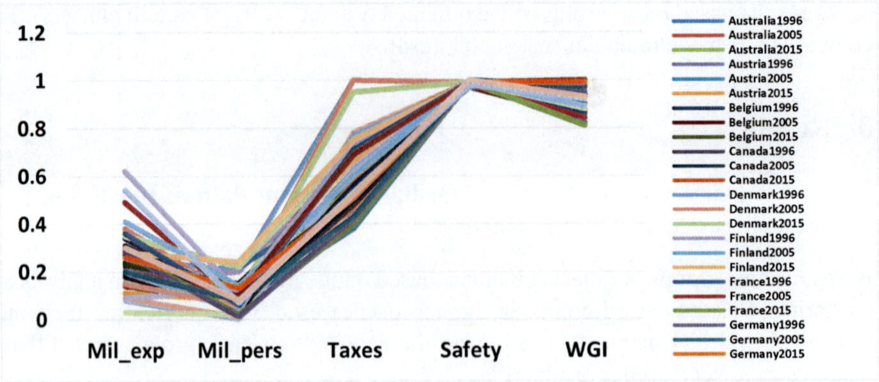

Fig. 4. Another example is combining objects into groups based on 2 methods.

A detailed description of the obtained division into the studied countries into groups is presented in [3].

3.2 Dynamic Analysis

There are 4 groups: "a–a–a" (not changing their development path), "a–b–b" (changing once), "a–b–a" (changing, but returning to the 1996 strategy), "a–b–c" (changing the strategy for each time period). An example of the results obtained is shown in Fig. 5.

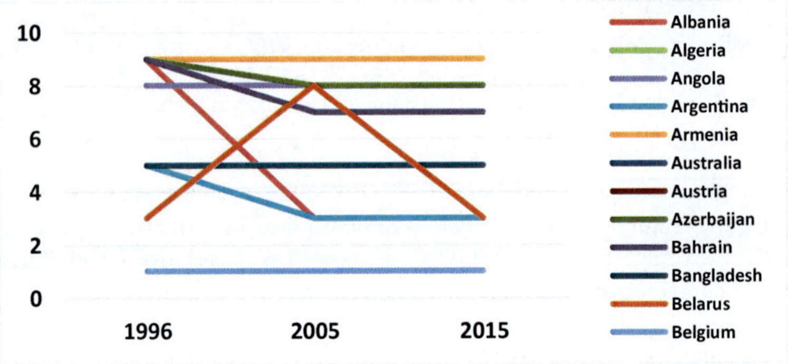

Fig. 5. Dynamic trajectories of development of state capacity in some countries.

From Fig. 5 it can be seen that countries such as Belgium and Armenia did not change their group for the entire period; Belarus changed once. A similar analysis was carried out for 150 countries.

4 Conclusion

This study investigates the trajectories of the development of state capacity in the countries of the world in the period 1996–2015. A basic system of indicators is defined, a comparison of 2 different methods (hierarchical clustering and ordinal-invariant pattern-clustering) is proposed to obtain stable results. The main emphasis is placed on the research methodology: comparing the results obtained using very different (according to the metrics and partitioning principles) data analysis methods: ordinal-invariant pattern clustering (adjusted for centroids) and hierarchical clustering.

Acknowledgements. The article was prepared within the framework of the Basic Research Program at the National Research University Higher School of Economics (HSE) and supported within the framework of a subsidy by the Russian Academic Excellence Project'5-100'.

References

1. Aleskerov, F., Nurmi, H.: A method for finding patterns of party support and electoral change: an analysis of British general and Finnish municipal elections. Math. Comput. Model. **48**(9–10), 1385–1395 (2008)
2. Akhremenko, A.S., Gorelskiy, I.E., Melville, A.Y.: How and why should we measure and compare state capacity of different Countries? Theoretical and methodological foundations. Polis. Polit. Stud. (RU), № 2, pp. 8–23 (2019). (in Russia)
3. Akhremenko, A.S., Myachin, A.L. Pattern Analysis and Clustering in the Study of State Capacity: "Adaptive Optics" for Political Science. Political Science (RU), pp. 112–139 (2019). (in Russia)
4. Charron, N.: Diverging Cohesion? Globalisation, state capacity and regional inequalities within and across European countries. Eur. Urban Regional Stud. **23**(3), 355–373 (2016)
5. Jain, A.K., Murty, M.N., Flynn, P.J.: Data clustering: a review. ACM Comput. Surv. (CSUR) **31**, 264–323 (1999)
6. Dubes, R., Jain, A.K.: Clustering techniques: the user's dilemma. Pattern Recogn. **8**, 247–260 (1976)
7. Kjær, M., Hansen, O.H., Thomsen, J.P.F.: Conceptualizing state capacity. Democracy, the State, and Administrative Reforms Research Report, vol. 6 (2002)
8. Inselberg, A.: Parallel coordinates: visual multidimensional geometry and its applications, vol. 20. Springer Science & Business Media (2009)
9. Mirkin B.G. Cluster analysis methods for decision-making support: overview: preprint WP7/2011/03. Moscow, 88 p. (2011). (in Russ.)
10. Myachin, A.L.: Pattern analysis in parallel coordinates based on pairwise comparison of parameters. Automat. Remote Control **80**, 112–123 (2019)
11. Myachin, A.L.: Determination of centroids to increase the accuracy of ordinal-invariant pattern clustering. Upravleniye bol'shimi sistemami (RU), № 78, pp. 6–22 (2019). (in Russia)
12. Xu, R., Wunsch, D.C.: Survey of clustering algorithms (2005)

A Micro Simulation Approach for a Sustainable Reduction Traffic Jam

Alejandra Valencia[1](✉), Cecilia Montt[1], Astrid M. Oddershede[2], Luis E. Quezada[2]

[1] Pontificia Universidad Católica de Valparaíso, Valparaíso, Chile
{alejandra.valencia,cecilia.montt}@pucv.cl
[2] Universidad de Santiago de Chile, Santiago, Chile
{astrid.oddershede,luis.quezada}@usach.cl

Abstract. Public transport represents an important traffic flow in many countries and urban traffic management has to face this situation. Chile is not an exception, particularly in the case of the most populated cities. To deal with this scenario, a simulation model has been built for a study case in a Chilean city to demonstrate how the problem can be approached. This case shows a serious conflict of public and private transport vehicles. These conflicts generate longer travel times between homes, work or study, higher vehicles operating costs and environmental impacts. The objective of the study is to propose a simulation model to produce and evaluate action plans to reduce traffic jam and the generated conflicts. The VISSIM^{TM} computational micro-simulation programme is used. The model simulates intersections where conflicts occur and it applies urban traffic management for an efficient use of roads. This computational micro-simulator uses a vehicle tracking model and lane change model plus other models that have been incorporated into this work. For the collection of information, field data were obtained such as vehicular flows, traffic light programming, speeds and measurement of queue lengths. Two models were simulated, where the best of them manages to mitigate the congestion problem and suggests changing the traffic, programming the traffic light from 120 s to 90 s and propose the use of a type short bus tracks for public transport. This result positively influence in speeds and queues, at 14% 18% respectively. Eventhough, the developed model solves a particular case, it can be tailored to other situations.

Keywords: Simulation · Traffic management · Public transport · Emissions

1 Introduction

1.1 A Subsection Sample

In many developing countries, traffic management becomes more relevant when a large number of public and private transport interact [1, 2]. Chile is not the exception, where in the main cities of the Country exists a large number of conflicts. The interaction of public buses and other vehicles, mainly cars, cause it, which is due to their differences in road behavior. In this complex reality, the use of modeling tools is an alternative for decision-making to manage the movement of the different modes of transport. They would allow

I. Dzitac et al. (Eds.): ICCCC 2020, AISC 1243, pp. 206–219, 2021.
https://doi.org/10.1007/978-3-030-53651-0_18

the estimation of benefits and costs of the measures that could be implemented at each site.

Simulation is one of the tools that has been used successfully in many areas, including the support for traffic management [3–5]. Regarding software systems, there is a wide range of microsimulators available in the market (AIMSUNTM, PARAMICSTM, TSISTM, VISSIMTM, etc.).

This research is based on a case study. It is the case of a road, where there is a complex intersection receiving a great flow of vehicles. In peak hours, there is a high degree of congestion (jam) in all of its arteries, which generates an increase in vehicular conflicts, longer travel times between households, work or study and higher operating costs of vehicles and environmental impacts. Therefore, the objective of the study is to propose a methodology to reduce congestion (jam) and the conflicts generated, simulating the intersection and applying urban traffic management for efficient use of roads.

The City Council Government would like to study the convenience of building an underpass to reduce the congestion. For doing this, we propose the utilization of the VISSIMTM microsimulation program. On the other hand, this work has to consider the government policy of giving a high priority to public transport, suggesting action plans such as a short bus track, redesigning bus stops based on bus flow and passenger demand and programming traffic lights using short cycle. Complementary to this, it is suggested to make efficient use of vehicles through the shared vehicle system in order to find a mobility plan for those who wish to use the private vehicle and make the use of public transport more attractive for different purposes that leads to reduce travel times, giving better use of the proposed road structure.

In the following sections, the importance of microsimulation, the characteristics of the study area, the methodology will be described, and the case study will be analyzed to subsequently present the results and conclusion.

2 Microsimulation

Microscopic traffic models are potentially useful for analyzing control strategies and making predictions of the operational behavior of the road network in different situations. Therefore, they constitute a fundamental tool for the analysis of management strategies, traffic control and vehicular traffic. They allow reproduce in detail the individual behavior of each vehicle and simulate the trajectories in time and space of each vehicle using models representing the interaction between them.

In this work, the VISSIMTM microsimulator is used, which uses a vehicle-tracking model and a lane change model, among others, that have been incorporated [6]. It describes the movement of a vehicle whose driver wants to drive faster than the speed presented by the preceding vehicles. If more than one lane is available, vehicles will tend to overtake, which is modeled by a rule based on lane change algorithms.

3 The Case Study Area Description

According to Government recommendations [7, 8], the road under study, named "Street I", has characteristics of being a road that has a high capacity for vehicle flows in both directions, with the total number of vehicles of approximately 6500 [veh/h], at peak hours (07:00–09:00 AM) and with a design speed between 50 and 80 [km/h].

On the other hand, there is another perpendicular road named "Street II". The intersection Street II-Street I adds a high flow of cars and public transport vehicles. In Street II the flow composition is different, where the circulation of taxis is relevant. There are also bus stops that facilitate the loading/unloading of passengers when they arrive to Street II, as well as the location of a Train Station nearby, which generates an important pedestrian flow. The intersection attract people traveling, like academic, students and workers, increasing the complexity.

In relation to the design features, the Street I axis connects North and South and intersects Street II. Both Streets I and II are bidirectional and irregular in terms of the number of tracks in certain sections. The Street I has 3 tracks, but when approaching the traffic light intersection reaches 5 tracks, with 2 tracks of exclusive turn towards Street II. This last street has 2 tracks in the East direction and 3 tracks in the direction of Street I with a median when approaching the intersection, used as a pedestrian rest, almost 2 m wide. Figure 1 shows a diagramme of the traffic flows, where the numbers correspond to the various directions of the vehicles movements. As an example, number 30 corresponds to the vehicles that turn from Street II to Street I. These numbers identify the movements in the simulation model.

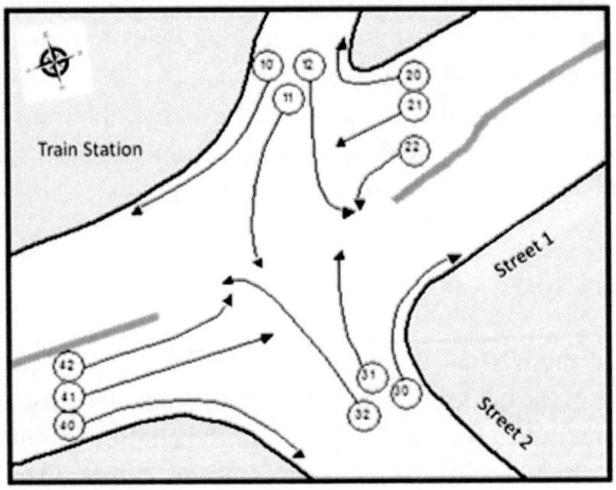

Fig. 1. Movements of the sector Street I and Street II

4 Methodology

For modeling the sector, it is necessary to carry out data collection in the field such as vehicle flows, traffic light programming, bus detention zones and the bus stops, bus and vehicle occupancy rates, speeds and queue lengths, in the current state. These data are used to specify the process of simulation of the current situation and the calibration of the model. The data collection during the calibration process for vehicle dimensions (specifically buses) and average speeds to model the traffic network was undertaken using the recommendations by the Government [7, 8].

Once the model is calibrated, a base situation (project proposed by the Government) is modeled, followed by an improved situation corresponding to the optimization of the base situation, the latter includes the proposals for shared vehicle analysis and the priority to public transport with only short bus stops tracks.

Finally, the result comparisons are obtained through the VISSIMTM microsimulator reports and conclusions are delivered.

5 Modeling and Calibration

This section explains the modeling and calibration procedure of the current situation in VISSIM 8.0 [9] of the Street I and Street II sector.

5.1 Input Data and Model

The simulation is carried out in the area depicted in Fig. 2(a). The flow of pedestrian from universities and the metro-train station is also included. The input data correspond to the information described in the methodology and the movements described in Fig. 1. The calibration was performed comparing the tail lengths and average speeds, when adjusting the parameters giving differences of less than 10%.

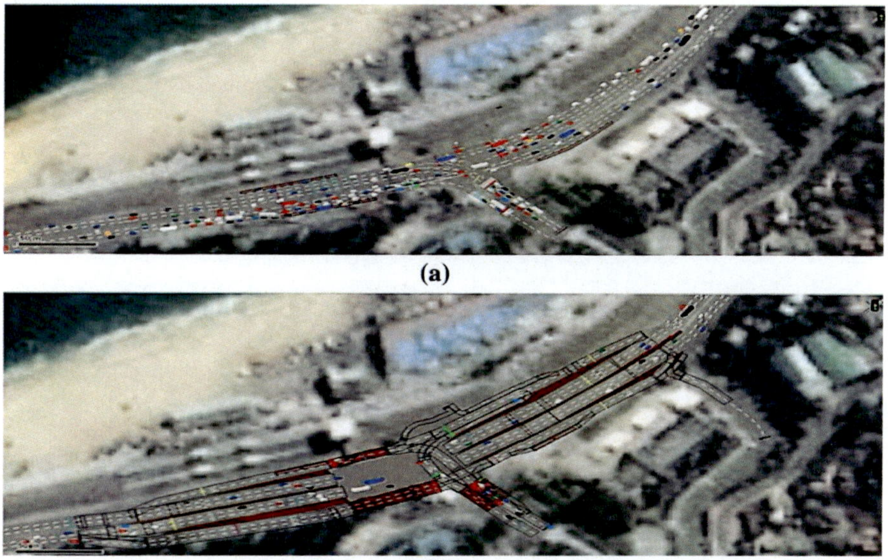

(a)

(b)

Fig. 2. **a**) Simulation area of the sector Street I-Street II. **b**) Proposed situation.

5.2 Results Analysis and Diagnosis

Figure 2(b) shows the VISSIM^{TM} simulation. A brief analysis with the VISSIM^{TM} microsimulator indicates that the intersection shows saturation due to the high demand for public transport interacting on bus stop and the strong presence of private light vehicles, driver behavior and lack of demarcations in the sector increasing inefficiencies for roads used. The above is verified with the results of performance indicators of detentions and queues, where in the first case the total detention rates exceed 1 det/vehicle and the queues reach a distance greater than 300 m in a congestion situation.

The traffic light that regulates the intersection is programmed in three times, which allows the passage of all the movements as shown in Fig. 3. It operates with a cycle of 142 s, 37 s of green for the access Street II and the remaining time is given priority to the flow coming from Street I. Therefore, in Street II there is the greatest saturation, where you can reach queues of more than 15 vehicles per track. This situation produces the total blockage of the axis, impeding the intersection upstream of Street II.

6 Modeling the Base and Improved Situation

This section presents the proposal of the government project as a base situation and as an improved situation, priority plans for public transport with short bus only tracks on Street I, traffic light optimization and use of the shared vehicle.

The project seeks to reduce directional conflicts and the impact of intimidation caused by the nature of the vehicular flow through the intervention of a underpass of 3 tracks per direction. This would cause that the greater volumes of vehicular flow have a continuous circulation, without being affected by the regularization, while other traffic volumes from Street II distribute their movements by the accesses regulated by the traffic light as observed in Fig. 3, simulated in VISSIM^{TM}.

Another situation that generates conflicts in the sector is the non-established stops to take and leave passengers, and the geometry, which prevent vehicles from maneuvering smoothly, thus decreasing traffic speeds [10].

Fig. 3. Government project of the sector Street I and Street II

With this proposal, a significant recovery of space for pedestrians is achieved, reducing not only congestion, gas emissions, idling effects and an increase in visual perception of safety.

However, the project does not contemplate an efficient operation of the use of bus stops, since their efficient use is given by the number of sites recommended according to the length of time of the buses in stop areas [11]. The buses currently have an average of permanence between 10–30 s and having a flow, greater than 180 [bus/h] per direction, theoretically justifies designing between 3 and 5 sites at bus stops with greater demand for passengers (Street I Axis); however, when the number of sites is 5, divided bus stops should be considered, since the efficiency is optimal at a maximum value of sites, if these are linearly arranged.

The improved situation has the same characteristics as the base situation, with the incorporation of traffic light optimization, implementation of short bus tracks only and shared vehicle (Fig. 2(b)).

6.1 Optimization of Traffic Light Programming

Regarding the traffic light, several traffic light cycles were modeled with VISSIM^{TM}, choosing a time that a short time cycle of 90 s with the same group of phases of the base situation, since it improves the performance of the intersection, especially in the access Street II. The assignment of a cycle like the one obtained showed that traffic light programming influences commercial speed if priority measures are implemented for public transport at traffic lights. Thus, as one might expect increases of up to 5% in the commercial speed if the traffic lights are programmed as established by [12] and an effective green ratio (green/cycle) equal to or greater than 0.6 that deliver better performance rates due to the decrease in public transport delays. Figure 4 shows the traffic light schedules for the base situation (120 s) and improve situation (90 s).

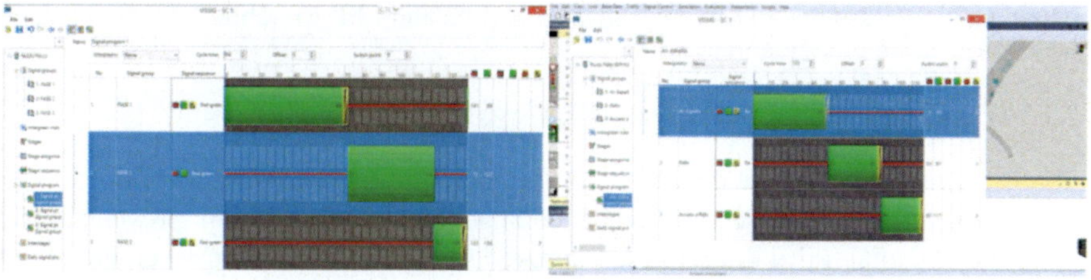

Fig. 4. Traffic light programming base and improved situation VISSIM^{TM}

6.2 Implementation of Short Tracks with Only Buses

These types of tracks are those in which traffic is allowed for only public transport, leaving private vehicles to the rest of the roads through fences or studs. These tracks include bus stops that vary their spacing according to the demand in each place [13]. The idea of these routes is that they work during peak hours, Monday through Friday from 7:30 a.m. to 10:00 a.m. and between 5:00 p.m. to 9:00 p.m., except holidays. The main advantage of this system is that it allows a large number of people to be transferred in a short time and at a much lower infrastructure cost than a meter.

Short only tracks with only bus must meet a minimum bus flow recommended according to some criteria such as [14] 60 [bus/h] per direction. On Street I, 359 [bus/h] circulate in each direction. Given the above, a redesign of the bus system must also be done, arbitrarily modeling 14 lines of higher frequency from 45 currently operating in bus stops as bus stop at Street I. The rest use underpass, without stopping indicated in the model.

6.3 Use of the Share Vehicle

Concerning to the case of the shared vehicle, only the input data (vehicle flow volume factors) are modified.

Taking into account that the use of the shared vehicle transforms possible drivers into possible passengers, it follows naturally that there is less participation of light vehicles within the network, for this reason, the volumes of the calibrated situation are reduced.

7 Results

The main focus of this section is that, through the simulation reports of the VISSIM^{TM} microsimulator, the general comparisons of some performance indicator are revealed.

The main results of the performance indicators of the case study are presented below, where (10) corresponds to private transport results and (30) to public transport results.

7.1 Travel Times

Figure 5 and Table 1 show the results of the travel time, where the travel times of the vehicles entering the intersection and the comparison of the base and improved situation show that the total travel times of the improved situation show an improvement of 24% for private transport and 42% for transport public regarding the base situation.

Table 1. Result of travel time VISSIM^{TM}

Origen-Destino	Base situation (min)			Improve situation (min)		
	Travtm(all)	Travtm(10)	Travtm(30)	Travtm(all)	Travtm(10)	Travtm(30)
Street 2 to Street 1 South (S)	8,9	8,2	12,6	6,99	7,5	7,2
Street 1 (S) to Street 1 North (N)	2,66	2,2	2,7	1,3	1,5	1,3
Street 2 to Street 1 (N)	0,59	0,57		0,6	0,6	
West to Street 1 (S)	8,1	8,2		6,7	6,2	
Street 1 (N) to Street 2	3,6	2,4	2,6	2,5	3,01	2,6
Street 1 (N) to Street 1(S)	4,2	1,3	4,6	3,1	0,97	3,2
Street 1 (N) to West	0,94	1,02		0,7	1,3,	

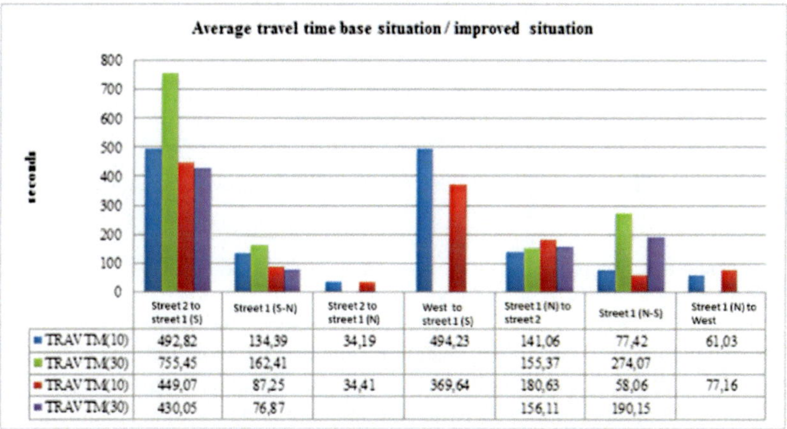

Fig. 5. Result of travel time base and improved situation

7.2 Queue

The total average queue lengths of the base situation are 47.58 m and the improved one has a total average tail of 38.99 m, showing a variation of 18%. This is shown in Table 2 and Fig. 6.

Table 2. Result of queue length VISSIM^{TM}

Queue length, base situation		Queue length, improve situation	
	Length (m)		Length (m)
South Access without and with underpass	37,31	South Access without and with underpass	2,8
Street 1 (S) with underpass	75,78	Street 1 (S) with underpass	79,02
Street 2	49,97	Street 2	45,92
North Access without and with underpass	86,6	North Access without and with underpass	66,82
Street 1 (N) with underpass	63,95	Street 1 (N) with underpass	37,51
West	2,79	West	0,04

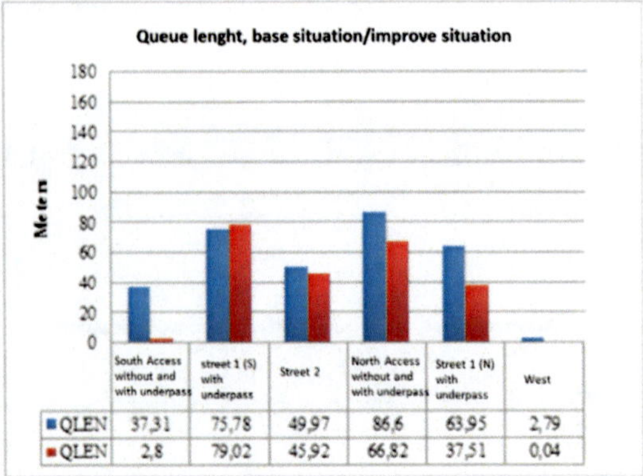

Fig. 6. Result of queue length base and improved situation

7.3 Intersection Stops

The number of stops is important, because from it derives the direct consumption of fuel and emissions of gases into the atmosphere. The results show that the detentions have an important improvement, mainly those of public transport with a decrease of 82.6% and 62% for private transport with respect to the base situation. This is shown in Table 3 and Fig. 7.

Table 3. Result of intersection stops VISSIM™

	Base situation (stops/veh)			Improve situation (stops/veh)		
	Stops(all)	Stops(10)	Stop(30)	Stops(all)	Stops(10)	Stop(30)
1	1	0,73	3,25	0,63	0,43	1,85
AVG	2,25	2	4,54	1,45	1,29	2,47
STDDEV	2,45	2,52	2,23	1,85	1,93	1,96
MIN	0,95	0,71	3,22	0,43	0,27	0,56
MAX	5,91	5,78	7,86	7,51	7,49	8,26

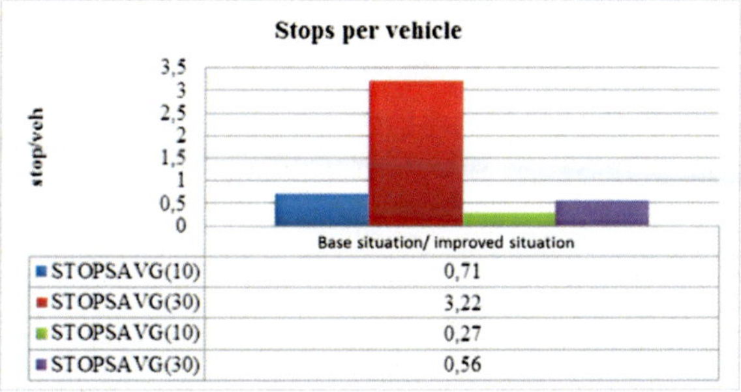

Fig. 7. Result of stops per vehicle in base and improved situation

7.4 CO_2 Emissions

One of the most important aspects today when evaluating projects, is the level of sustainability that offers at the optimization of processes, therefore, the proposals for this case are not alien to the requirements established regarding the environmental impact.

The results show that the emissions generated both in the base situation and in the improved situation fell from 1263 [gCO_2/km] to 686 [gCO_2/km]. While it is evident that the government proposal is an alternative solution that decreases several impacts of congestion, it does not manage to avoid congestion at the intersection accesses, since buses currently make free use of all stops and for that reason there are greater gas emissions to the atmosphere. Therefore, the emission reduction is 54.3% when incorporating the project.

7.5 Fuel Consumption

The level of emissions also implies lower fuel consumption from 18 [km/L] to 9.8 [km/L] which translates into a 50.2% decrease in the improved situation with respect to the base situation.

7.6 Average Speeds

The speeds have an increase with respect to the base situation, not only for private transport of 4 km/h, but also for public transport of 7 km/h. The variation between the base and improved situation translates into a 14% increase in travel speeds. This is shown in Table 4 and Fig. 8.

Table 4. Result of average speeds VISSIM^{TM}

	Base situation (km/h)			Improve situation (km/h)		
	Average speed(all)	Average speed(10)	Average speed 30)	Average speed(all)	Average speed(10)	Average speed(30)
1	34,52	39,91	15,90	39,22	45,38	20,66
AVG	30,64	35,28	13,56	35,30	39,88	21,98
STDDEV	6,93	8,27	3,37	5,63	7,10	6,90
MIN	20,30	22,90	8,69	20,30	21,87	10,84
MAX	34,66	39,91	15,90	42,32	49,25	37,86

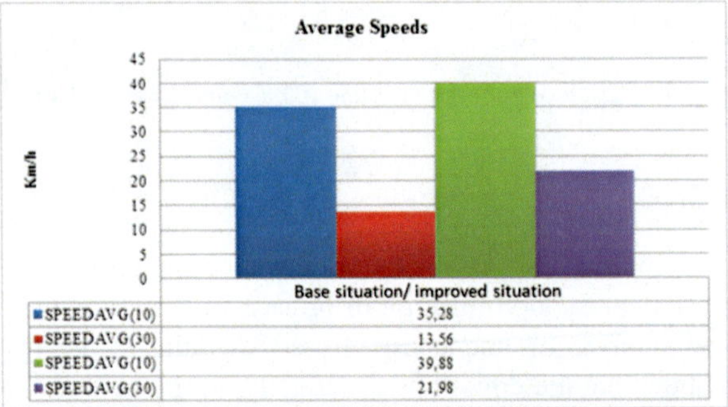

Fig. 8. Result of average speeds in base and improved situation

7.7 Delays

The speeds have an increase with respect to the base situation, not only for private transport of 4 km/h, but also for public transport of 7 km/h. The variation between the base and improved situation translates into a 14% increase in travel speeds.

The speeds have an increase with respect to the base situation, not only for private transport of 4 km/h, but also for public transport of 7 km/h. The variation between the base and improved situation led to an increase of 14% in travel speeds.

On the other hand, the hypothesis raised about the participation of the shared vehicle helps to reduce the presence of private vehicles, favoring a whole 38.3% increase in bus travel speed. This is shown in Table 5 and Fig. 9.

Table 5. Result of average delays VISSIM™

	Base situation (seconds)			Improve situation (seconds)		
	Delays(all)	Delays(10)	Delays(30)	Delays(all)	Delays(10)	Delays(30)
1	43,04	33,14	130,99	30,08	22,70	77,32
AVG	57,95	46,73	171,56	40,30	33,25	91,25
STDDEV	27,71	25,26	64,69	20,11	19,25	50,82
MIN	42,75	33,14	130,99	16,11	11,81	20,52
MAX	99,43	84,60	267,19	98,23	88,16	216,61

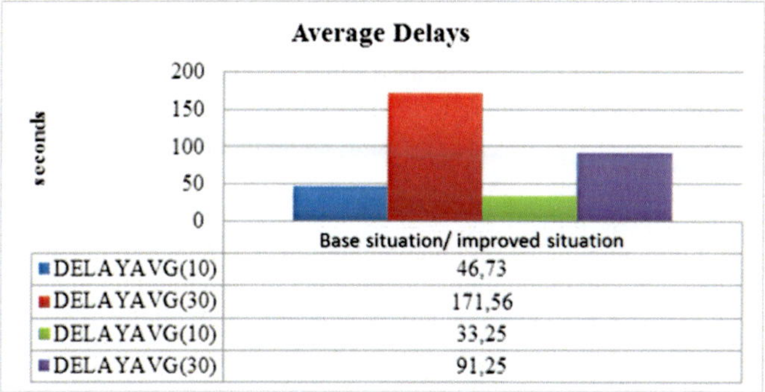

Fig. 9. Result of average delays in base and improved situation

8 Conclusions

Regarding the experience of using PTV VISSIM 8.0™, a friendly interface with great versatility and fluidity of work in modeling is rescued, but there are difficulties in certain aspects related to the reports and their subsequent interpretation. For example, there is no a distinction between the delay in bus stops and passengers, and in terms of speeds, those of public transport route are reported, but mention the commercial speed of these are not mentioned.

Another observation regarding the reports, are the indicators of urban traffic performance, specifically referring to the degrees of saturation and the capabilities of the road devices (sections of track, bus stops and intersections), these are also not explicitly specified.

Regarding the modeling, this work indicates in the first instance that the intersection under study obviously requires an intervention in the design of its road infrastructure. The Government proposal is advisable to be carried it out, since it reduces not only the congestion of the sector, but also positively influences others such as pollution. However, not considering the redesign of the system of public transport lines would cause a serious problem in terms of the proposed design, since the number of sites required for bus stop (supply) and the high presence of passengers (demand) would not have the adequate response capacity, with congestion occurring at the departures to on city (at morning peak hour) and as a course towards the other city (at evening peak hour).

In the area of the proposals presented about, considering the use of only short bus lanes in bus stop, the use of the shared vehicle in a minimum 30% and a traffic cycle of 90 s with respect to the base situation produce a mitigation the problem of congestion.

On the other hand and in a complementary way, it is of interest to use variable message signaling regarding the hours of use of these proposed tracks, in order to adequately inform the user of their presence.

The results obtained would have better performance if observations are considered in regard the design proposed by the Government by redesigning stops based on bus flow and demand according to some of the recommended criteria (such as TCRP recommendations [15].

Further Research

It is important to mention that the methodology used in this study case can be extended to other situations of traffic management. To do this, it is necessary to taylor the model to other cases. The results of this study would be the base line for building of a piece of software for future applications.

References

1. Hajinasaba, H., Persson, D.: On the use of on-line services in transport simulation. Transp. Res. Procedia **21**(2017), 208–215 (2016)
2. dos Santos Lopes, H., da Silva Lima, R., Leal, F., de Carvalho Nelson, A.: Scenario analysis of Brazilian soybean exports via discrete event simulation applied to soybean transportation: the case of Mato Grosso State. Res. Transp. Bus. Manage. **25**, 66–75 (2017)
3. Hajinasab, B., Davidsson, P., Holmgrena, J., Perssona, J.: On the use of on-line services in transport simulation. In: International Symposium of Transport Simulation (ISTS 2016 Conference), 23–25 June (2016)
4. Brügmann, J., Schreckenberg, M., Luther, W.: A verifiable simulation model for real-world microscopic traffic simulations. Simul. Model. Pract. Theory **48**, 58–92 (2014)
5. Simoni, M., Claudel, C.: A fast simulation algorithm for multiple moving bottlenecks and applications in urban freight traffic management. Transp. Res. Part B **104**, 238–255 (2017)
6. Wiedemann, R., Reiter, U.: Microscopic Traffic Simulation: The Simulation System MISSION, Background and Actual State. Proj. ICARUS Final Report, Brussels, CEC 2 (1992)
7. Ministry of National Planning (MIDEPLAN). Recommendation for the Design of Urban Road Space, Santiago of Chile (2009). (in Spanish)
8. SECTU. Methodology for Urban Road Projects, Ministry of National Planning (MIDEPLAN), Santiago of Chile (1988). (in Spanish)
9. VISSIM PTV: User Manual Vissim 7.0. Germany (2015)
10. Service of Housing and Urbanism (SERVIU), Engineering Design, Improvement of the Intersection España Aveniu/Pelle Street, City of Valparaíso, 3rd edn. Valparaíso, Chile (2013). (in Spanish)
11. TCRP: Guidelines for the Location and Design of Bus Stops. Report 19. Texas Transportation Institute. National Academy Press. Washington, D.C. (1996)
12. Gibson, J., Fernández, R.: Effect of an intersection downstream on the capacity of a bus stop with multiple sites. Apuntes de Ingeniería **19**(4), 31–40 (1996)

13. Camus, C.: Analysis of the Theory and Practice about Measures to Assign Priorities to Public Transport, Final Project for the Degree of Civil Engineering, Universidad de Los Andes, Chile (2014). (in Spanish)
14. Thompson, I.: Buses over Segregates Lines, Nueva Sociedad, No 212 (2007). ISSN 0251-3552
15. TCRP Transit Capacity and Quality of Service Manual, 2nd edn. Report 100. Transportation Research Board, Washington, D.C. (2003)

Empirical Versus Analytical Solutions to Full Fuzzy Linear Programming

Bogdana Stanojević[1](✉) and Milan Stanojević[2]

[1] Mathematical Institute of the Serbian Academy of Sciences and Arts, Kneza Mihaila 36, 11000 Belgrade, Serbia
bgdnpop@mi.sanu.ac.rs

[2] Faculty of Organizational Sciences, University of Belgrade, Jove Ilića 154, 11000 Belgrade, Serbia
milans@fon.bg.ac.rs

Abstract. We approach the full fuzzy linear programming by grounding the definition of the optimal solution in the extension principle framework. Employing a Monte Carlo simulation, we compare an empirically derived solution to the solutions yielded by approaches proposed in the literature. We also propose a model able to numerically describe the membership function of the fuzzy set of feasible objective values. At the same time, the decreasing (increasing) side of this membership function represents the right (left) side of the membership function of the fuzzy set containing the maximal (minimal) objective values. Our aim is to provide decision-makers with relevant information on the extreme values that the objective function can reach under uncertain given constraints.

Keywords: Full fuzzy linear programming · Extension principle · Fuzzy numbers · Monte Carlo simulation

1 Introduction

Mathematical optimization is an important topic of operations researches and is widely used in making good decisions. Including fuzzy concepts in mathematical models the researchers extended even more its applicability. In this study we analyze possible solutions to linear programming problems with both fuzzy coefficients and decision variables. Such problems arise when a real life system is modeled as an optimization problem under a certain kind of uncertainty. Li et al. [12] described a method able to transform heterogeneous information into trapezoidal fuzzy numbers, and Morente-Molinera et al. [15] used the sentiment analysis and fuzzy linguistic modeling procedures to organize the unstructured information to properly work with it.

Highlighting the limits and achievements of fuzzy approaches from the literature and comparing the use of quantitative and qualitative scales in measuring the decisions, Dubois [7] provided critical overview of the role of fuzzy sets in the field of decision analysis. One of the open questions formulated in [7] is

I. Dzitac et al. (Eds.): ICCCC 2020, AISC 1243, pp. 220–233, 2021.
https://doi.org/10.1007/978-3-030-53651-0_19

related to developing a methodology able to validate the fuzzy decision analysis approaches.

Ghanbari et al. [10] surveyed the models and solutions provided in the literature to fuzzy linear programming problems. One section was devoted to solving full fuzzy linear programming problems, including full fuzzy transportation problems. Their survey presented the practical applications of this class of problems and discussed some limitations of the existing solving methods.

Baykasoglu and Subulan [1] carried out a research on fuzzy efficient solutions to full fuzzy reverse logistics network design problem with fuzzy decision variables. Their study took into consideration different levels of uncertainty and the risk-averse attitude of the decision maker. Later on, Baykasoglu and Subulan [2] proposed a constrained fuzzy arithmetic approach for solving a wide variety of fuzzy transportation problems. They compared their solution approach with the state of the art approaches from the literature.

Pérez-Cañedo and Concepción-Morales [16] proposed a solution approach to derive a unique optimal fuzzy value to a full fuzzy linear programming problem with inequality constraints containing unrestricted LR fuzzy parameters and decision variables. Later on, in [17] they used the lexicographic optimization to rank LR-type intuitionistic fuzzy numbers and introduced a method to derive solutions to full intuitionistic fuzzy linear programming problems with unique optimal values.

Stanojević et al. [19] applied the interval expectation to trapezoidal fuzzy numbers and used it to transform the fuzzy linear programming problem into an interval optimization problem. An order relation was employed to rank the obtained intervals; and a parametric model was used to handle the acceptance degree of the violated fuzzy constraints. Finally, the Pareto optimal solutions to the parametric bi-objective linear programming problem were analyzed.

There are many studies in the recent literature announcing the usefulness of fuzzy linear programming models in their fields, and the possibility to use them to replace the classic linear models. For instance, Zhang et al. [22] proposed soft consensus cost models for group decision making, and discussed their economic interpretations. Emphasizing that fuzzy information is widely used to describe the uncertain preferences of the decision-makers when crisp values fail to represent the real viewpoints, the authors planed to include interval preferences in their linear models.

Liu and Kao [14] proposed a solution approach to a fuzzy transportation problem with crisp decision variables based on the extension principle. In order to simplify their approach they imposed a restrictive constraint of total supply less or equal to the total demand, thus loosing its generality. However, their approach is significant and can be extended to a more general one, namely to solve a full fuzzy linear programming problem. Liu [13] solved a fractional transportation problem with fuzzy parameters using the same solution concept based on the extension principle.

Ezzati et al. [9] applied fuzzy arithmetic and derived the triangular fuzzy value of the objective function with respect to the triangular fuzzy values of the decision variables and parameters. Then, he constructed a 3-objective crisp problem and solved it by a lexicographic method. We include one of the examples solved in [9] in our experiments; compare their solution to the empirical solution obtained by our Monte Carlo simulation; and draw some conclusions.

Bhardwaj and Kumar [4] pointed out a shortcoming that arose in [9] when fuzzy inequality constraints were transformed into equalities.

Kumar et al. [11] employed a ranking function (for the fuzzy number values of the objective function) and a component-wise comparison of the left and right hand sides of the constraints to transform the full fuzzy problem into a deterministic one. They finally solved one single objective linear programming problem deriving optimal values for all components of all decision variables.

Das et al. [6] introduced a new method to solve fully fuzzy linear programming problem with trapezoidal fuzzy numbers. Their method was based on solving a mathematical model derived from the multiple objective linear programming problem and lexicographic ordering method. They illustrated the applicability of their approach by solving real life problems as production planning and diet problems.

The main contribution of this paper is three-fold: (i) we define the membership functions of the fuzzy set solution components to a full fuzzy linear program; (ii) using a Monte Carlo simulation we compare the empirical solutions based on the new introduced definition to the analytical solutions provided in the literature; and (iii) we propose a model able to numerically describe the membership function of the fuzzy set of the feasible objective values.

Our presentation further includes: Sect. 2 that presents notation and terminology; Sec. 3 that introduces our advances in fuzzy linear optimization through the extension principle, and the new optimization model to depict the feasible objective values; Sect. 4 that reports our numerical results, and Sect. 5 that contains our concluding remarks.

2 Preliminaries

Zadeh [20] introduced the fuzzy sets as collection of elements with certain membership degrees. With respect to a universe X, a fuzzy subset $\widetilde{A}$ in X is generally defined by $\widetilde{A} = \left\{ \left(x, \mu_{\widetilde{A}}(x) \right) | x \in X \right\}$, where $\mu_{\widetilde{A}}$ is the membership function of $\widetilde{A}$, $\mu_{\widetilde{A}}(x)$ represents the membership degree of x in $\widetilde{A}$, $\mu_{\widetilde{A}}(x) \in [0,1]$.

The set of elements of the universe X whose membership degree in $\widetilde{A}$ is greater than α, $\alpha \in [0,1]$, is called the α-level set (or α-cut) of the fuzzy subset $\widetilde{A}$. It is formalized by

$$\left[\widetilde{A} \right]_{\alpha} = \left\{ x \in R | \mu_{\widetilde{A}}(x) \geq \alpha \right\}.$$

The set $\left\{ x \in R | \mu_{\widetilde{A}}(x) > 0 \right\}$ is called the support of the fuzzy subset $\widetilde{A}$.

The special fuzzy subsets of the real numbers universe $\mathbb{R}$ that are convex and normalized, and whose membership function is upper semi-continuous and has the functional value 1 at least at one element are called fuzzy numbers. They were introduced in [21] together with their elementary arithmetic operations. Any classic arithmetic operator between real numbers can be extend to a fuzzy operator between fuzzy numbers with the help of the extension principle.

A triangular fuzzy number $\widetilde{A} = \left(a^1, a^2, a^3\right)$, with $a^1 \leq a^2 \leq a^3$ is an especial fuzzy set with the membership function defined as follows:

$$\mu_{\widetilde{A}}\left(x\right) = \begin{cases} \left(x - a^1\right) / \left(a^2 - a^1\right), & a^1 \leq x < a^2, \\ \left(a^3 - x\right) / \left(a^3 - a^2\right), & a^2 \leq x \leq a^3, \\ 0, & \text{otherwise.} \end{cases}$$

For each $\alpha \in (0, 1]$ the α-cut of a triangular fuzzy number $\widetilde{A} = \left(a^1, a^2, a^3\right)$ is the interval

$$\left[\widetilde{A}\right]_\alpha = \left[\alpha a^2 + (1 - \alpha)\, a^1, \alpha a^2 + (1 - \alpha)\, a^3\right]. \tag{1}$$

An LR flat fuzzy number [8] is a quadruple $(m, n, \alpha, \beta)_{LR}$, $\alpha, \beta > 0$ whose components define its corresponding membership function as follows:

$$\mu_{\tilde{A}}\left(x\right) = \begin{cases} L\left((m - x)/\alpha\right), & x \leq m, \\ 1, & m < x < n, \\ R\left((x - n)/\beta\right), & n \leq x, \\ 0, & \text{otherwise,} \end{cases}$$

where L and R are reference non-increasing functions defined on the interval $[0, \infty)$, taking values from the interval $[0, 1]$, and fulfilling the double equality $L\left(0\right) = R\left(0\right) = 1$.

Bellman and Zadeh [3] formulated the extension principle widely used to aggregate the fuzzy subsets. Applying the extension principle, the fuzzy subset $\widetilde{B}$ of the universe Y that is intended to be the aggregation of the fuzzy subsets $\widetilde{A}_1, \widetilde{A}_2, \ldots, \widetilde{A}_r$ over their universes $X_1, X_2, \ldots, X_r$ through the function f that is a mapping of the Cartesian product $X_1 \times X_2 \times \cdots \times X_r$ to the universe Y is defined through its membership function as

$$\mu_{\widetilde{B}}(y) = \begin{cases} \sup\limits_{(x_1, \ldots, x_r) \in f^{-1}(y)} \left(\min\left\{\mu_{\widetilde{A}_1}\left(x_1\right), \ldots, \mu_{\widetilde{A}_r}\left(x_r\right)\right\}\right), & f^{-1}\left(y\right) \neq \varnothing, \\ 0, & \text{otherwise.} \end{cases}$$

In Sect. 3 we apply the extension principle to define a solution to full fuzzy linear programming (FF-LP) problems. Zimmermann [23] and [24] has already formulated solutions to fuzzy mathematical programming problems based on the extension principle. However, he solved another class of problems, namely vector optimization problems involving fuzzy goals and fuzzy constraints.

The extension principle was also used to define the elementary arithmetic operations over the set of fuzzy numbers. We recall the definitions for addition, subtraction and multiplication of triangular fuzzy numbers since we use them in the sequel.

Given two triangular fuzzy numbers $\widetilde{A} = (a_1, a_2, a_3)$ and $\widetilde{B} = (b_1, b_2, b_3)$, the basic arithmetic operations are defined as follows:

- addition: $\widetilde{A} + \widetilde{B} = (a_1 + b_1, a_2 + b_2, a_3 + b_3)$;
- subtraction: $\widetilde{A} - \widetilde{B} = (a_1 - b_3, a_2 - b_2, a_3 - b_1)$;
- multiplication: the exact result of $\widetilde{A} \cdot \widetilde{B}$ is commonly approximated by the triangular fuzzy number (c_1, c_2, c_3), where $c_2 = a_2 b_2$ and

$$c_1 = \min\left\{a_1b_1, a_1b_3, a_3b_1, a_3b_3\right\}, c_3 = \max\left\{a_1b_1, a_1b_3, a_3b_1, a_3b_3\right\}.$$

3 The Optimization via the Extension Principle

Without loss of generality an optimization problem consists in finding maximum of a real valued objective function f over a feasible set X. A formalized model is given by

$$\begin{array}{ll} \max & f(x), \\ \text{subject to} & \\ & x \in X, \end{array} \tag{2}$$

where, x is the vector decision variable and $f: X \to R$.

In a full fuzzy linear programming problem the objective function is linear; the feasible set is defined with the help of linear constraints; and uses fuzzy numbers for the both coefficients and decision variables. For a maximization problem, a formalized model is given below.

$$\begin{array}{lll} \max & \widetilde{f}(\widetilde{x}) = \sum\limits_{j=1}^{n} \widetilde{c}_j \widetilde{x}_j, & \\ \text{subject to} & & \\ & \sum\limits_{j=\overline{1,n}} \widetilde{a}_{ij} \widetilde{x}_j \preceq \widetilde{b}_i, & i = \overline{1,m}, \\ & \widetilde{x}_j \succeq 0, & j = \overline{1,n}. \end{array} \tag{3}$$

In order to define an optimal solution to Problem (3), we employ the crisp linear programming problem

$$\begin{array}{lll} \max & \sum\limits_{j=1}^{n} c_j x_j, & \\ \text{subject to} & & \\ & \sum\limits_{j=\overline{1,n}} a_{ij} x_j \leq b_i, & i = \overline{1,m}, \\ & x_j \geq 0, & j = \overline{1,n}, \end{array} \tag{4}$$

where the real numbers a_{ij}, b_i, c_j and x_j, $i = 1, \ldots, m$, $j = 1, \ldots, n$ belong to the supports of the fuzzy numbers $\widetilde{a}_{ij}$, $\widetilde{b}_i$, $\widetilde{c}_j$ and $\widetilde{x}_j$ respectively.

3.1 The Membership Functions of the Fuzzy Set Solution

Let us denote by $X_{a,b}$ the feasible set of Problem (4). Let $P_{(\widetilde{a},\widetilde{b},\widetilde{c})}(a,b,c)$ denote the aggregated membership level of all parameters,

$$P_{(\widetilde{a},\widetilde{b},\widetilde{c})}(a,b,c) = \min\left\{\mu_{\widetilde{a}}(a), \mu_{\widetilde{b}}(b), \mu_{\widetilde{c}}(c)\right\}, \tag{5}$$

where

$$\begin{aligned} \mu_{\widetilde{a}}(a) &= \min\left\{\mu_{\widetilde{a}_{ij}}(a_{ij}) \,|i = \overline{1,m}, j = \overline{1,n}\right\}, \\ \mu_{\widetilde{b}}(b) &= \min\left\{\mu_{\widetilde{b}_i}(b_i) \,|i = \overline{1,m},\right\}, \\ \mu_{\widetilde{c}}(c) &= \min\left\{\mu_{\widetilde{c}_j}(c_j) \,|j = \overline{1,n}\right\}. \end{aligned}$$

We follow Liu [13], and apply the extension principle to define the membership function of the fuzzy optimal value

$$\mu_{\widetilde{f}}(z) = \begin{cases} \max\limits_{(a,b,c)|z=\max\limits_{x\in X_{a,b}} c^T x} \left(P_{(\widetilde{a},\widetilde{b},\widetilde{c})}(a,b,c)\right), & \exists a,b,c|z = \max\limits_{x\in X_{a,b}} c^T x, \\ 0, & \text{otherwise.} \end{cases}$$

Liu [13] used such membership function to describe the fuzzy solution value to a transportation problem with fuzzy numbers and crisp decision variables. In addition, we use the extension principle to define each fuzzy optimal solution component as seen in Formula (6).

$$\mu_{\widetilde{x}}(x) = \begin{cases} \max\limits_{(a,b,c)|c^T x=\max\limits_{y\in X_{a,b}} c^T y} \left(P_{(\widetilde{a},\widetilde{b},\widetilde{c})}(a,b,c)\right), & \exists a,b,c|c^T x = \max\limits_{y\in X_{a,b}} c^T y, \\ 0, & \text{otherwise.} \end{cases} \tag{6}$$

3.2 The Mathematical Model

Keeping in mind above definitions we now introduce the mathematical model (7) that computes the membership degree of each feasible objective value z. It is clear that the decreasing piece of the so determined membership function describes the fuzzy set solution to a maximization problem (3). Model (7)

$$\begin{aligned} &\max \quad \alpha \\ &\text{subject to} \\ &\qquad \mu_{\widetilde{a}_{ij}}(a_{ij}) \geq \alpha,\ i = \overline{1,m},\ j = \overline{1,n} \\ &\qquad \mu_{\widetilde{b}_i}(b_i) \geq \alpha, \quad i = \overline{1,m}, \\ &\qquad \mu_{\widetilde{c}_j}(c_j) \geq \alpha, \quad j = \overline{1,n}, \\ &\qquad c^T x = z, \\ &\qquad x \in X_{a.b}, \end{aligned} \tag{7}$$

is non-linear and maximizes α with respect to variables x, a, b, c and α.

Model (7) consists in maximizing $P_{(\widetilde{a},\widetilde{b},\widetilde{c})}(a,b,c)$ over the feasible combinations of the parameters (a,b,c) able to derive a vector $x \in X_{a.b}$ such that $c^T x = z$, separately for fixed values z of the objective function. Indeed, for a value $\alpha = P_{(\widetilde{a},\widetilde{b},\widetilde{c})}(a,b,c)$ we have

$$\mu_{\widetilde{a}}(a) \geq \alpha,\ \mu_{\widetilde{b}}(b) \geq \alpha,\ \mu_{\widetilde{c}}(c) \geq \alpha,$$

with at least one equality among the inequalities. Therefore, the maximization of α determines the highest level of membership for the given crisp value z.

Considering that all parameters are expressed by triangular fuzzy numbers we use the definition of α-cut intervals (1) to rewrite the constraint system of (7). Since

$$\begin{array}{ll} \mu_{\widetilde{a}_{ij}}(a_{ij}) \geq \alpha \Leftrightarrow a_{ij} \in [\widetilde{a}_{ij}]_\alpha, & i = \overline{1,m}, j = \overline{1,n} \\ \mu_{\widetilde{b}_i}(b_i) \geq \alpha \Leftrightarrow b_i \in \left[\widetilde{b}_i\right]_\alpha, & i = \overline{1,m}, \\ \mu_{\widetilde{c}_j}(c_j) \geq \alpha \Leftrightarrow c_j \in [\widetilde{c}_j]_\alpha, & j = \overline{1,n}, \end{array}$$

we derive the specific Model (8) whose first three sets of constraints are linear.

$$\begin{array}{lll} \max & \alpha & \\ \text{subject to} & & \\ & \alpha a_{ij}^2 + (1-\alpha) a_{ij}^1 \leq a_{ij} \leq \alpha a_{ij}^2 + (1-\alpha) a_{ij}^3, & i = \overline{1,m}, j = \overline{1,n} \\ & \alpha b_i^2 + (1-\alpha) b_i^1 \leq b_i \leq \alpha b_i^2 + (1-\alpha) b_i^3, & i = \overline{1,m}, \\ & \alpha c_j^2 + (1-\alpha) c_j^1 \leq c_j \leq \alpha c_j^2 + (1-\alpha) c_j^3, & j = \overline{1,n}, \\ & c^T x = z, & \\ & x \in X_{a.b}. & \end{array} \tag{8}$$

Figure 1 shows a fuzzy set of the feasible objective values obtained by solving Model (8). The envelope of optimal objective values obtained by a dual optimization as suggested in Liu [13] can be also seen in Fig. 1.

3.3 The Monte Carlo Simulation

Monte Carlo algorithms are widely used to provide numerical results by a random sampling of certain parameters. Buckley and Jowers's book [5] on Monte Carlo methods in fuzzy optimization focused exclusively on generating random fuzzy numbers to be used in evaluating the objective functions. Our approach generates random crisp values for the parameters, finds crisp optimal values, and uses them in constructing the fuzzy set solution value to the original full fuzzy linear problem.

The definitions of the membership functions given in Sect. 3.1 are difficult to be directly used in practice. An analytical definition would be the most convenient, but such a definition is also hard (if not impossible) to be obtained. Model (7) introduced in Sect. 3.2 provides the upper bounds on the maximal values of the objective function with respect to their membership degrees, but lower bounds on those maximal values are still unknown.

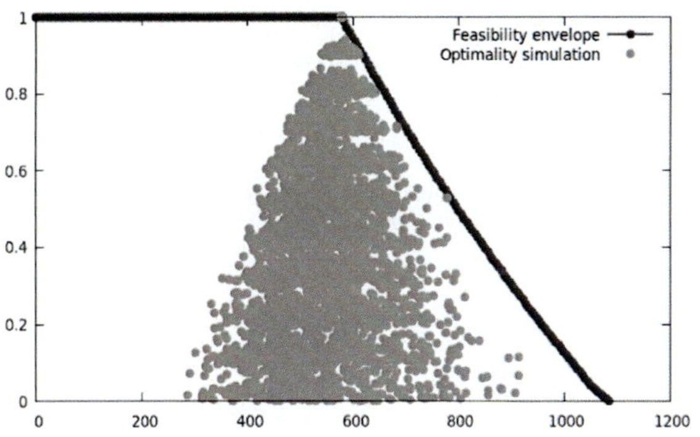

Fig. 1. Fuzzy set of feasible objective values obtained by solving Model (8); and the results of a Monte Carlo simulation that describe the fuzzy set of optimal solution values

---

Algorithm 1. The Monte Carlo simulation-based algorithm to disclose the shapes of the membership functions of the fuzzy set solutions to Problem (3)

Input: a natural number p; a sequence $\alpha_1, \alpha_2, \ldots, \alpha_p$ of equidistant values from $[0,1]$; and the membership functions of the fuzzy sets of the coefficients $\widetilde{a}$, $\widetilde{b}$ and $\widetilde{c}$.

1: Set $L = \varnothing$.

2: **for** $k = \overline{1,p}$ **do**

3: Randomly generate $a_{ij} \in [\widetilde{a}_{ij}]_{\alpha_k}$, $b_i \in \left[\widetilde{b}_i\right]_{\alpha_k}$, $c_j \in [\widetilde{c}_j]_{\alpha_k}$, $i = \overline{1,m}$, $j = \overline{1,n}$.

4: Solve Problem (4) with the generated coefficients.

5: Set $L = L \cup \left\{\left(x^k, z^k, P_{(\widetilde{a},\widetilde{b},\widetilde{c})}(a,b,c)\right)\right\}$, where z^k is the optimal value and x^k is the optimal solution to Problem (4); and $P_{(\widetilde{a},\widetilde{b},\widetilde{c})}(a,b,c)$ is defined as in (5).

6: **end for**

Output: the list L.

Under additional assumptions, Liu and Kao [14] described the lower bound on the maximal values of the objective function of a full fuzzy transportation problem. They made use of the dual of the crisp transportation problem to transform a two level min-max model to a two level max-max problem that was finally solved as an one-level max problem.

We aim to disclose the shapes of the membership functions of the fuzzy sets solution to Problem (3) by a Monte Carlo simulation which generates random values for the coefficients needed in obtaining optimal solutions to Problem (4). In order to obtain more accurate shapes, for fixed values of $\alpha \in [0,1]$, the values of the coefficients are randomly chosen from the α-cut interval of their corresponding fuzzy numbers.

Algorithm 1 describes the steps needed for the simulation.

The third component of the triple

$$l_k = \left(x^k, z^k, P_{(\widetilde{a},\widetilde{b},\widetilde{c})}(a,b,c)\right),$$

which is an element of the output list L of Algorithm 1, namely $P_{(\widetilde{a},\widetilde{b},\widetilde{c})}(a,b,c)$, represents a minorant of the membership degree of both x^k and z^k.

After running the algorithm we group the elements $l_k \in L$ with respect to their second component as follows: we split the interval $\left[\min_{l_k \in L} z^k, \max_{l_k \in L} z^k\right]$ in q sub-intervals I_1, I_2,..., I_q of the same length $\frac{1}{q}\left(\max_{l_k \in L} z^k - \min_{l_k \in L} z^k\right)$, and compute the values

$$f_j = M\left(\left\{z^k | l_k \in L, z^k \in I_j\right\}\right), j = 1, \ldots, q,$$

where $M(S)$ represents the mean value of the elements belonging to the set S. The membership degree of the value f_j is obtained as maximum of the values of the third component of those elements $l_k \in L$ that have the second component in I_j. In a similar way, grouping the elements of L in q sub-intervals of the same length, with respect to the values of the components of the vector x^k, we compute the membership values that correspond to the mean values of the components belonging to the same sub-interval.

This simulation provides an empirical computation for the membership functions of the solution and solution value to Problem (3) when solving Model (7) fails to obtain a correct solution due to its non-linearity. Model (7) is replaced by the quadratic Model (8) when triangular fuzzy numbers are used to describe the components of the parameters (a,b,c). In this case, and also in the case of the trapezoidal fuzzy numbers used to describe the parameters, the non-linearity appears in last two constraints $c^T x = z$ and $x \in X(a,b)$, and it is very likely one to be able to solve the model correctly. On the other side, if LR flat fuzzy numbers [8] are used to describe the parameters of Problem (3), the non-linearity appears in all constraints, and may create difficulties in finding a global optimal solution.

4 Numerical Example

We recall the following example from the literature ([9,11]).

$$\begin{array}{ll} \max & \widetilde{c}^T \widetilde{x}, \\ \text{subject to} & \\ & \widetilde{a}\widetilde{x} = \widetilde{b}, \\ & \widetilde{x} \succeq 0, \end{array} \tag{9}$$

where the fuzzy number values of the coefficients are

$$\widetilde{a} = \begin{bmatrix} (8;10;13) & (10;11;13) & (9;12;13) & (11;15;17) \\ (12;14;16) & (14;18;19) & (14;17;20) & (13;14;18) \end{bmatrix}, \tag{10}$$

$$\widetilde{b} = \begin{bmatrix} (271.75;411.75;573.75) \\ (385.5;539.5;759.5) \end{bmatrix}, \widetilde{c} = \begin{bmatrix} (10;15;17) \\ (10;16;20) \\ (10;14;17) \\ (10;12;14) \end{bmatrix}. \tag{11}$$

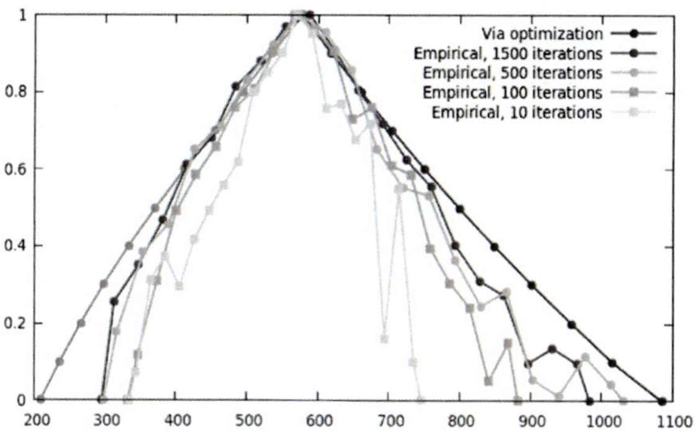

Fig. 2. The results of several Monte Carlo simulations with different number of iterations; and the envelope's approximation obtained via the optimization model

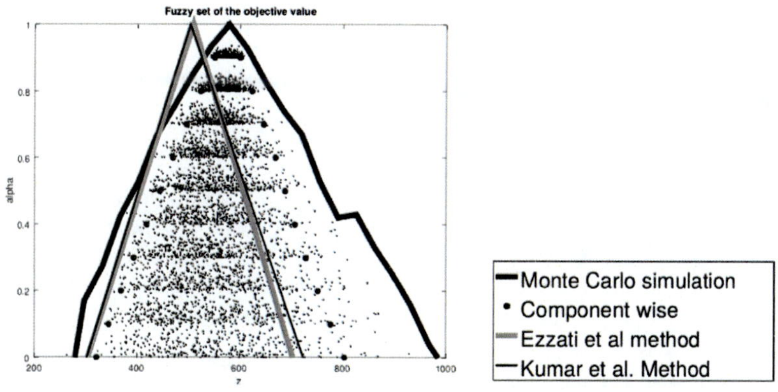

Fig. 3. Fuzzy set of optimal objective values to Problem (9) defined with the values of the parameters given in (10) and (11)

In short, the solution approach proposed in Kumar et al. [11] used a ranking function to transform the fuzzy number values of the objective function into crisp values; and a component-wise comparison of the left and right hand sides of the fuzzy constraints to transform them into deterministic ones. The obtained crisp linear problem was solved, and the optimal values for all components of all decision variables were derived. The fuzzy number value of the objective function was finally computed using the fuzzy number values of the decision variables and parameters.

The solution approach proposed in Ezzati et al. [9] first formally computed the components of the triangular fuzzy value of the objective function with respect to the triangular fuzzy values of the decision variables and parameters. Second, it constructed a three-objective crisp problem and solved it by a lexicographic method.

The component-wise method (described in Stanojević et al. [18] for solving full fuzzy linear fractional programming problems, but applied here to the linear case) was based on solving two crisp linear programming problems for several fixed α levels. The parameters for these crisp problems were the left and right endpoints respectively of the α-cut intervals. Their optimal solutions were used to construct the membership functions of the fuzzy numbers representing the final solution.

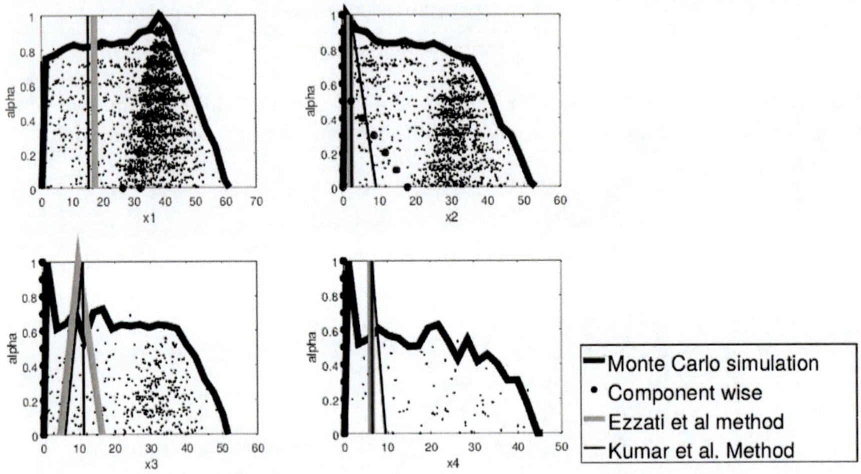

Fig. 4. Fuzzy sets of optimal values of the decision variables to Problem (9) defined with the values of the parameters given in (10) and (11)

Figures 3 and 4 show our empirical results obtained by a Monte Carlo simulation compared to the results found in the literature. Increasing the number of iterations, the simulation provides more accurate results (Fig. 2).

Aiming to find the maximal values that the objective function can reach we must pay attention to the right sides of the fuzzy numbers. It is clearly shown in Fig. 3 that both solutions from the literature are far from the desired values no matter how the membership degree is chosen.

Table 1. Components of the fuzzy sets representing the optimal values of the objective function

Method	Fuzzy number $\widetilde{z}^*$
Component-wise [18]	$(321.25, 578.04, 803.63)$
Ezzati et al. [9]	$(304.58; 509.79; 704.37)$
Kumar et al. [11]	$(301.83; 503.23; 724.15)$
MC simulation	$(279.37; 579.32; 985.13)$

Table 2. Components of the fuzzy sets representing the optimal values of the decision variables

Method	Comp.-wise [18]	Ezzati et al. [9]	Kumar et al. [11]	MC simulation
$\widetilde{x}_1^*$	$(32.12, 38.54, 42.39)$	$(17.27; 17.27; 17.27)$	$(15.28; 15.28; 15.28)$	$(0; 38.28; 61.24)$
$\widetilde{x}_2^*$	$(0, 0, 17.78)$	$(2.16; 2.16; 2.16)$	$(2.40; 2.40; 9.10)$	$(0; 1.32; 53.08)$
$\widetilde{x}_3^*$	$(0, 0, 0, 0)$	$(4.64; 9.97; 16.36)$	$(6.00; 11.25; 11.25)$	$(0; 1.28; 51.54)$
$\widetilde{x}_4^*$	$(0, 0, 0, 0)$	$(6.36; 6.36; 6.36)$	$(6.49; 6.49; 9.49)$	$(0; 1.14; 45.60)$

Tables 1 and 2 report the same solutions by their components as triangular fuzzy numbers. For the Monte Carlo simulation we approximate the results providing the extreme values, i.e. the minimum, the value with maximal amplitude and the maximum. For our analysis and graphic representations we run the component-wise method and reported the obtained solution, but used the numerical values reported in Ezzati et al. [9] for other two methods [9] and [11].

5 Conclusion

In this study we addressed the full fuzzy linear programming, and formalized the definition of the optimal solution via Zadeh's extension principle. We conducted a Monte Carlo simulation, and compared the solution derived in this way to the solutions yielded by approaches proposed in the literature. Concluding that there is a wide gap between our empirical results and the results provided by the solution approaches from the literature, we proposed a model that is able to describe numerically the membership function of the fuzzy set of feasible objective values that follows the extension principle. Our goal was to provide decision-makers with more relevant information on the extreme values that the objective function can reach under uncertain given constraints.

In our future research we will conduct more experiments on a wider class of mathematical programming problems (e.g. full fuzzy linear fractional programming problems) aiming to prove that the simulation we proposed can be used as a basic method to test the validity of any solution approach to full fuzzy programming problems.

Acknowledgments. This work was supported by the Serbian Ministry of Education, Science and Technological Development through Mathematical Institute of the Serbian Academy of Sciences and Arts and Faculty of Organisational Sciences of the University of Belgrade.

References

1. Baykasoglu, A., Subulan, K.: An analysis of fully fuzzy linear programming with fuzzy decision variables through logistics network design problem. Knowl. Based Syst. **90**, 165–184 (2015). https://doi.org/10.1016/j.knosys.2015.09.020

2. Baykasoglu, A., Subulan, K.: Constrained fuzzy arithmetic approach to fuzzy transportation problems with fuzzy decision variables. Expert Syst. Appl. **81**, 193–222 (2017). https://doi.org/10.1016/j.eswa.2017.03.040
3. Bellman, R.E., Zadeh, L.A.: Decision-making in a fuzzy environment. Manage. Sci. **17**(4), B—141–B–164 (1970). https://doi.org/10.1287/mnsc.17.4.B141
4. Bhardwaj, B., Kumar, A.: A note on 'a new algorithm to solve fully fuzzy linear programming problems using the molp problem'. Appl. Math. Model. **39**(19), 5982–5985 (2015). https://doi.org/10.1016/j.apm.2014.07.033
5. Buckley, J.J., Jowers, L.J.: Monte Carlo Methods in Fuzzy Optimization. Springer, Heidelberg (2008)
6. Das, S.K., Mandal, T., Edalatpanah, S.A.: A mathematical model for solving fully fuzzy linear programming problem with trapezoidal fuzzy numbers. Appl. Intell. **46**, 509–519 (2017). https://doi.org/10.1007/s10489-016-0779-x
7. Dubois, D.: The role of fuzzy sets in decision sciences: old techniques and new directions. Fuzzy Sets Syst. **184**(1), 3–28 (2011). https://doi.org/10.1016/j.fss.2011.06.003
8. Dubois, D., Prade, H.: Fuzzy sets and systems: theory and applications. In: Mathematics in Science and Engineering (1980)
9. Ezzati, R., Khorram, V., Enayati, R.: A new algorithm to solve fully fuzzy linear programming problems using the molp problem. Appl. Math. Model. **39**(12), 3183–3193 (2015). https://doi.org/10.1016/j.apm.2013.03.014
10. Ghanbari, R., Ghorbani-Moghadam, K., De Baets, B.: Fuzzy linear programming problems: models and solutions. Soft Comput. (2019). https://doi.org/10.1007/s00500-019-04519-w
11. Kumar, A., Kaur, J., Singh, P.: A new method for solving fully fuzzy linear programming problems. Appl. Math. Model. **35**(2), 817–823 (2011). https://doi.org/10.1016/j.apm.2010.07.037
12. Li, G., Kou, G., Peng, Y.: A group decision making model for integrating heterogeneous information. IEEE Trans. Syst. Man Cybern. Syst. **48**(6), 982–992 (2018). https://doi.org/10.1016/j.ejor.2019.03.009
13. Liu, S.-T.: Fractional transportation problem with fuzzy parameters. Soft Comput. **20**(3), 3629–3636 (2016). https://doi.org/10.1007/s00500-015-1722-5
14. Liu, S.-T., Kao, C.: Solving fuzzy transportation problems based on extension principle. Eur. J. Oper. Res. **153**(3), 661–674 (2004). https://doi.org/10.1016/S0377-2217(02)00731-2
15. Morente-Molinera, J.A., Kou, G., Pang, C., Cabrerizo, F.J., Herrera-Viedma, E.: An automatic procedure to create fuzzy ontologies from users' opinions using sentiment analysis procedures and multi-granular fuzzy linguistic modelling methods. Inf. Sci. **476**, 222–238 (2019). https://doi.org/10.1016/j.ins.2018.10.022
16. Pérez-Cañedo, B., Concepción-Morales, E.R.: A method to find the unique optimal fuzzy value of fully fuzzy linear programming problems with inequality constraints having unrestricted l-r fuzzy parameters and decision variables. Expert Syst. Appl. **123**, 256–269 (2019). https://doi.org/10.1016/j.eswa.2019.01.041
17. Pérez-Cañedo, B., Concepción-Morales, E.R.: On LR-type fully intuitionistic fuzzy linear programming with inequality constraints: solutions with unique optimal values. Expert Syst. Appl. **128**, 246–255 (2019). https://doi.org/10.1016/j.eswa.2019.03.035
18. Stanojević, B., Dzitac, I., Dzitac, S.: On the ratio of fuzzy numbers - exact membership function computation and applications to decision making. Technol. Econ. Dev. Econ. **21**(5), 815–832 (2015). https://doi.org/10.3846/20294913.2015.1093563

19. Stanojević, B., Dzitac, S., Dzitac, I.: Solution approach to a special class of full fuzzy linear programming problems. Procedia Comput. Sci. **162**, 260–266 (2019). https://doi.org/10.1016/j.procs.2019.11.283
20. Zadeh, L.A.: Fuzzy sets. Inf. Control **8**(3), 338–353 (1965). https://doi.org/10.1016/S0019-9958(65)90241-X
21. Zadeh, L.A.: The concept of a linguistic variable and its application to approximate reasoning I. Inf. Control **8**(3), 199–249 (1975). https://doi.org/10.1016/0020-0255(75)90036-5
22. Zhang, H., Kou, G., Peng, Y.: Soft consensus cost models for group decision making and economic interpretations. Eur. J. Oper. Res. **277**(3), 964–980 (2019). https://doi.org/10.1016/j.ejor.2019.03.009
23. Zimmermann, H.-J.: Fuzzy programming and linear programming with several objective functions. Fuzzy Sets Syst. **1**(1), 45–55 (1978). https://doi.org/10.1016/0165-0114(78)90031-3
24. Zimmermann, H.-J.: Applications of fuzzy set theory to mathematical programming. Inf. Sci. **36**(1), 29–58 (1985). https://doi.org/10.1016/0020-0255(85)90025-8

Critical Analysis of Faults in Operation of Energy Systems Using Fuzzy Logic

Victorita Radulescu(✉)

University Politehnica of Bucharest, Splaiul Independentei 313, 060042 Bucharest, Romania
vradul4@gmail.com

Abstract. Nowadays, establishing an efficient regime of functioning for hydraulic installations represents a complex problem requiring a large amount of calculation. This paper presents a multi-criteria method using the Fuzzy logic to improve the functioning of a pumping station, by minimizing the electric energy consumption and maximizing its efficiency in critical regimes. This method supposes the minimization of the maximum flow rate pumped, by establishing an efficient time interval between the starting and stopping of the installation. It has a significant effect on the energetic efficiency during functioning. The numerical model includes the flow rate consumption correlated with its optimum parameters, by planning a proper time functioning. A new model non-deterministic is introduced, associated with a Fuzzy controller system improved with a model of inference algorithm. This model is used to reduce the consumed flow rate with the help of the linear optimization and transition from a branched network analyzed as a neural network, to an annular one. The neural network is structured on five input variables and 9 hidden layers (seven as sub-input and two as output). The schematic structure of the implemented neural network is presented associated with its main objective and improved functioning. The Fuzzy numerical model is tested with the Matlab software, using a permanent function for the input and output variables. Some of the numerical results, conclusions, and references are finally mentioned.

Keywords: Fuzzy control · Neural networks · Optimization methods

1 Introduction

Fuzzy systems represent one of the modern methods of command and control for the management and automation for all kinds of systems, including the hydraulic machines. This method is applied to the most diverse installations and equipment, mainly due to its possibility of introducing, in a simple manner, of different types of non-linear conditions. The problem of improving efficiency represents a multi-criteria complex model, requiring a large amount of calculation. By using the Fuzzy method the optimization and operation of one hydraulic pumping system, aims to reduce the operating costs, minimizing the electric energy consumption, and maximizing its efficiency during the functioning. There are situations when the energy and economic optimization solutions do not coincide. A more general approach to these issues allows the formulation of a

I. Dzitac et al. (Eds.): ICCCC 2020, AISC 1243, pp. 234–244, 2021.
https://doi.org/10.1007/978-3-030-53651-0_20

mixed problem when the minimized power consumption is associated with the reduction of the number of units in operation. For increasing the efficiency of the pumping installations and its operating characteristics represented by the pressure drop and flow rate (which are nonlinear), it is necessary to impose additional restrictions for assuring the maximum transported flow.

By applying the Fuzzy logistic method for this problem the objectives are:

- Reducing the maximum pumped flow rate for the same structure, but in such manner that, at the same time, will provide the same hourly consumption.
- Energetic optimization of the installation operation, by functioning with the maximum efficiency, simultaneously with a decrease in the energy costs.

2 Fuzzy Systems Applied at Command of the Pumping System

A Fuzzy system is applied at the command of a pumping system structured on several pumps with constant rotation, powering a network system with variable flow demand.

The objectives of the command system are:

- Assurance of a random flow rate to consumers, with minimum energy consumed
- Increased the reliability of the aggregates, avoiding the cavitation phenomenon
- Ensuring a rational policy of reparations

For supplying the consumers in proper conditions can be considered two variants of exit from the pumping station:

- Ensuring a constant pressure at the discharge of the pumping station
- Assuring an increased pressure correlated to the flow rate required by consumers

In the present paper was adopted the second solution, closer to the estimated optimum performance. The considered linguistic input variables are the aggregate flow rate, the interval between the stopping and starting of the aggregate, the total number of operating hours since the last overhaul, and the safe operating pressure.

The operation with variable speed requires the introduction of a new variable physical and linguistic, the rotation speed [1]. In the Fuzzy analyzed processes were chosen simple functions, of type triangular or trapezoidal.

2.1 Fuzzy Method for the Fluid Flow Rate

The flow rate is an independent parameter and the Fuzzy system can be created only by association with the efficiency of the pumping system. It starts from the curve $\eta = \eta(Q_p)$, in which, for different values of the efficiency are determined for the flow rate, different domains [2]. If the physical variable coincides with the linguistic variable, can be defined, based on the correlation with the optimum value of the flow rate, the following seven areas:

- Very Small (VS), the flow rate is far away from the optimal value, much smaller
- Small (S) - rates comparable to the optimal value, but a little smaller
- Good (G) - around the optimum value
- Large (L) - comparable to the optimum flow rates, but a little higher
- High (H) - rates comparable to the optimal value, but a little higher
- Very large (VH) - far away from the optimum value, much higher.

In Fig. 1 can be seen how the domain of the flow rate has been divided into five areas, according to the linguistic definitions, associated with each specific field [3].

In these conditions there are defined some base-rules:

1. Flow rate - Too small - TS; necessary some supplementary fluid; open the tap
2. Flow rate is about the correct value, then no changes are necessary, just right-R
3. Flow rate is too high and closes the valve, Too high-TH

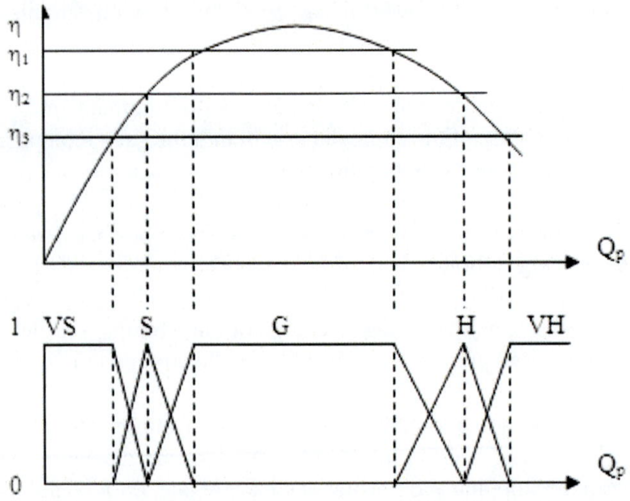

Fig. 1. Fuzzy functions for the input and output flow rate associated with the efficiency.

The intersection points are at 0.54 and 0.48. In these conditions, the flow rate intersects the mentioned areas consisting only of rules 2 and 3. The area of the right triangle is 0.1458 and the area TH is 0.1296. The output is determined by calculating the point at which a fulcrum can balance the two triangles. Solving the functional equations simultaneously $D_1 = 0.2859$ and $D_2 = 0.2541$.

2.2 Fuzzy Method for the Interval Between Stop-Start of the Aggregates

Applying the Fuzzy method for the time interval between switch off and start is dependent on the large currents necessary for the electric engine, which produces its delay. The non-permanent phenomena from the hydraulic circuits affect also this time interval. Maneuvers of on-off correspond to some Boolean variables and a changeover control, generating instability and shocks. In Fig. 2 is shown the application of the Fuzzy method for the range of time maneuver, using functions as semi-trapezoidal and triangular areas. The initial evolution of the temperature function of time T(t) is linear. Considering the physical and linguistic variable, the time between stopping and powering can be defined in three areas:

- Too Short (TS) - a too short time interval, inadmissible, between two maneuvers
- Little Short (LS) - a short interval between maneuvers, but acceptable
- Correct (C) - sufficient time between maneuvers

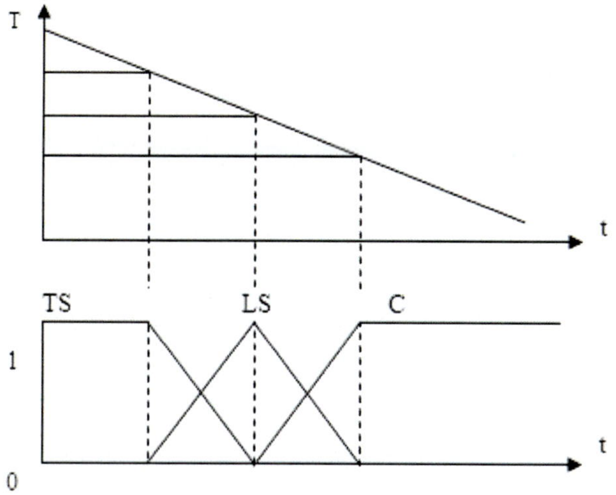

Fig. 2. Fuzzy functions for the time interval between two maneuvers.

Limits for this time interval depend on T-the winding temperature of the electric motor, and its design characteristics. In Fig. 3 are presented two versions of Matlab windows for sequences between stopping and starting: 1 - immediate start (minimum time interval) and, 2 - a break between the two intervals of maneuver.

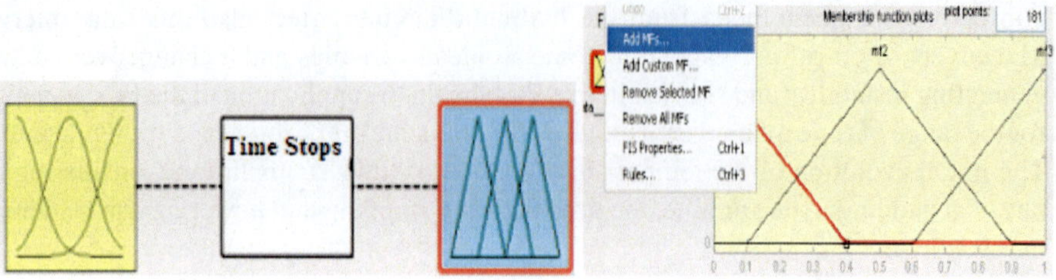

Variant 1 Variant 2

Fig. 3. Matlab windows in both cases.

2.3 Fuzzy Method for the Wear of the Aggregates

In a pumping station with several units may appear several politics of exploitation, but they can be divided into two extreme situations [4].

Politics of the service aggregate, which establishes an aggregate that starts working the first and stops the last. Therefore he has the maximum wear, the second which started has smaller wear, etc. In terms of efficiency, this policy requires that the first aggregate has the highest yield, the second lower than the first, and the last the smallest, with the lowest yield and most rarely used. After an operating period, the service unit must be repaired or replaced, and its place is taken by the second aggregate in the starting order. This policy of pumping assures permanent functioning without ceasing the operation during the repairs; for example the supply stations for drinking water.

The equalization policy of the wear rate is based on the idea of starting or maintaining in operation the aggregates so, their wear is about equal. A demand for a supplementary flow rate starts with the pump with the lowest number of the operating hours. Thus, all units achieve the maximum degree of wear at approximately the same time. To repair the station this one must be completely stopped; as an example an application is the irrigation seasonal [5]. The Fuzzy method in this case, is more difficult to be expressed because the wear variation is nonlinear. By considering the physical and linguistic variables for the wear degree, the areas may be defined as:

- Reduced (S) - aggregate with low wear
- Medium (M) - average degree of wear
- High (H) - advanced state of wear

The Fuzzy method, assuming a linear variation of the wear, is applied in Fig. 4. The colored portions are places where operate only one qualifier.

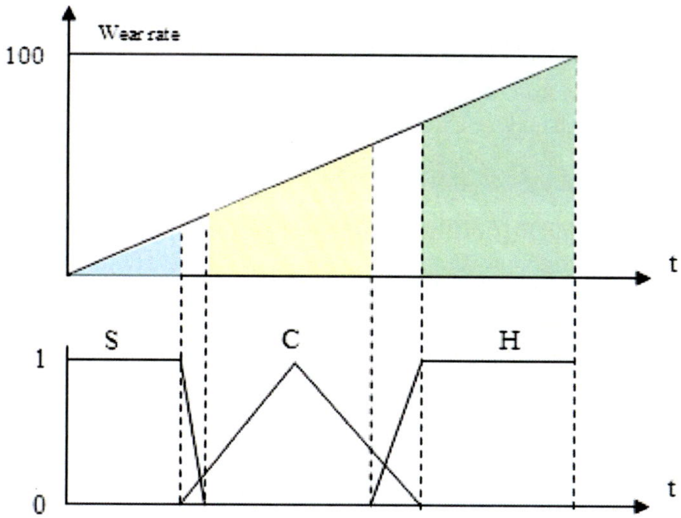

Fig. 4. Fuzzy function for the wear rate.

Applying the Fuzzy method at this objective may have different basic criteria structured on comparing the number of functioning hours, the energy used and the flow, or the volume pumped [5]. At the request of an additional flowing rate the strategy of the sacrificed aggregate requires to be stopped the unit with the smallest degree of wear, while the policy of equalization of the wear rate imposes the unit with the highest wear rate [7]. The relationship between the four input variables involves a four-dimensional matrix, which will be represented by many normal matrices.

In Table 1 is mentioned a normal matrix for the variables Q - flowing rate, and T - time between on-off. The output is U - the command of starting the aggregate.

From linguistically have been applied these rules of interference:

– If T - time since the last stop is TS, and Q is TS then for any request of the flowing rate from TS to H the answer is N (no), considering that the aggregate can not support a new start.mm

Table 1. Table with rules of interferences.

Output U		Q				
		TS	S	C	L	H
T	TS	N	N	N	N	N
	LS	P	P	N	N	N
	G	P	P	P	P	P

- If T is LS, and Q is TS or S, then U is P (positive), is considered that the load of the unit will be small and of short duration.
- If T is LS and Q is from R to H, the answer is N, considering that the heat load of the engine will be high and for a long time.
- If T is G and Q is from TS to H, answer is P and no reason to refuse the startup.

3 Improving the Functioning of the Hydraulic Installations

The improvement of the energy consumed by an installation may concern in [6]:

- Determination of the optimal flowing rate of the pumping aggregates, based on the instantaneous yield of the machines.
- Determination of the optimum pumped volume in a certain period, based on consumer demand.

For the analyzed installation, the Glina pumping station, from the south part of the Bucharest, the obtained results are set in Table 2.

The main solutions to the problems previously mentioned are:

- For reducing the maximum flow pumped, by maintaining the consumer demand and the hourly structure, some supplementary tanks are necessary for alimentation
- For optimizing the produced energy, without any new investment requires the utilization of the aggregates, in conditions as close to their maximum efficiency. This means some steps for the pressure and adaptation at the installation structure
- To increase the energy with new investments is the simplest way and can be achieved by replacing the outdated equipment with new ones, more efficient
- Replacement of the branched network with an annular one
- Economical efficiency of the installation functioning

Table 2. Obtained results for the Glina pumping station

Nr.	Q_p [^3_m/s]	t_i [h]	H_{pi} [m]	η_ι	V_i [m^3]	H_iV_i [m^4]	H_iV_i/η_ι [m^4]
1	1	3	20	0.91	10800	216000	237362
2	1.3	1	20.5	0,92	4670	95940	104283
3	1.5	1	21.5	0.93	5400	116100	124839
4	0.8	2	19.2	0.9	5760	110592	122880
Total	4,6	7	–	–	26640	538632	589364

In these conditions, the obtained average values may be expressed as:

$$H_{p,av} = \frac{\sum_i H_{pi} Q_{pi} t_i}{\sum_i Q_{pi} t_i} = 20,22\,\mathrm{m} \tag{1}$$

$$\eta_i = \frac{\sum_i H_{pi} Q_{pi} t_i}{\sum_i \frac{H_{pi} Q_{pi} t_i}{\eta_i}} = 0,914 \tag{2}$$

$$E = \frac{1}{367} \sum_i \frac{H_{pi} Q_{pi} t_i}{\eta_i} = 1606\,\text{kWh} \tag{3}$$

Where: $H_{p,\,av}$ – the average pumping head, Q_p – the flowing rate, t – the time of functioning, η – installation efficiency, and E – the consumed energy.

4 Analyze the Pumping Stage by Neural Network

It was assumed one unidirectional multi-layers network, with total connections to the consumers between two layers the output and input, accomplished by a second hidden layer. The input layer of the neural networks receives different input variables.

The variable components are determined by Fourier series, keeping the significant harmonics, for T - the time of functioning

$$S\_T_w(t) = \bar{T}_w + \sum_{i=1}^{N} \left[a_{1i} \cos\left(\frac{2\pi it}{365}\right) + a_{2i} \sin\left(\frac{2\pi it}{365}\right) \right] \tag{4}$$

Here: $\bar{T}_w$ – average multi-annual recorded values, a_{1i}, a_{2i} – coefficients, i – the harmonic, with $i = 1, \ldots, N$, and t – is the range of the day from the year, from 1 to 365. Based on some preliminary studies the structure of the series (4) was tested having as input, seven variables, Fig. 5.

The crossover structure of the correlation coefficients between the residual series of the water flow rate and time intervals is analyzed. Secondary correlations with other parameters such as efficiency, ware rate, etc. were also considered but associated with the average parameters, considered as sliding values, respectively the influence of the recorded parameters in the preceding days, $F\_\overline{T_w(j, j-1)}$, $F\_\overline{T_w(j, j-2)}$, $F\_\overline{T_w(j, j-3)}$, $F\_\overline{T_a(j, j-1)}$, $F\_\overline{T_a(j, j-2)}$, $F\_\overline{T_a(j, j-3)}$.

The residual values were considered up to the value $j-3$ for Q and T. Contact points W_{ij}, W_{jk} are adjusted by the retro-propagation method, with a gradient algorithm, to minimize the mean square error (ASE - Average Square Error) between the output of the neural networks F_T_w and the relevant values observed as input F_T_a. To control the overload of the neural network the data series was separated into two areas:

- Even days, considered the values for synchronization, used for the calculation of the parameters W_{ij}, W_{jk}, consisting of 9420 data.
- Odd days, values used for the validation and testing of the neural network on data that are not involved in the estimation of the parameters.

Taking into account the utilization of the sigmoid function, the data were converted into the domain (0, 1). Consequently, for example, a value $t = 10$ represents the average

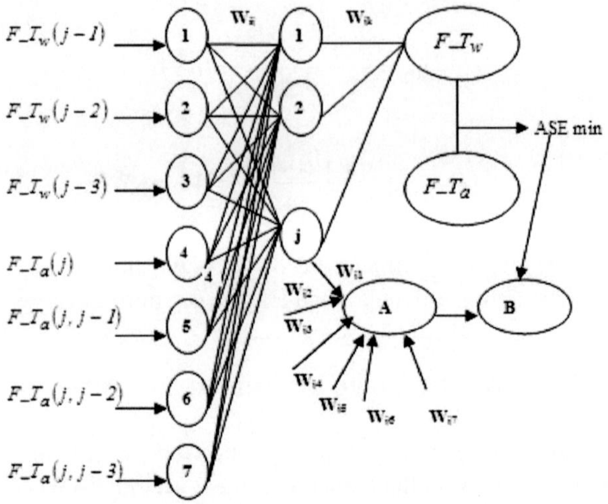

Fig. 5. The architecture of the neural networks multi level tested.

values recorded on 10 March, during the period 2010–2019. Each component can be graphically determined as $S\_Q_w(t)$, $S\_T_w(t)$, and all other components. For their multi-annual calculation was chosen $R^2 = 0.98$ for Q and $R^2 = 0.94$ for T.

$$S\_Q_a(t) = 12.311 - 2.3637 \cdot \sin\left(\frac{2\pi t}{365}\right) - 6.9353 \cdot \cos\left(\frac{2\pi t}{365}\right) \tag{5}$$

$$S\_T_w(t) = 14.153 - 2.4939 \cdot \sin\left(\frac{2\pi t}{365}\right) - 8.427 \cdot \cos\left(\frac{2\pi t}{365}\right) \tag{6}$$

Residual series of both characteristics are presented in Table 3, where $j = 1 - 9420$, taking into account the self-correcting coefficients, for different timing offsets.

Table 3. Characteristics of the residual series

	Min (m^3/s)	Max (m^3/s)	Av (m^3/s)	R L_d1	R L_d2	R L_d3	R L_d4	R L_d5
$F\_T_a(t)$	14.8	8.7	0.2	0.76	0.54	0.4	0.31	0.26
$F\_T_w(t)$	7.2	6.8	0.1	0.95	0.85	0.74	0.64	0.55
$F\_T_w(t)/F\_T_a(t)$				0.7	0.78	0.76	0.69	0.61

Here L_d1, L_d2, etc. represent de delay for one day, for two days, etc. In terms of the cross-correlation can be noted that the flow rate noted j-day, is better correlated with the T registered 2–3 h ago, meaning $j - 2$, $j - 3$. We could say that the flow rate has inertia in its variation. That was the reason why it was created another series, based on the sliding averages, reported to a few days ago. Under these conditions, the correlation coefficients between $F\_T_W(j)$ and $F\_T_a(j, j-1)$, $F\_T_a(j, \ldots, j-2)$, $F\_T_a(j, \ldots, j-3)$ are better improved: 0.79, 0.84, 0.86.

The transition from a branched network to an annular network assures minimization of the pressure losses and maximizes the produced electric power because the flow is distributed more rational on different routes. In these conditions, an annular network is more economical.

5 Economic Efficient Functioning

The network previously described proves that the largest share of expenditure is due to the expenses related to the electricity consumption and transport of the pumped volume, including here the staff costs, investments, etc. Under these circumstances, the revenues from the supplied water must be higher than the current expenses as to be economic. The formula of validation the economy becomes:

$$\frac{\sum_i V_P(i) \cdot E_c \cdot P_e}{V_c \cdot P_W} + \frac{1}{P_w \cdot \eta_V}\left(\frac{C_P}{V_{Tp}} + \frac{C_c}{V_C}\right) \le 1 \tag{7}$$

Where: V_p - the pumped volume, E_c - energy consumption for 1 m^3 of pumped water, P_e - price for electric energy, V_c - volume received by consumers, P_w - the price of water delivered to the consumers, η_V - volumetric efficiency, C_p - costs related to the total pumped volume - V_{Tp}, and C_c - costs with staff, investments.

It can be seen from the Eq. (7) that for ensuring the economy, from the first term can act only over the price of the delivered water, and from the second term to assure the efficiency may be increased the pumped volume, for the minimum price of the electric energy.

By considering also the V_{loss} – the volume lost through the pipes, and V_R – the volume possible to be stored into supplementary tanks, the condition of estimating the economy may be written

$$\sum_i \left(\frac{V_P(i) + V_R(i)}{V_c}\right)\frac{E_c \cdot P_e}{P_W} + \frac{1}{P_w}\left(\frac{C_P}{V_{Tp}} + \frac{C_c}{V_C}\right) \le 1 \tag{8}$$

The problem of the economic improvement can be also treated by Fuzzy method, as a matter of risk analysis of the project's investment, considering R - the rate of a possible return having as input the variables D - the time of recovering the investment, DF - the effective time of functioning, and Eta - the efficiency concerning the flow rate, during the operation, from the pumping system. The interference is considered as having three inputs, one output, and more rules. The recovery period is considered of D = 4 years, and the universe of the discourse is covered by two linguistic terms R (right) and W (wrong), after being mentioned the membership functions.

In this case, the interference steps consist in the determination of the strong interference, considered D = 4 years, analysis of the activated rules, determining the degrees of the membership, the establishment of the functions of the membership for the conclusion of each rule, and setting the membership function for the output.

6 Conclusions

It was analyzed by the Fuzzy method a command system of pumping, considering three of the main factors which affect the most: the fluid flowing rate, the interval between stopping and starting, and the wear of the aggregates. Briefly was pointed the main conditions for improving the functioning of the hydraulic aggregates, correlated with their main parameters. Based on these observations the pumping system was analyzed with the neural methods, as a branching network and mentioned some possibilities to transform it into an annular network. It was assumed one unidirectional multi-layers network, with total connections to the consumers, between two layers, the output and input, accomplished by a second hidden layer. The input layer of the neural networks was assumed that receives different input variables. Some rules of the economic parameters, some remarks considering the economic optimization of such complex systems are also mentioned, in both approaches, by an analytic method, and by the Fuzzy method.

References

1. Harpreet, S., Madan M. G., Meitzler T., et al.: Real-life applications of fuzzy logic. Adv. Fuzzy Syst. **1**(1), 112–115 (2013). http://dx.doi.org/10.1155/2013/581879
2. Liu, F., Zhang, G., Lu, J.: A novel fuzzy neural network for unsupervised domain adaptation in heterogeneous scenarios. In: Proceedings of International Conference on Fuzzy Systems, New Orleans, USA, pp. 154–160 (2019)
3. Jurado, S., Nebot, A., Mugica, F.: Optimal selection in fuzzy inductive reasoning for smart grid applications, In: Proceedings of International Conference on Fuzzy Systems, New Orleans, USA, pp. 154–160 (2019)
4. Domoto, E., Shibata, J., Fukushima, M., Okuhara K.,: Multivariate analysis of text data from image information acquired by wearable device. In: Proceedings of SCIS & ISIS, Toyama, Japan, pp. 272–280 (2018)
5. Angleberger, T., Dellinger, P.: Fuzzy. Harry N. Adams, New York (2016). 272 p. ISBN 1419721224
6. Radulescu, V.: New method in estimation of the turbulent drag coefficient for two-phase of fluid flow through pipelines. In: Proceedings ASME International Mechanical Engineering Congress and Exposition (IMECE), vol. 7 (2017). https://doi.org/10.1115/imece2017-702362-s2.0-85040926027
7. Ross, T.: Fuzzy Logic with Engineering Applications, 3rd edn. Wiley, Hoboken (2010). ISBN-13: 978-0470743768

Fuzzy-Logic Based Diagnosis for High Voltage Equipment Predictive Maintenance

Mihaela Frigura-Iliasa[1], Attila Simo[1,2(✉)], Simona Dzitac[3], Flaviu Mihai Frigura-Iliasa[1,2], and Felicia Ioana Baloi[1]

[1] Politehnica University Timisoara, 300223 Timisoara, Romania
attila.simo@upt.ro
[2] Romanian Academy of Scientists, 300220 Timisoara, Romania
[3] University of Oradea, 410087 Oradea, Romania

Abstract. This paper presents a fuzzy-logic algorithm for predictive maintenance (Industry 4.0) and at the same time for future design improving, applicable for high voltage equipment (switches, surge-arresters, etc.). Starting from this algorithm a software tool can be developed tool for high voltage equipment maintenance. It is an example of implementing advanced mathematical and software solutions for the maintenance of operational high voltage switching equipment (and other high voltage devices, too). For testing and validating the algorithm experimental data (and also experimental setups) were operated from the industrial environment (manufacturers and users of that equipment) and used for conceiving a monitoring and diagnosing digital based procedure, both for a more efficient design of that equipment as well as for efficiently assess the technical state of their main HV contacts. The results obtained are encouraging and recommend the use of the algorithm on a larger scale.

Keywords: Fuzzy logic · Predictive maintenance · High voltage devices

1 Introduction

The fuzzy logic is a dedicated class of artificial intelligence having a recent history and application. In 1965, Lotfy Zadeh a computer scientist, proposed the fuzzy logic theory [2, 3]. He argued that human thinking is often fuzzy, vague, or imprecise in nature, and, therefore cannot be represented by yes (1) or no (0). The fuzzy control is also nonlinear and adaptive in nature, which gives it a very robust performance under parameter variations. In modern control techniques, uncertainty and vagueness have an important position, also supports the non-linear design techniques [4, 5].

A fuzzy logic-based algorithm (which could be integrated on a dedicated controller) adjusts the system input to get a desired output by just looking at the output without any requirement of mathematical model for the controlled system that are now being exploited power systems applications. For this reason, a fuzzy logic-based algorithm is different from a classical control one and it is possible to get desired control actions for complex, uncertain, and non-linear systems by using the fuzzy logic controller (FLC)

I. Dzitac et al. (Eds.): ICCCC 2020, AISC 1243, pp. 245–253, 2021.
https://doi.org/10.1007/978-3-030-53651-0_21

without the requirement of their mathematical models and parameter estimation [6]. Although FL deals with imprecise "IF … THEN …" rules, the information is processed in sound mathematical theory, which has progressed (seen advances) in recent years. Each fuzzy set is defined by a linguistic variable (low, large) which is again defined by a multi valued membership function (MF).

An MF varies between 0 and 1 and can have a range of shapes. The logical operations on fuzzy sets and crisp sets are and (intersection), or (reunion), not, sum, prod. The IF part of the rule is defined as antecedent and THEN part is defined as consequent. The corresponding fuzzy set of each rule is defined by a linguistic variable.

The basic steps in fuzzy inference systems are:

- fuzzification of input variables;
- application of fuzzy operator in the IF part of the rule;
- implication from the antecedent to the consequent part of the rule;
- aggregation of the consequents across the rule;
- defuzzification.

The linguistic variables used for the inputs and output are Negative Very Big (NVB), Negative Big (NB), Negative Medium (NM), Negative Small (NS), ZERO, Positive Small (PS), Positive Medium (PM), Positive Big (PB), Positive Very Big (PVB).

Fuzzyfication defines the correlation between a crisp value and one or more linguistic values.

The conversion of fuzzy outputs in a fuzzy system into crisp output is defined as defuzzyfication. The fuzzy output in constructed by superimposing the outputs of individual rule. The most usual method of defuzzification is the center of area method (also called centroid) which involves integration covering the outer envelope.

2 Fuzzy Logic Analysis

The inference rules are made with linguistic variables to help describe the behavior of the process.

In case of a complex description of that process, it is necessary to start with a brief description which will be completed during the implementation of the control. First step is the determination of the variation area for the parameters of the process during the transition process, taking into account the reach of the function points.

To describe the dynamic behavior of the process are necessarily as many indications as we can find, like the number of the time values and their approximately dimension. These specifications are indispensable for the determination of the sample time and for the definition of membership function.

Starting with the description of the controlled process it can be chose the structure of the regulator, if required. First, we must specify the number and the nature of the input values. For example, one of the most used input value is the adjustment error, meaning the difference between the prescription value and the adjusted value. Often the input values are in number of two.

For a large number of input values, the inference rules are voluminous and hard to define. In that case, we can use a structure of control with many regulators placed in cascade, or a more complex structure, if the process claims it. In some situations, at this level it can be mandatory to go at the initial level to finish the description of the process.

Next stage is the determination of the fuzzification rule, meaning the definition of the membership functions for the input and the output values. At this level, we will establish the number of the fuzzy terms for every value, the shape and the distribution of the membership function.

Normally, three or four fuzzy terms are enough. A larger number of fuzzy terms will be used only if it is needed to obtain some accentuated nonlinear effects. A big number of fuzzy terms increase the number of inference rules and the run time.

To obtain a linear characteristic for the fuzzy regulator (RF), we use a triangular membership function with uniform distribution. A problem appears only when we chose the scale factors (C1 and C2) between the normalized field $-1 \leq x \leq 1$ and the corresponding physical value. As shown in Fig. 2, these scale factors are corresponding to some reaction coefficient which are determined with the help of the dynamic behavior of the control system [7].

Obviously, this scale factors aren't known before, and it is indicated that the implementation to be made based on the structure shown in Fig. 1. So, during the tests on the equipment it can be modified this scale factors until are find properly values for the transition process. The change of the dimension and the distribution of the membership function aren't mandatory in this case.

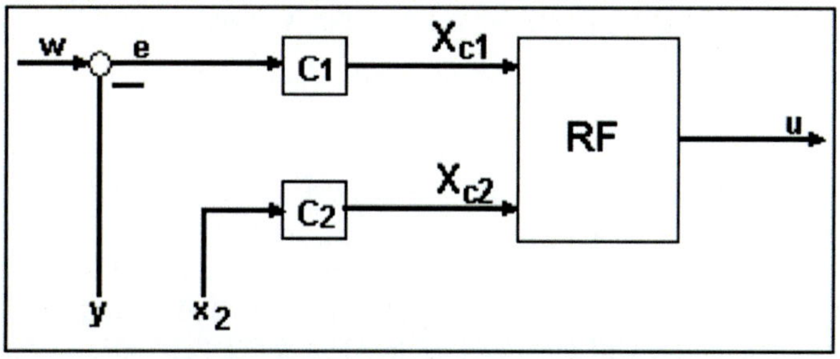

Fig. 1. Fuzzy regulator with two scale factors.

For the output value aren't problems with the definition of the scale factors. The normalized field corresponds with the maximum tolerable variation of the command signal.

Starting with the definition of the process using linguistic variables and membership function it can be established the inference rules. This are in the most of cases incomplete. The rules can't be in contradiction. A representation with inference matrix makes the verification more simple.

During the tests on the equipment it can be also mandatory to modify the inference rules. Generally, it attempts to reduce their number or to add complementary rules.

The selection of the inference rules is made by taking into account the implementation mode of the regulator, with the help of a program or the equipment, when the running time is defining.

If the control system is described with a model of the process, the tests on the installation are replaced with simulation on the PC. All the adjustments can be made on the simulation. At the implementation on the equipment are made only minor adjustments.

3 Technical Analysis

In Fig. 2, we describe a simple assessment method for the status of the main contacts (located inside the high voltage power switching devices), which is mostly recommended for the offline scheduled maintenance, based on the measurement of the working contact resistance.

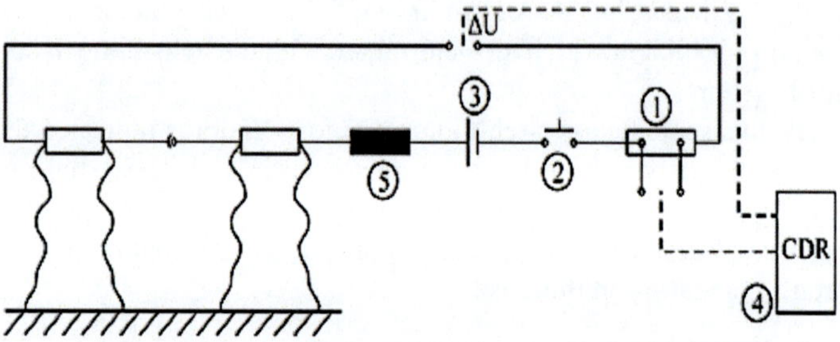

1 – coaxial shunt, 2 – monopolar contactor, 3 – 12 V, 66 Ah battery, 4 – CDR (Contact Disturbance Recorder), 5 – limiting resistance.

Fig. 2. Main diagram of the measuring system.

When applying this method, a 100–1000 A current is injected through the contact. The voltage drop U_2 between the contacts and the current I passing on the closed contacts of the equipment are recorded. A CDR unit, (Contact Disturbance Recorder) dedicated interface records, computes, compares, send to dispatcher and stores all data received by the measurements performed according to elements of Fig. 2. It will count the number of commutations N performed by the equipment between two consecutive measurements. All data recorded from the scheduled measurements of the contact resistance, R_{NC}. after these N connections, is used to estimate the relative error $\delta_j(N)$, related to the value of the same parameter, saved at the end of the latest measurement, R_{CO}. It could be connected to certain PLCs, also.

$$\delta_j(N) = 1 - \frac{R_{CO}}{R_{CN}} \tag{1}$$

The evolution of this parameter after N commutation cycles (between two consecutive measurements) $\varepsilon_j(N)$, is obtained by:

$$\varepsilon_j(N) = \delta_j(N) - \delta_j(N-1) \tag{2}$$

When $\delta_j(N) < \delta_j(N-1)$, resulting from (1) we will notice that $R_{cN-1} > R_{cN}$, because the contact resistance decreases. The cause of this incident (contact resistance diminution) in service is the change of position, size and number of the real microscopic contact areas, as consequence of the self-polishing action.

Locating $\delta_j(N)$ and $\varepsilon_j(N)$ on the fuzzy multitudes from Fig. 3 and relating them to fuzzy values SP, MP, LP, briefly described in Fig. 4, we can proceed to the calculation of the final fuzzy multitude output functions:

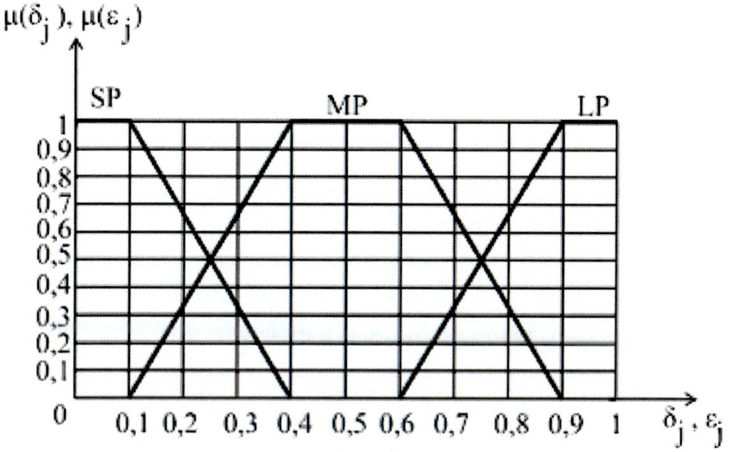

Fig. 3. Fuzzy belonging and multitude functions.

ε_j \ δ_j	SP	MP	LP
SP	SP	MP	MP
MP	MP	MP	LP
LP	LP	LP	LP

Fig. 4. The diagram of the fuzzy rules.

$$\mu_j(SP) = \underset{x_i}{MAX}[MIN(\mu(x_i))] \tag{3}$$

$$\mu_j(MP) = \underset{x_i}{MAX}[MIN(\mu(x_i))] \tag{4}$$

$$\mu_j(LP) = \underset{x_i}{MAX}[MIN(\mu(x_i))] \tag{5}$$

The estimation of the $\mu_f(SP)$, $\mu_f(MP)$ and $\mu_f(LP)$ functions is obtained considering Fig. 4 and the decision diagram of fuzzy multitudes and belonging functions obtained from Fig. 3.

Next operation consists in computing the output of the belonging function μ, based on the algorithm mostly described in Fig. 5 by applying this relation:

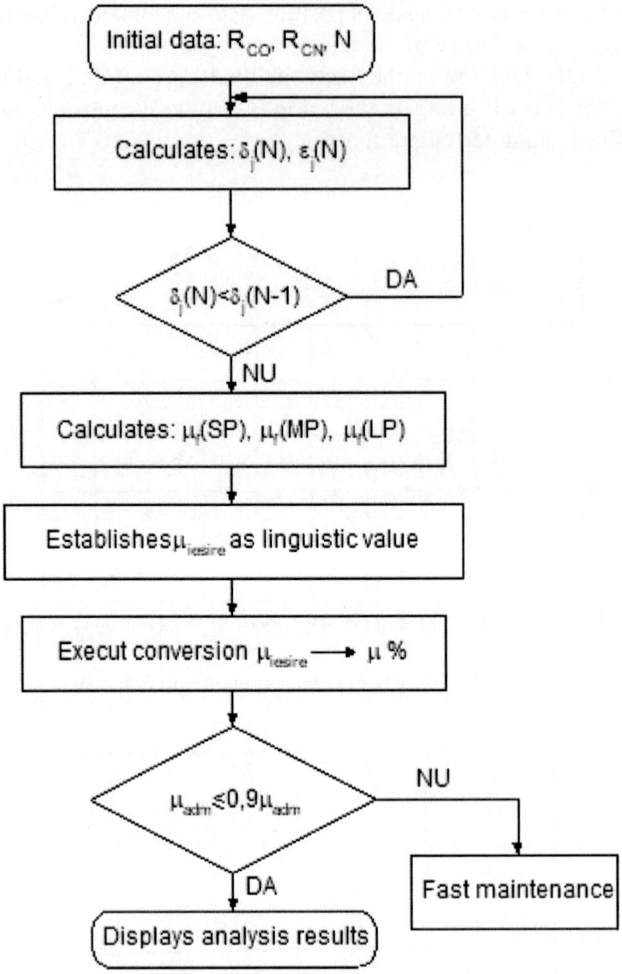

Fig. 5. Dedicated fuzzy algorithm.

$$\mu_{output} = MAX\,[\mu_f(SP),\ \mu_f(MP),\ \mu_f(LP) \qquad (6)$$

When converting the linguistic value μ into a corresponding numerical value μ, in %, (which is representing the natural aging process of the high voltage main contacts in percentages) and comparing this one with the admissible value of main electric contacts fatigue, μ_{dam}, we can determine the exact time moment of the next assessment test and measurement on that piece of equipment [7].

It could be also used on the design stage of that piece of equipment.

4 Implementation of the Measurement System

One of the major causes for service perturbations and even economic issues during the active state of the whole high voltage power system, is the fault of certain main contacts. Continuous monitoring consists in measuring and automatically comparing certain parameters with referential values of a process or equipment. In our case it is based on methods of diagnosis and analysis of internal faults [5]. The decisional process in this situation corresponds to an extremely hazardous process, depending on a lot of parameters. The specific of these command operations is shown in Fig. 6.

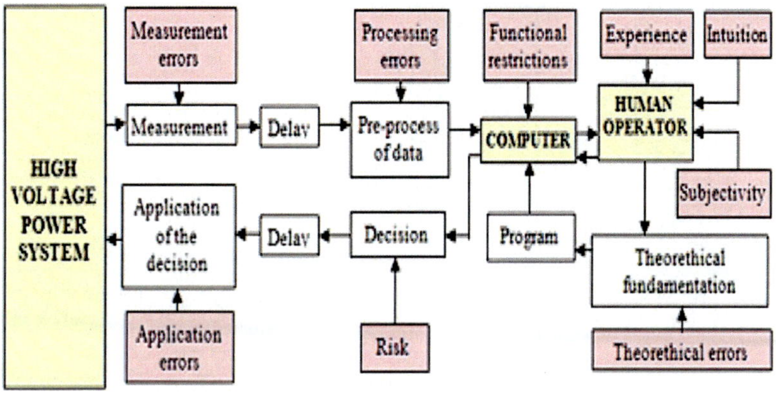

Fig. 6. Principles of decision in case of hazard on high voltage areas.

With no artificial intelligence or expert system involved, it is still based on the interaction between the human operator and the machine. Future progresses in this field will lead to a total automation of this procedure. All control operations and the tests performed by the manufacturer of a high voltage switching device could also be performed by the user of that switching device (located inside a high voltage power system), therefore they are very well defined and widely applied. The expert ideas about the methods used for monitoring and assessment of the technical status of a high voltage power equipment are different in each situation. All these methods could have a certain importance only when they are avoiding operational incidents, or increasing the maintenance time schedule of that device, and, also if they provide an economical advantage. The answer to such a problem is a continuous monitoring and diagnosis operation of the technical status for that electrical device, by involving some special systems for monitoring and diagnosis [8, 9].

All hazardous factors described below are increasing the risks associated with one decision taken based on incomplete and/or wrong information. After the decisional act, there are also two major risk factors: the delay in the application of the decision and all errors in the application issues of that decision. The intervention of errors in information and, generally, all the errors and delays in processing the information, imposed the use of fuzzy techniques for taking the right decision [10, 11]. A continuous and permanent operation of contact status monitoring is possible to be executed online, when involving a method based on using optic fiber, connected at the manufacturing moment on that insulating switch and, also, connected to an interface for introducing, computing and analysing data. It consists in verifying the thermal images provided and informing the decision operating staff concerning the variation of the contact resistance R_c in order to prevent any faults. According to the stationary over temperature of the contact, τ_c ($\tau_c = \theta_c - \theta_{amb}$) and the total intensity of the permanent service current, I, as function of the nature of the superficial material from the contact pieces (copper or silver), we can calculate a voltage drop, U_c.

The block diagram for the monitoring and diagnosis equipment is described in Fig. 7.

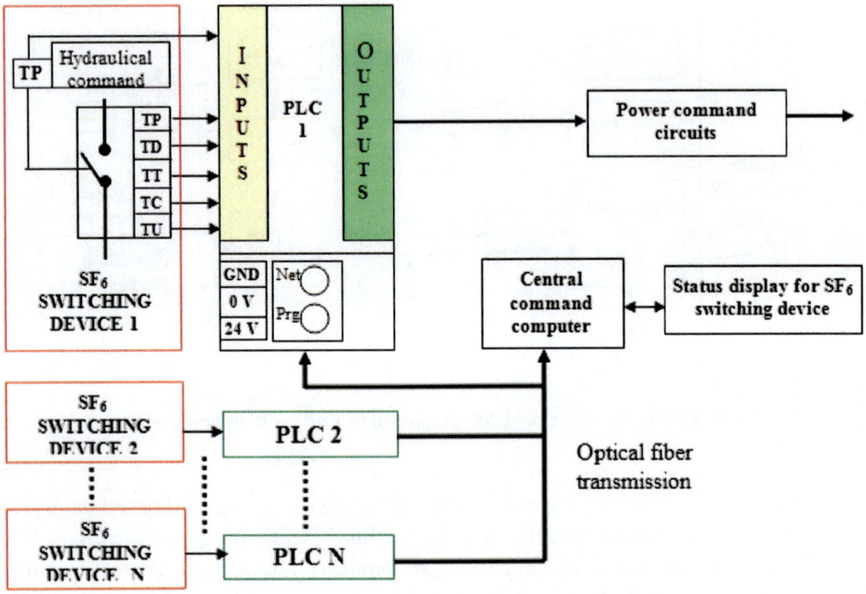

Fig. 7. Block scheme of the monitoring and diagnosis system.

The total number and the final types of sensors, their location and the choice of the measuring methods for the monitored parameters are influencing decisively the solution for this problem.

All data about the SF6 equipment integrated inside a power system could be obtained by means of direct measurements (as quantitative data). All these measurements are involving the risk of error occurrence (accidentally or systematically). The information resulting from these measurements could be already old in the moment of system introduction and the final status of the switching device could be different at that precise moment. We have also some errors caused by simple information processing. The human operator who is deciding could be affected by normal subjectivity; the hardware and the software involved could present some technical restrictions and/or different errors.

The consequences of all these incidents could be limited if the fault causing component is disconnected far before the beginning of that fault's occurrence, and, that's why it is important to determine all symptoms of a certain fault. The monitoring and diagnosis systems will allow the determination of the maintenance scheduling, by analyzing the technical status of all the equipment and switching devices belonging to that power system. The economical task of any switching device owner is to reduce at a minimum the maintenance cost.

5 Conclusions

All the methods presented in this article allow a high quality diagnosis for the technical status of the commutation pieces belonging to a high voltage capsulated switch, providing accurate information concerning: the electric contact resistance, the fatigue status of its main electric contacts and the best moment for performing any predictive maintenance operations for this piece of equipment.

This offline monitoring procedure and system, dedicated mostly for the contact resistance of capsulated switching devices, integrated inside high voltage power stations, briefly described in this paper, allows the prevention of their faults, or their accidental breaking, increasing the service efficiency, reducing the periods of unavailability, and, of course, generally increasing the reliability of all these pieces of equipment.

In case of extending the application, as well as the diagnosing methods, it could be applied for large number of high voltage devices and we can record large data bases concerning all parameters involved, not only the contact resistance. It also provides a final report concerning the tested equipment inside that database, thus developing the knowledge about their work, in order to optimally predict the duration of safe service.

References

1. Bimal Bose, K.: Power Electronics and Motor Drives: Advances and Trends, Knoxville, Tennessee (2006)
2. Zadeh, L.A.: Fuzzy sets. Inf. Control **8**, 338–353 (1965)
3. Zadeh, L.A.: Outline of a new approach to the analysis of complex systems and decision processes. IEEE Trans. Syst. Man Cybern. **SMC-3**(7), 28–44 (1973)
4. Fonseca, J., Afonso, J.L., Matins, J.S., Couto, C.: Evaluation of neural networks and fuzzy logic techniques applied to the control of electrical machines. In: Proceedings of the 5th UK Mechatronics Forum, International Conference, Portugal, vol. 2, pp 15–20 (1996)
5. Eminoglu, I., Altas, I.H.: The effects of the number of rules on the output of a fuzzy logic controller employed to a PM D.C. motor. Comput. Electr. Eng. **24**, 245–261 (1998)
6. Alexandridis, A.P., Siettos, C.I., Sarimveis, H.K., Boudouvis, A.G., Bafas, G.V.: Modelling of nonlinear process dynamics using Kohonen's neural networks, fuzzy systems and Chebyshev series. Comput. Chem. Eng. **26**, 479–486 (2002)
7. Uddin, M.N., Rahman, M.A.: High-speed control of IPMSM drives using improved fuzzy logic algorithms. IEEE Trans. Ind. Electron. **54**(1), 190–199 (2007)
8. Doru, V., Dan, S.F.: Monitoring of the power quality on the wholesale power market in Romania. In: Proceedings of the 9th WSEAS International Conference on Electric Power Systems, High Voltages, Electric Machines, Genova, Italy, 17–19 October 2009, pp. 59–64 (2009). ISBN 978-960-474-130-4, ISSN 1790-5117
9. Simona, D., Felea, I., Dzitac, I., Vesselenyi, T.: An application of neuro-fuzzy modellingto prediction of some incidence in an electrical energy distribution center. Int. J. Comput. Commun. Control IJCCC **III**, 287–292 (2008). ISSN 1841-9836. Supplementary Issue
10. Sorin, M., Nicola, O.V., Doru, V., Ciprian, S.: Theoretical and experimental determination of equivalent parameters of three-phase induction motor windings in case of power electronic converters supply. WSEAS Trans. Syst. **8**(10), 1115–1134 (2009). ISSN 1109-2777
11. Sorin, N., Simona, D.: Neutrosophic TOPSIS: a general view. In: 6th International Conference on Computers Communications and Control (ICCCC), Oradea, Romania, 10–14 May 2016, pp. 250–253. Electronic ISBN 978-1-5090-1735-5. https://doi.org/10.1109/icccc.2016.7496769

Decision Making and Support Systems

Making a Multi-criteria Analysis Model for Choosing an ERP for SMEs in a KM World

Ramona Lacurezeanu(✉) and Vasile Paul Bresfelean

Babes-Bolyai University, Cluj-Napoca, Romania
{ramona.lacurezeanu,paul.bresfelean}@econ.ubbcluj.ro

Abstract. There has never been such a rich offer in terms of ERP solutions that customers can choose from because the problem of knowledge has never been emphasized in the past. An ERP system in line with the company's business is metamorphosed into a knowledge management tool that improves information transfer and generates "knowledge". On the other hand, major decisions are rarely simple, and the best alternative can befall only after careful deliberation. In the present research we endeavor to provide a model of multiple-criteria decision-making (MCDM) in the selection of ERP applications. We illustrate and apply a MCDM technique, namely analytical hierarchy process (AHP) to assist SMEs to select the most appropriate ERP. We formulate an AHP decision model and apply it to a hypothetical case study to demonstrate the feasibility of the choice to select the most appropriate ERP software for a specific SME. We believe that our work could become a didactic source of inspiration for teaching decision making techniques to young people and students. The application of the proposed model indicates that it can be applied to improve decision-making processes and condense the time interval needed for ERP selection. Our model can be considered good practice and identified with the know-what and know-who component of a KM model.

Keywords: ERP · AHP · MCDM · SME

1 Introduction

The rapid global dynamics of ideas linked to the knowledge-based organization is beneficial from the point of view of possible applications, but, however, is of an almost confusing abundance and diversity, while a writings' systematization was still to be expected, as noted by [20] quoted in [12]. "At the present time, there is a range of technological and managerial solutions sufficient to make possible the operationalization of the concept of knowledge-based organization; their usage implies discernment of the choice, consistency in the effort of learning and receptivity in perceiving and assimilating their advantages" [12]. In the world of Knowledge Management (KM) technology has a double characteristic: it has progressed from a small number of processes to all processes, and what functions today may not function as well in a year or two. The digital transformation we comment on today affects everything by breaching the entire range

I. Dzitac et al. (Eds.): ICCCC 2020, AISC 1243, pp. 257–273, 2021.
https://doi.org/10.1007/978-3-030-53651-0_22

of operations. Technology is now part of all business operations, because every process is automated. In other words, when mentioning the technology invasion we start from customers' experience all the way to their involvement in the development of products and problems related to UX, UI, and usability. The synergy effects between Enterprise Resource Planning (ERP) and KM consequently include a more efficient implementation of methodologies, reduced implementation costs, increased user satisfaction, as well as strategic business achieved through state-of-the-art ERP applications [33]. The organizational implementation of an ERP system has been documented as a complex process, in which change management, communication and stakeholder involvement are as important as technical implementation [30] quoted in [33]. In this paper we join those who consider that KM and the ERP core will send adequate knowledge (goods, funds) to the right individuals (position), at the right time. We accept the idea that in a KM world the correct implementation of an ERP system will support companies to better achieve their goals [29]. For a long time, around 20 years, ERP solutions have largely been the focus of corporations and large companies. The new small and medium-sized enterprises' (SMEs) approach in recent years in accessing, implementing and using ERP applications, has created an increasing demand in business areas for ERP software. In spite of the growing aspiration to implement this type of information systems in organizations, the fear of failure leads puts the implementation of the project to a halt or delays it. In other words, considering the high failure rate to an estimate of up to 61% [15], it is essential for organizations to identify the success factors of the implementation project [7, 45]. In today's highly competitive environment, the globalization of the market for each sector is gradually increasing. As a consequence, the number of potential suppliers increases and also the number of criteria when choosing the right supplier. The evaluation and selection of a certain supplier could provide an organization quality products and/or services at a convenient price, convenient quantities and at a convenient schedule. This process can be extremely complex, for the reason that it combines a wide variety of varying factors, which can be diversified based on the nature of the products and services to be purchased. Due to its importance in terms of organization, there should be a systematic way of selecting an appropriate supplier. [9, 28, 37, 39, 50] quoted by [1]. Selecting the best supplier amongst the diversified options has become crucial to achieving customer satisfaction. Even though there is an increase in the number of ERP providers, the number of problems reported regarding the lack of quality has also increased earnestly. Multiple-criteria decision-making (MCDM) methods are an efficient way of evaluating and selecting a supplier. This process results in a broad set of criteria and sub-criteria to consider and leads to numerous different relationships between these criteria and sub-criteria, which should be carefully evaluated in order to make a decision. These methods support decision makers while providing an unblemished comprehension of the problem. These instruments take into account the views of several decision makers and evaluate each one's assessments. In the last decade, there has been an increase in MCDM methods' implementation in order to support decision makers. One of the essential reasons for this growth is the acceptance of the results provided by these methods. Furthermore, MCDM methods are more convenient to implement for larger and more complex issues [31, 35] quoted by [1]. From a phenomenological point of view, the Romanian organizational environment could not make an exception from the contemporary global tendency of orientation towards knowledge, however, in the absence of a strategic option, it manifested itself non-systematically, being located mainly in an informal level; interpersonal information intermediation, informal knowledge transactions within and between organizations, ad hoc professional assistance, local improvisations in knowledge management are facts that plainly confirm such a verdict [12]. The ERP software market was inaugurated in Romania by large companies in early 2000s. These packages have been adopted either in close connection with the "parent companies" or by copying them in large companies with domestic capital. After discerning upon the importance of SME in GDP configuration and after the adoption at EU level of measures to encourage their development, ERP market and vendors took them into consideration. Furthermore, the development of cloud solutions and on the other hand, the emergence of open source solutions have enlarged the number of ERP providers but also the technical problems due to implementation and maintenance. But there was also the issue of training the workforce who can fully operate the functions of "such an integrated" software. Nonetheless all these problems are materialized in costs, time and performance. Taking into account the specificity of the Romanian ERP market, in the present paper we intend to propose a framework for selecting an ERP system:

- to determine the best ERP software for a SME that meets its needs and expectations using the Analytic Hierarchy Process (AHP),
- set out a set of criteria with a focus on supporting business objectives and business strategies,
- describe a numerical example used for ERP selection for an enterprise, implemented in a widespread software Microsoft Excel.

We demonstrate just how with zero additional costs, but with the involvement of SME's existing staff and their digital skills, a rational methodology can be used based on MCDM theories to select the optimal ERP solution for Romanian SMEs. In the following sections we endeavor to justify this demarche, as we consider it a novelty in national researches. Therefore, this paper exercises the AHP model developed by [46, 47] for ERP selection and evaluation in SME firms in which the goal being pursued has multiple, often conflicting attributes. With this in mind, Sect. 2 presents an abbreviated review of relevant literature on the approaches used in ERP. The 3rd section focuses on the steps of AHP method with the calculations based on the given criteria. In Sect. 4 we present the particularities of the Romanian emerging market related to ERP implementation in SMEs and regarding the practical use of an AHP model through Microsoft Excel, thus demonstrating that our approach is an original one, unique in the indigenous research literature and practice. As a final point, the findings are shown from a managerial perspective in the conclusion part, Sect. 5.

2 Literature Review ERP for SMEs

2.1 The Role of SMEs in a Country's Economy

Throughout the world the SMEs sector is considered the determining factor of a competitive economy, the main source of economic growth and new jobs. SMEs are the most widespread organizational form of business. It is attributable to the fact that the development of this sector augments prosperity and the standard of living. Small and medium-sized enterprises play an essential role in EU economy, accounting for over 99% of European businesses and securing two thirds of jobs from the private sector. SMEs embrace different definitions, one of the reasons derives from the most obvious and indisputable characteristic of SMEs, namely diversity [10, 18, 53, 54, 56]. According to Article 2 of the Annex from Recommendation 2003/361/EC (integrated in the Romanian legislation by OG no. 27/2006, M.Of. No. 88/31.01.2006, approved by Law no. 175/2006, M. Of. no. 438/22.05.2006), "The category of Micro, Small and Medium-sized Enterprises (SMEs) is made up of companies that employ less than 250 people and have a net annual turnover of up to 50 million Euro and/or have total assets of up to 43 million Euro". Based on dynamism, flexibility, adaptability to the needs of the economy and society, on the modern entrepreneurial spirit and on the capacity for innovation, SMEs play a leading role in the development of the Romanian economy within the European Union and the global economy [44]. Having the status of specific forms of organizing the economic activity, SMEs are directly involved in transforming the socio-economic environment compelled by the technological explosion of the last decades. Internet and e-commerce continuous development have determined these enterprises to aspire to implement various classes of software packages on a much larger scale, not only as complementary elements, but as a foundation for their organizational structure. As regards the conditions, ERP software implementation to assist the management is a normal consequence, imposed by the growth of the economic environment, required to transform the leadership style, but also the technological one [4].

The information included in this paper was equally gathered from primary and secondary sources. It was collected from various Internet sources, research papers, journals, surveys, in addition to the other necessary information accumulated from the below sources.

2.2 About ERP Systems for SMEs

Over the last two decades, software selection has become a dynamic area of research attributable to its complex nature. Researchers concentrated more on the selecting specific software products and less on the general models and methods for solving this selection problem. Software selection represents a multi-criteria decision problem that refers to electing a software product from a lot of software products based on various criteria evaluation. The most appropriate methods that can be applied to solve software's selection problem are the multi-attribute decision methods in which the set of variants is a finite, non-empty set, while the set of criteria, and possibly sub-criteria, is clearly established, finite and measurable [42]. Starting with this assumptions, in the following paragraphs we leaned on studying the ways in which SMEs select the most appropriate ERP. After examining several articles from the research literature, we noticed a global concern on this topic: purchase and implementation of an ERP system by SMEs. Researchers from all over the world are concerned about the same subject - identify the success factors of ERP implementation in SMEs. We were able to assess and compare studies from different areas (Europe, North America, Asia) both qualitative and quantitative, but limited to a single country; we remarked more complex researches that took into account regional economic factors but also the cultural factor, which tried to demonstrate that there were not such wide differences between an ERP system implementation for a SME from China versus USA, or the United Kingdom. One thing is already validly expressed and accepted: Most ERPs are designed and developed having in mind the wealthiest companies – since these were the first to be able to pay the high prices. The principal targets for the ERP market are large and middle-size companies from developed countries, because most SMEs are unable to invest large financial or time resources while also facing restrictive planning functionalities in their daily operations [42]. There are voices that consider that owing to the reduction of the need to implement an ERP in large companies (corporations, banks, multinationals, big 4) ERP vendors have currently begun to focus more on SMEs. Obviously, a coherent argument considers that globalization, partnerships, value networks, and the huge information flow between SMEs today, is increasingly determining SMEs to adopt ERPs. The risks of adopting an ERP are present, based on SMEs' limited resources and specific features that make their ERP needs (modules) very different from the needs of large companies [19]. Other concerns regarding the implementation of an ERP in a SME were stated by [28] in their study. "SMEs have less resources and competencies about complex ERP systems compared to larger companies. Thus, ERP projects have proven to be risky and costly for SMEs. SMEs can easily be an easy prey for experienced vendors and consultants and end up with a system far from what they expect". ERP market is currently expanding, and these systems are more and more implemented, unfortunately they are still unknown to many SMEs to a certain degree. Although many of these enterprises are aware about the benefits of a modern ERP system, they still have doubts to purchase one, this being attributable to financial and risk reasons [42]. The findings from this study [37] demonstrate that even though SMEs managers are aware of their goals when implementing an ERP system, there is a lack of documented ERP business cases. The authors argue for the need to develop business cases and suggest a minimum list of steps to assist SMEs. SMEs differ considerably from large organizations; hence, an increasing number of such enterprises implement ERPs. SMEs encounter more challenges than large enterprises in ERP implementation due to SMEs' major disadvantage - lack of human and financial resources. Staff unavailability in such enterprises may even cause production to stop in order to facilitate training [9]. The process of ERP selection is a multipart issue encompassing various parts and variables, due to the its unstructured type decision-making characteristic (made by Turban and Simon) quoted by [39]. A good ERP selection process is crucial for reducing the waste of funds (up to several million dollars), because economic studies indicate [56] that an ERP implementation in a SME could cost from fifty thousand dollars to a few million dollars. Most ERP selection methodologies have as basic principles the information on development companies, technical structure and software capabilities, costs related to software, provided services and support. In the first stage, the needs of the organization are identified by comprehending the processes with the purpose of improving them. [9] found 42 criteria for ERP selection and evaluation, which are quantitatively higher than in the research literature. The outcomes of this study sustain many previous researches; nevertheless, there are differences between this study and similar studies because of the prevalent conditions from different countries and organizations. A comprehensive review of literature on ERP in SMEs (77 articles) which exemplifies the status of research in this area was made by [19]. A part includes an analysis of ERP acquisition, phase that consist of the process of ERP package and vendor selection best fitting in organization's requirements. The authors enunciate several issues such as: factors affecting selection, selection criteria, in-house developed systems, and different acquisition issues. After studying the research literature on the topics revealed up to this point, we came to the idea that at the present time there are numerous ERP software products and each of them has different characteristics. Their cost can vary from tens of thousands to several million Euros, therefore acquiring a new ERP solution can be a major expense for an enterprise [19]. An unsuccessful selection of an ERP system of this magnitude can set in motion erroneous strategic decisions with subsequent economic losses for the organization. We took into account the views of practitioners, who provide ERP solutions for SMEs but also those who have begun the implementation of such software. The vast majority believes that a sensitive but weighty issue is how to choose the right ERP software. There are simple but very vague answers, such as companies simply answering five simple questions related to: number of employees, industry, option of implementation, business complexity and the expected growth of employees. We can infer the following: globalization has led to the advent of disruptive technologies that SMEs must use to expand or even stay afloat. The appropriate resources or lack of them can make the difference between survival and dissolution for a small business. We strongly believe that SMEs need to invest in ERP solutions in order to benefit from integrated analysis, rapid implementation and best practices for different business processes, including financial, production, human resources, procurement and supply chain management. Based on this outline we move on to the next section related to the AHP multi-decision analysis model.

3 Literature Review AHP

The AHP method, a multi-criteria decision method belonging to the multi-attribute decision methods class, was first defined by the prominent Saaty, in 1980 in the work "The Analytic Hierarchy Process", with several re-editions and reinterpretations to this day [46, 47]. For the last 30–35 years AHP has been thoroughly tested by thousands of organizations around the globe and it seems to work. Therefore, there are large numbers of case studies that describe how large organizations used AHP for their strategic decisions to attain superior outcomes [50, 55]. AHP has been attracting the interest of many researchers in point of fact by simplicity which is characterized by the pair-wise comparison of the alternatives according to specific criteria [54] and because it allows decision makers to benefit from a specific and mathematical decision support tool. Its application not only supports and qualifies the decisions, but also enables decision makers to justify their choices, in addition to simulate potential results [9, 54]. From [55] we can retain "The structure of an AHP model is a model of an inverted tree. There is a single purpose at the top of the tree that represents the purpose of the problem of decision making. A hundred percent decision weight is at this point. Just below the goal is a leaf point indicating the criteria, both qualitative and quantitative. Goal Weight should be divided between rating points based on rating". In the case study discussed in this study [32, 55]," the analytic hierarchy process (AHP) was used because is an application to an evaluation process concerning a pre-feasibility study of strategic solutions in the logistics and transport field offers some features that can facilitate the inclusion of the above-mentioned aspects in the decision aiding process". As some researchers mentioned in their work [29] this method has proved one of the most applied and is mentioned in most manuals and guides [9, 15, 16, 32, 50]. Researches on the AHP method applied in ERP systems' evaluation and selection by SMEs can be found in Romanian research articles [21, 22, 25] stipulating that this method is: a) Interactive (through which a single or a group of decision-makers transmit their preferences to the analyst, and opinions and results can be debated or discussed), b) Emerged from theories on human behavior, including those related to the process of thinking, logic, intuition, experience and learning theories. We consider that a well-defined and easy to understand structure of the AHP method's algorithm for the problem of multi-criteria decision is found in [41]: "We consider the n decision criteria C_1, C_2, ..., C_n as leaf nodes of a simple hierarchy, which constitute the decomposition of a requirement (objective). The AHP method (algorithm) has three steps:

1) Compare the pairs of decision alternatives according to each decision criterion, in order to rank them in relation to the corresponding factor,
2) Compare the pairs of decision criteria; a relative hierarchy of them is obtained,
3) Create the performance matrix and calculate the scores of alternatives for all decision criteria using the hierarchy of alternatives obtained in step 1) and the criteria hierarchy from step 2)" [41]. A crucial element of this algorithm consists of constructing the comparison matrices for ranking the alternatives with regard to each criterion. Every pair of alternatives in the set $\{A_1, A_2, A_3, \ldots, A_m\}$ is compared according to each criterion C_k from the decision criteria set $\{C_1, C_2, \ldots, C_n\}$, obtaining the comparison matrices $\{D(k), 1 \leq k \leq n\}$. The process has two sub-steps: (1a) constructs the raw comparison matrices D(k) and (1b) normalizes them [38, 40, 41, 47]. The method of comparing a pair of alternatives (Pairwise comparison) seeks to establish hierarchies among decision alternatives. It should be noted that if alternative A_1 was compared with alternative A_7, we would no longer compare alternative A_7 with A_1, for the reason that its importance is 1/the importance of alternative A_1 versus A_7. In addition, the pairwise comparison matrix should satisfy the consistency property: if alternative A_i was preferred to alternative A_j and alternative A_j was preferred to alternative A_k, then alternative A_i would be preferred to alternative A_k. A comparison example from the grid completed by Saaty. We consider the construction of these matrices as being very important for the reason that the consistency of the model depends on their accuracy. [47] recommended using consistency index (CI) and consistency ration (CR) to check for the consistency associated with the comparison matrix. The model is presented in numerous articles [38, 46]. The relationship between CI and RCI is defined as the Consistency Ratio (CR): CR = CI/RCI. The Binary comparisons are sufficiently coherent with one another if: - CR < 5% for n = 3; - CR < 9% for n = 4; - CR < 10% for n > 4, where n represents the number of criteria [2]. The description of all AHP steps is performed in the following studies [3, 11, 13, 36]. An example of an AHP method application is described in the following sections.

4 ERP in Romania: Selection Criteria and AHP Model

4.1 ERP for Romanian SMEs

A natural question arose: Why talk about ERP solutions for SMEs in an emerging country like Romania? The answer is neither simple nor problematic. In the last decade, there has been a noticeable rise in the number of SMEs willing to take on this technology and integrate it into their businesses. In addition, major ERP providers have created modules expressly for SMEs, for example SAP Business One for SMEs (SAPB1). Although SAPB1 lacks some of the rich features of other products, it compensates with lower prices, rapid implementation and TCO. The idea of using an ERP in Romanian SMEs has been argued since 2005 and we bring to mind that since then, attention has been drawn to the fact that the development of small and medium-sized enterprises is a necessity that need not be further disputed. Information technologies are definitely contributing to the growth of SMEs performance. The salvation solution was called "information integration" or ERP [17]. According to the practitioners, those who offer implementation software, the large ERP distribution companies, pointed out: there are 560,000 companies in Romania, of which 90–95% are SMEs, according to data from the Trade Register (2016). When SMEs take the first steps towards adopting an ERP system, they have to take into consideration a number of factors. Among all these, the phase of selecting the best ERP solution is an important process and should be completed with great care. With all this in mind, we have chosen to propose the topic of this chapter. In this chapter an AHP decision model is formulated and applied to a hypothetical case study to demonstrate the feasibility of his choice for selecting the most appropriate ERP software for a specific Romanian SME. Based on the literature review up to this point we have summarized the following: nowadays ERP represents a mandatory software for any SME. The procurement and implementation processes are expensive, both financially and humanly. Since there are plenty of ERP solutions on the market, the decision to choose and purchase such a software package as being the best variant can be based on methods and methodologies from the MCMD and MCAD (multiple-criteria decision analysis) areas. While the AHP might not be the most recent approach, it is a qualified process for collaborative decision making, is user-friendly and reasonably priced therefore it can be employed for both small and large decisions. In the following paragraphs we formulate the current criteria carried on the Romanian ERP market: on the one hand criteria listed by software vendors and on the other hand criteria requested by Romanian SMEs. We apply these criteria on an AHP model and implement it in Excel, one of the most popular and user-friendly spreadsheet software. This implementation supports the assertion that a suitable software is needed when working with AHP because despite the fact AHP is simple in concept, the math is time consuming. We chose to carry out this case study because we found the following:

- It represents a hot topic (resulting from the articles on this topic published in the last decade and cited in the bibliography section),
- Most studies focus on the implementation of popular ERP systems,
- Studies capture the perspectives of both sellers and consumers, but there are a number of fissures regarding the criterion of local and regional economic development.

In 2015 Eurostat assigned Romania a penetration rate of business solutions (ERP category) of only 20–21%, which indicated that a considerable number of companies did not use an integrated management system and relied on simple accounting and spreadsheets solutions. Three years later in 2018, the Ministry of Business Environment, Trade and Entrepreneurship officially announced that 48.31% of SMEs had planned to adopt new technologies, in order to become more competitive [23, 27]. In the light of all that has been presented to this point, the evolution of Romanian SMEs sector is not out of the ordinary, but is part of the general tendency chronicled in Europe. Although the Romanian market has not yet reached maturity, it enjoys consistent offers from well-known software companies (Microsoft, SAP, Oracle, Scala), as well as from Romanian companies (Crisoft, Transart, Bit Software, TotalSoft, WizRom, etc.) who have a tendency to offer products specifically tailored for SMEs. It is well known that success does not only come from the best hardware or software. People and their commitment are the elements who make the difference. In today's business world, individualism is a declining concept, and the concept of group has become a part of everyday life. Nowadays, nearly every project has to be completed in groups, and very few are done by individuals. For that reason, we consider the collaborative approach in which group members can create and participate in different types of decision sessions through an iDS platform according to [6]. To this it is also necessary to incorporate the concept of Cloud. The idea of iDS that caught our attention refers to a proposed set of tools that consists of: a) discussion list (a forum-like tool for discussions), b) voting (a tool that allows grading or expressing the agreements over a set of issues), c) electronic brainstorming based on the Issue Based Information System (IBIS) approach, d) Mind Map and e) Categorization. Furthermore, we also believe that the Cloud represents the future of ERP. Most of the major ERP manufacturers, such as Microsoft, Oracle, SAP have been making significant investments in Cloud Business. The benefits are known to be cheaper to deploy and faster to deploy, therefore most ERP providers use these facts to increase sales of cloud-based ERPs particularly for SMEs.

4.2 A Hypothetical Example of the Romanian's SME Firms

Taking into consideration the idea outlined in [14] that selection of IT&C tools must be regarded as a multi-attribute decision making, we commenced by presenting an appropriate model for selecting the most suitable ERP for an SME in Romania. Thus, we consider that together with the general selection criteria grouped into suitability criteria, the quality of implementation and the quality of delivery, some authors have developed sets of criteria for filtering (with or without experts) the software products from a long initial list. Moreover, from the mentioned research article we apprehend that there are methodologies for evaluating and selecting software through the use of MADM but also systems of experimental experts proposed especially by dominant software selection companies. Unquestionably, all these require the acceptance of a service for a fee, but more precisely we want to talk about the limited resources of a SME, and through our model we guarantee that an SME should not spend more and can use the people of the company. We believe that through the national education system at high school and faculty level, Romanian employees have already attained the necessary Excel knowledge and even the knowledge and skills to assimilate the AHP decision-making method. In this project of selecting the most suitable ERP for a SME wishing to implement it, we adapted a model based on [52]: we make a general presentation of the AHP method and the implementation steps in Excel. Based on the literature review, a model can be considered as a simplified exposition of reality. By simplifying the assumption, it is possible to develop a model of the problem that is simple enough to understand and analyze and yet provides a good overview of the real problem. We personalize a model adapted from the research literature:

A) We present the AHP general method – A team of 3–5 people from SMEs will create the selection criteria (those that will become the alternatives of the AHP method), after answering a set of questions; they will be chosen from a different department, depending on the SME organization chart, so as to be representative. It will be possible to generate the list of qualified software programs by choosing from the top companies that offer ERP solutions for SMEs for the Romanian market.
B) Implementation steps in a spreadsheet software (Excel) – The AHP model will be individualized (customized) to the concrete situation

The present article does not incorporate this questionnaire, but on the other hand we have one in research. Every person will be chosen from a different department, depending on SME's organizational chart, so that they will be representative. It will be possible to generate the list of qualified software programs, either by choosing the top companies that offer ERP solutions for SMEs on the Romanian market, or a specialized company could be employed to make the selection, in which case an additional cost will be generated. In this present study, we assume that we use the idea of unanimity of the decisions of the group of experts (unanimous decisions), that occurs when all agree without reservation. These are easier for trivial matters, but very difficult for important and/or higher-pressure situations. Groups using the Consensus-Oriented Decision Making (CODM) model experience increased cohesion and commitment and stronger relationships as a result of their successful cooperation. By incorporating the principles of collaboration, inclusion, empathy and openness, the CODM process encourages common ownership of group decisions. It combines the best practices in professional facilitation, mediation and non-violent communication. The AHP model will be personalized to the specificity of each situation (as presented in Subsect. 4.1). We demonstrate that with zero additional costs a rational methodology based on the MADM theories for selecting the best ERP software can be applied even for Romanian SMEs, with the involvement of existing staff and using the digital skills of the SMEs' employees. In our paper we chose to acknowledge the point of view expressed by Deloitte & Touche [24, 26, 49]. The detail that differentiated this study from others is that it contained consecutive interviews with companies involved in two ERP software implementations. They first implemented an ERP and had certain criteria, with a certain hierarchy of importance. Over the years they had another implementation in which they used all the criteria from the first implementation, but with another hierarchy of importance (Table 1). An important conclusion from those who acknowledged this study but also from our research is that the buyers who are in the second round of ERP implementation have gained important experience and knowledge, which lacks for those who purchase an ERP system for the very first time. Based on this logic, second-time rankings are an enhanced guide to what really matters when examining the options of an ERP system. The list is presented in Table 1. Following a brainstorming by the authors of the article (on the one hand based on their experience of lecturers teaching Decision Support Systems (DSS) & ERP, on the other hand, after discussions with hundreds of students from business specializations, on this topic, namely the selection criteria of an ERP) we went on with the following criteria: Functionality (C1), Ease of use named by us usability (C2), Ease of implementation (C3), Quality of documentation (C4), Price of software (C5), Growth potential named by us scalability (C6), Ability to fit to business (C7), Developer's track record of performance (C8), Level of reseller support (C9).

Table 1. Comparing criteria for selecting an ERP system, (based on 49)

1st-time buyers	2nd-time buyers
1) Price of software	1) Level of reseller support
2) Ease of implementation	2) Developer's track record of performance
3) Ease of use	3) Ability to fit to business
4) Ability to fit to business	4) Growth potential
5) Functionality	5) Price of software
6) Ability to work with existing hardware	6) Quality of documentation
7) Growth potential	7) Functionality
8) Level of reseller support	8) Ease of use
9) Quality of documentation	9) Ease of implementation
Which criteria really matter for selecting an ERP system?	

Here, a hypothetical example based on opinion of [13, 36] is presented to illustrate how the AHP process can be used. Each criterion will be evaluated compared to the others. In this step we set the 9 criteria, set out above and noted from C1 to C9. For example: functionality is more important than ability to fit to business with 5 points. Based on the evaluations of previous step, we build the pairwise comparison matrix according to the following rule:

$$a_{ij} > 0, a_{ji} = \frac{1}{a_{ij}}, a_{ii} = 1 \tag{1}$$

The value of the (i, j) position of the pairwise comparison matrix is determined using Saaty's scale (1, 3, 5, 7, 9), the inverse value of the assigned number is assigned to the (j, i) position. In order to obtain the weight of each criterion, the following instructions are used: First, the sum of each column is calculated, then the normalization of the matrix by dividing the content of each cell by the Sum of its column, and finally calculation of the average of the rows [13, 36] (Figs. 1 and 2).

Criteria	C1	C2	C3	C4	C5	C6	C7	C8	C9
C1	1.00	3.00	3.00	3.00	3.00	3.00	5.00	7.00	7.00
C2	1/3	1.00	3.00	3.00	3.00	3.00	5.00	7.00	7.00
C3	1/3	0.33	1.00	3.00	3.00	3.00	3.00	5.00	7.00
C4	1/3	0.33	0.33	1.00	3.00	3.00	3.00	5.00	7.00
C5	1/3	0.33	0.33	0.33	1.00	3.00	3.00	3.00	5.00
C6	1/3	0.33	0.33	0.33	0.33	1.00	3.00	3.00	5.00
C7	1/5	0.20	0.33	0.33	0.33	1/3	1.00	3.00	5.00
C8	1/7	0.14	0.20	0.20	0.33	1/3	0.33	1.00	5.00
C9	1/7	0.14	0.14	0.14	0.20	1/5	0.20	0.20	1.00
	3.15	5.82	8.68	11.34	14.20	16.87	23.53	34 1/5	49.00

Fig. 1. Criteria

Normalized pairwise comparison matrix										Calculation of criteria's weight	
	C1	C2	C3	C4	C5	C6	C7	C8	C9	Sum of lines	Weight(avg of line sums)
C1	0.32	0.52	0.35	0.26	0.21	0.18	0.21	0.20	0.14	2.39	0.27
C2	0.11	0.17	0.35	0.26	0.21	0.18	0.21	0.20	0.14	1.84	0.20
C3	0.11	0.06	0.12	0.26	0.21	0.18	0.13	0.15	0.14	1.35	0.15
C4	0.11	0.06	0.04	0.09	0.21	0.18	0.13	0.15	0.14	1.10	0.12
C5	0.11	0.06	0.04	0.03	0.07	0.18	0.13	0.09	0.10	0.80	0.09
C6	0.11	0.06	0.04	0.03	0.02	0.06	0.13	0.09	0.10	0.63	0.07
C7	0.06	0.03	0.04	0.03	0.02	0.02	0.04	0.09	0.10	0.44	0.05
C8	0.05	0.02	0.02	0.02	0.02	0.02	0.01	0.03	0.10	0.30	0.03
C9	0.05	0.02	0.02	0.01	0.01	0.01	0.01	0.01	0.02	0.16	0.02
										weight	

Fig. 2. Normalized pairwise comparison matrix

Deduction: C1 is the most important criterion with a 0.27 weight, then comes C2 criterion with 0.20 weight, the third criterion is the C3 with a 0,15 weight, and so all until C9 criterion with 0.02 weight. Consistency check: Before proceeding to the next step of the AHP method, it is essential to make sure that we did not make any absurd comparisons and that criteria weights are indeed consistent, that's why this step is mostly important, to check the system consistency it is necessary to follow the steps below:

Calculate Weight sums vector:

$$\{\mathrm{Ws}\} = \{\mathrm{M}\}.\{\mathrm{W}\} \tag{2}$$

Find the Consistency vector:

$$\{\mathrm{Cv}\} = \{\mathrm{Ws}\}.\left\{\frac{1}{W}\right\} \tag{3}$$

Average Consistency vector, this measure is termed $\lambda_{\max}$

Determinate Consistency Index:

$$CI = \frac{(\lambda_{max} - n)}{(n-1)} \tag{4}$$

Calculate Consistency ratio:

$$CR = \frac{CI}{RI} \tag{5}$$

The RI or Random Index is obtained from the following data, the first row represents the number of criteria (n) (Source Saaty, 1977):

n 123 4 5 6 7 8 9 10

RI000.580.91.121.241.321.411.451.49

The final test is: If CR < 0,1: then the matrix is consistent and judgments can be considered coherent. If CR > 0,1: then the matrix results are inconsistent and were exempted for the further analysis. The next step is to check the results of the previous example (Figs. 3 and 4).

First determine the Weight sums vector Ws

	C1	C2	C3	C4	C5	C6	C7	C8	C9			
C1	1.00	3.00	3.00	3.00	3.00	3.00	5.00	7.00	7.00		0.27	2.77
C2	1/3	1.00	3.00	3.00	3.00	3.00	5.00	7.00	7.00		0.20	2.18
C3	1/3	0.33	1.00	3.00	3.00	3.00	3.00	5.00	7.00		0.15	1.58
C4	1/3	0.33	0.33	1.00	3.00	3.00	3.00	5.00	7.00	X	0.12	1.24
C5	1/3	0.33	0.33	0.33	1.00	3.00	3.00	3.00	5.00		0.09	0.88
C6	1/3	0.33	0.33	0.33	0.33	1.00	3.00	3.00	5.00		0.07	0.68
C7	1/5	0.20	0.33	0.33	0.33	0.33	1.00	3.00	5.00		0.05	0.47
C8	1/7	0.14	0.20	0.20	0.33	0.33	0.33	1.00	5.00		0.03	0.31
C9	1/7	0.14	0.14	0.14	0.20	0.20	0.20	0.20	1.00		0.02	0.17

Fig. 3. Determine the weight sums vector Ws

Find the Consistency vector:

2.77		3.76	10.42	2.77		3.57	9.89
2.18		4.90	10.70	2.18		5.00	10.92
1.58		6.67	10.58	1.58		6.67	10.56
1.24	X	8.22	10.20	1.24	X	7.69	9.55
0.88		11.30	9.96	0.88		1.11	0.98
0.68		14.27	9.73	0.68		16.67	11.37
0.47		20.40	9.69	0.47		33.33	15.83
0.31		30.08	9.40	0.31		33.33	10.42
0.17		56.38	9.69	0.17		50.00	8.59
			90.37				
						λ_{max}	10.04

Fig. 4. Find the consistency vector

After finding λ_{max} value 10.04, we determinate CI and CR: CI = (10.04 − 9)/8 = 0,13, then CR = 0.13/1.45 (Source Saaty 1977, above table) = 0.09. CR < 0.1 The consistency ratio is acceptable; the system is consistent. We can move forward. Alternatives weights: This is the final step of AHP. The goal is to get the weight of each ERP solution in order to choose the best ERP system, for each criterion, the matrix of ERP solution will be created, there are 9 criteria, that is why 9 matrixes are obtained. Following the same steps as before, the ERP weights are determined as shown below: example for C1 criteria (Fig. 5). We used the same model, the same steps for the remaining 8 criteria from 2 to 9. The results (weights) were combined in the matrix (Fig. 6).

Alternative's weight for Criteria C1

	ERP1	ERP2	ERP3
ERP1	1.00	3.00	3.00
ERP2	1/3	1.00	3.00
ERP3	1/3	1/3	1.00
Sum	1.67	4.33	7.00

Normalized pairwise comparison matrix / *Calculation of criteria's weight*

			Sum of line	Weight
0.60	0.69	0.43	1.72	0.57
0.20	0.23	0.43	0.86	0.29
0.20	0.08	0.14	0.42	0.14

C1 criteria's weight

ERP1	0.57
ERP2	0.29
ERP3	0.14

Fig. 5. Example for C1 criteria

	C1	C2	C3	C4	C5	C6	C7	C8	C9
ERP1	0.57	0.61	0.14	0.10	0.61	0.57	0.10	0.61	0.14
ERP2	0.29	0.26	0.24	0.21	0.26	0.29	0.21	0.26	0.24
ERP3	0.14	0.13	0.62	0.69	0.13	0.14	0.69	0.13	0.62

weight of criteria
0.28
0.20
0.15
0.13
0.09
0.06
0.03
0.03
0.02

Fig. 6. Combined matrix

B106 fx =B93*B96+C93*B97+D93*B98+E93*B99+F93*B100+G93*B101+H93*B102+I93*B103+J93*B104

	A	B
105	ERP1	0.4331
106	ERP2	0.4695
107	ERP3	0.3086

Fig. 7. Final results

In order to obtain the last result, the transpose of the previous matrix was multiplied by the determined weight vector. We finally established the weight of each ERP: 0.4331 for ERP1, 0.4695 for ERP2 and 0.3086 for ERP3 (Fig. 7). Based on the 9 criteria and the data we gathered, and by applying the AHP method, we can conclude that ERP2 is a more suitable solution for the company in comparison with ERP1 and ERP3.

5 Conclusions

SMEs need to invest in technology to be successful in the digital economy and seamlessly integrate processes within the organization, which will help them pull the complexities out of the business and focus on growth. The implementation of an integrated ERP system represents a strategic decision from the top management of an organization and it must be well established and supplemented by a coherent plan of activities, outlined with terms and responsibilities in order to allow an easy monitoring and evaluation. SMEs, nor only from our country but from all over the world, have been considering ERP systems for many years. There were implementations that were considered successful while others have failed. In every implementation project, an important step was the one that established the evaluation criteria. The main contributions of this paper consist in:

a. Elaboration and implementation of a decision model with several criteria for the selection of ERP providers,
b. Group-based decision making for comparing ERP attributes and alternatives,
c. Efficient use of the AHP concept to facilitate the model solution for Romanian SMEs,
d. The proposed method will help managers to weigh ERP alternatives before actually implementing them, which in turn will save money and time (crucial for any SME).

The studies we analyzed found and combined various criteria, starting from some very simple ones (such as cost, vendor reputation, assuring maintenance and training), to some very refined ones (such as scalability, adapting to national legislation, adapting to the hardware and software components existing in the organization). Some researchers have taken into consideration classes of criteria with sub-criteria, starting from 3, 5, 7, 9 criteria and going up to 42 criteria. The retrieved criteria were framed in various methods and methodologies including AHP, Electre, etc. In our study, the analytic hierarchy process is used to solve the ERP selection problem for a SME. By using the model proposed by AHP the time and effort spent in making the right decision can be significantly reduced. At the same time, we are aware of the limitations of this approach but on the other hand we consider it as very important for the practice in our country, for the specialized national research literature but also for other research areas. This work is conceived as a primary step towards a smart decision support system that integrates the capabilities of expert systems (ES), fuzzy logic and additional criteria such as evolution to the cloud, the sector in which SME operates, but also specialized training for those who will use the ERP system in their activity. In other words, we will attempt to integrate in the criteria the way in which SMEs consider knowledge management. In our future research we will consider a method of determining the number of members of the expert group, possibly Wason selection group type. In addition, we believe improved psychophysiological tools could be interesting to measure electrodermal activity, skin temperature and body movement.

The boundaries of the paper that we will try to surpass in our future research derive from the fact that the ERP technology in the cloud has not been discoursed noticeably, although there are some providers for ERP systems in the cloud on the Romanian market. We reason this limitation on the one hand because there are cloud ERP packages, with licensing possibilities even for a minimum of 2 users, at particularly low costs, on the other hand because the Romanian market is still reluctant to the cloud.

Acknowledgments. We would like express our appreciations to the esteemed Professors Florin Gheorghe FILIP and Ioan DZITAC whose observations and suggestions during ICCCC2020 conference aided us to bring our study in this form and to pursue it further in the future.

Our paper is dedicated to the memory of Professor Stefan NITCHI, who established the Business Information Systems department, for his guidance on the path of DSS.

References

1. Akcan, S., Güldeş, M.: Integrated multicriteria decision-making methods to solve supplier selection problem: a case study in a hospital (2019). https://www.ncbi.nlm.nih.gov/pmc/articles/PMC6811789/
2. Application of Analytical Hierarchy Process (AHP) Technique To Evaluate and Selecting Suppliers in an Effective Supply Chain. https://pdfs.semanticscholar.org/2377/ac12fe82cd6d20e5816cb1a7b6d6ea3bd9e2.pdf
3. Aşchilean, I. et al.: Choosing the optimal technology to rehabilitate the pipes in water distribution systems using the AHP method. https://www.sciencedirect.com/
4. Berar, S.: Model conceptual al unui sistem software pentru managementul IMM-urilor virtuale. Revista Informatica Economica **3**(15) (2000)
5. Boer, L., Wegen, L.L.M.: Practice and promise of formal supplier selection: a study of four empirical cases. J. Purchasing Supply Manage. **9**(3), 109–118 (2003)
6. Candea, C., Filip, F.G.: Towards intelligent collaborative decision support platforms. Stud. Inform. Control **25**(2) (2016). http://www.sic.ici.ro
7. Cândido, C.J.F., Santos, S.P.: Strategy implementation: what is the failure rate? J. Manage. Organ. **21**(2), 237–262 (2015). https://www.researchgate.net/publication/264004530_Strategy_implementation_What_is_the_failure_rate
8. Chamodrakas, I.S.: Multiple criteria decision making methods for business-to-business electronic commerce. Thesis summary (2010). https://pdfs.semanticscholar.org/8cfd/fcd…527d.pdf
9. Chehrehpak, M., Alizadeh: A multi-criteria decision making model for ERP selection: an empirical case in industrial organizations of Iran. In: Conference on Industrial Engineering (2016)
10. Ciubotariu, M.S.: The role of small and medium enterprises in the modern economy and the importance of IFRS application for SMEs (2013). http://www.annals.seap.usv.ro/
11. Cruz-Cunha, M.M., et al.: ERP selection using an AHP-based decision support system. Inf. Resour. Manage. J. **29**(4), 65–81 (2016)
12. Dragomirescu, H: Organizatii bazate pe cunoastere. www.academiaromana.ro
13. Eldrandaly, K.: GIS software selection: a multi-criteria decision making approach. Appl. GIS **3**(5), 1–17 (2007)
14. Filip, F.G.: A decision-making perspective for designing and building information systems, vol. 7, pp. 264–272 (2012). ISSN 1841-9836
15. Ghosh, R.: A comprehensive study on ERP failures stressing on reluctance to change as a cause of failure. J. Mark. Manage. **3**(1), 123–134 (2012)
16. Goepel, K.D.: Implementing the analytic hierarchy process as a standard method for multi-criteria decision making in corporate enterprises – a new AHP excel template with multiple inputs. In: Proceedings of the International Symposium on the Analytic Hierarchy Process, Kuala Lumpur (2013)

17. Grama, A, Fotache, D.: Integrarea informaţională a IMM-urilor româneşti, analele ştiinţifice ale Universităţii "Alexandru Ioan Cuza" din Iaşi, Tomul LII/LIII Ştiinţe Economice (2006)
18. Habersaat, M., Schönenberger, A., Weber, W.: Les PME en Suisse et en Europe. SECO, Berne (2000)
19. Haddara, M., Zach, O.: ERP systems in SMEs: a literature review. In: Proceedings of the 44th Hawaii International Conference on System Sciences (2011)
20. Hendriks, P.H.J.: Do smarter systems make for smarter organizations? Decis. Support Syst. **27**, 197–211 (1999)
21. http://old.fonduri-ue.ro/res/filepicker_users/cd25a597fd-62/Documente_Suport/Studii/0_Studii_Instrumente_Structurale/Pag.3_ACB/19_Analiza_Multicriteriala.pdf
22. http://www.asecib.ase.ro/simpozion/2009/full_papers/pdf/58_Jantea_ro.pdf
23. http://www.business24.ro/
24. https://www.brinel.ro/article/top-10-criterii-de-selectie-a-unui-erp/
25. https://www.intechopen.com/predownload/51007
26. https://www.serrasoftware.ro/
27. https://www.startupcafe.ro/afaceri/software-afaceri-gestionare.htm
28. Hustada, E., Olsenb, D.H.: Critical issues across the ERP life cycle in small-and-medium sized enterprises: experiences from a multiple case study. Procedia Technol. 179–188 (2013). https://www.sciencedirect.com/
29. Yuena, K., Yangeng, W., Qun, Z.: ZhouQun: a mode of combined ERP and KMS knowledge management system construction. Phys. Procedia **25**, 2228–2234 (2012)
30. Kemppainen, I.: Change management perspectives in an ERP implementation. In Proceedings of the Twelfth European Conference on Information Systems (2004)
31. Koganti, V.K., Menikonda, N., Anbuudayasankar, S.P., Krishnaraj, T., Athhukuri, R.K., Vastav, M.S.: GRAHP TOP model for supplier selection in Supply Chain: a hybrid MCDM approach. Decis. Sci. Lett. **8**, 65–80 (2019)
32. Kosacka-Olejnik, M., Werner-Lewandowska, K., et al.: Which option of sustainable mobility should be chosen? – AHP method application for sustainable decision making in transport. Acta Technica Napocensis Appl. Math. Mech. Eng. **61** (2018)
33. Leknes, J., Munkvold, B.E.: The role of knowledge management in ERP implementation: a case study. In: Aker Kvaerner, ECIS, Proceedings, vol. 25 (2006). http://aisel.aisnet.org/ecis2006/25
34. Mandal, A., Deshmukh, S.G.: Vendor selection using interpretive structural modelling (ISM). Int. J. Oper. Prod. Manage. **14**(6), 52–59 (1994)
35. Mian, S.H., Al-Ahmari, A.: Comparative analysis of different digitization systems and selection of best alternative. J. Intell. Manuf. **30**(5), 1–29 (2017)
36. Motaki, N., Kamach, O.: ERP selection: a step-by-step application of AHP Method (2017). https://www.researchgate.net/publication/320477208
37. Mukwasi, C.M., Seymour, L.F: The growing trend of small to medium-sized enterprises adopting enterprise resource planning systems: an analysis of business cases in Zimbabwe and South Africa. JETEMS J. **5**(7), 138–145 (2014)
38. Nesticò, A., Somma, P.: Comparative analysis of multi-criteria methods for the enhancement of historical buildings. Sustainability. https://www.mdpi.com/journal/sustainability
39. Pacheco-Comer, A. González-Castolo, J.C.: An empirical study in selecting Enterprise Resource Planning Systems: the relation between some of the variables involve on it. Size Investment (2012). https://www.sciencedirect.com/science/article/pii/S2212017312002617
40. Boonyaprasit, P., Yang, S.: How SMEs make their decision for choosing an optimal ERP provider by using AHP method. https://pdfs.semanticscholar.org/619b/64e510a07fb3428c20917f3f67285af789d7.pdf
41. Prejmerean, V.: Sisteme pentru fundamentarea deciziilor (Course 5 Decisions), p. 13. http://www.cs.ubbcluj.ro/~per/Dss/Dss_5.pdf

42. Rădulesc, D.M.,: Selecţia produselor software – o problemă complexă de decizie multi-criterială. Revista Română de Informatică şi Automatică **25**(1), 51–61 (2015). https://rria.ici.ro/wp-content/uploads/2015/03/08-art.6-delia-radulescu-OK-8.pdf
43. Rahman, S.: Erp adoption in small and medium sized enterprises in Bangladesh (2015). https://www.academia.edu/....edium_Sized_Enterprises_in_Bangladesh
44. Rolul întreprinderilor mici şi mijlocii în dezvoltarea economiei româneşti - dezbatere sub egida AOSR (2015). http://www.marketwatch.ro/
45. Rouhani, S., Hosseini, S., Zanjani, M.S.: The role of knowledge management processes in ERP implementation success. **7**(3) (2017). https://www.igi-global.com/
46. Saaty, T.L: The analytic hierarchy process-what it is and how it is used. Mah. Model. **9**(3–5), 161–176 (1987). https://www.sciencedirect.com/science/article/pii/0270025587904738
47. Saaty, T.L.: The Analytic Hierarchy Process. McGrew Hill, New York. International, Translated and Revised Editions, Paperback (1996, 2000), vol. 9. RWS Publications (1980)
48. Sarkis, J., Talluri, S.: A model for strategic supplier selection. J. Supply Chain Manage. **38**(1), 18–28 (2002)
49. Scherer, J.: Top ERP selection criteria (2018). https://blog.gobiztech.com/top-erp-selection-criteria
50. Snider, B., et al.: ERP implementation at SMEs: analysis of five Canadian cases. http://citeseerx.ist.psu.edu/
51. Sonmez, M.: Review and Critique of Supplier Selection Process and Practices. Loughborough University, Loughborough (2006)
52. Thao, S.: Enterprise resource planning software selection for a small company located in Mid-Western Wisconsin (2002). https://minds.wisconsin.edu/bitstream/handle/1793/40695/2002thaos.pdf?sequence=1
53. Torres, O.: Les PME, Éditions Flammarion, Collection Dominos, Paris (1999)
54. Vargas, R.V.: Using the analytic hierarchy process (AHP) to select and prioritize projects in a portfolio (2010). https://www.pmi.org/learning/library/analytic-hierarchy-process-prioritize-projects
55. Widiantal, M.M.D., et al.: Comparison of multi-criteria decision support methods (AHP, TOPSIS, SAW & PROMENTHEE) for employee placement. J. Phys. Conf. Ser. **953**, 012116 (2018)
56. Wtterwulghe, R.: La PME, une entreprise humaine. De Boeck & Larcier, Bruxelles (1998)

A Multi-Objective Model for Devices Procurement with Application in Health Care

Constanţa Zoie Rădulescu[1](✉), Marius Rădulescu[2], Lidia Băjenaru[1], and Adriana Alexandru[1]

[1] National Institute for Research and Development in Informatics, 011455 Bucharest, Romania
{zoie.radulescu,lidia.bajenaru,adriana.alexandru}@ici.ro

[2] "Gheorghe Mihoc-Caius Iacob" Institute of Mathematical Statistics and Applied Mathematics, Calea 13 Septembrie No. 13, 050711 Bucharest, Romania
mradulescu.csmro@yahoo.com

Abstract. Managers need to make informed choices about what to buy in order to meet priority needs and to avoid wasting limited resources. The procurement decision is a very difficult task since there exists a great variety of brands, vendors and equipment performances. In the present paper, we have developed a decision process for equipment procurement in which are used, in combination, a Multi-Criteria subjective weighting method SWARA (Step-wise Weight Assessment Ratio Analysis) for equipment evaluation weights, an adaptation of SAW (Simple Additive Weighting) for equipment performance and a new Multi-Objective optimization model for equipment procurement. The Multi-Objective model considers several types of equipment, their costs and their performances. The model aims to be an aiding in the decision process of equipment procurement.

Managers of health care systems need to find in their choices a compromise between the cost of procurement, brands' reputation, vendors' reputation and equipment performance. A numerical example for medical equipment procurement, based on sensors, is studied.

Keywords: Equipment procurement · Equipment performance · Multi-Objective model · Weights · Medical equipment · Health care

1 Introduction

The procurement is a complex decision-making problem that is very important for the firm's profitability. It has received considerable attention in last years. Decision techniques are an important part of the procurement problem and contribute to the development of decision support applications. Decision-making is a specific form of information processing that aims at setting-up an action plan under specific circumstances [1].

The equipment procurement, within an organization, takes into account several factors. It is a particular decision problem in which are involved the manager or the group of decision makers who define the problem of purchasing equipment and select the expert or the group of experts who evaluate the vendors and the equipment according to a

I. Dzitac et al. (Eds.): ICCCC 2020, AISC 1243, pp. 274–283, 2021.
https://doi.org/10.1007/978-3-030-53651-0_23

set of criteria. Experts (decision makers) may have different evaluation abilities (which are represented by coefficients of authority) and criteria may have different importance (which are represented by coefficients of importance or weights). These weights considered in the evaluation can be obtained using a multi-criteria weighting method (cf. [2, 3]).

In the decision problem can be considered the equipment to purchased, the equipment brands, the cost, quality and performance of the equipment that is evaluated. The criteria of credibility and reputation of the vendors that sell the equipment are taken into account. The procurement decision problem can be a structured, semi structured and unstructured problem. It may be deterministic or having uncertainty. From the point of view of the number of people involved, the problem can be an individual decision problem or a group decision problem. The methods and techniques used to solve this decision problem include Multi-Criteria Decision Making, Artificial Intelligence and Mathematical Programming [4].

In the paper [5] is presented a review of papers for vendor selection in healthcare industry. Here is explained how to use the MARCOS method to select sustainable vendors in the private medical sector. A case study of a sustainable vendor selection for the healthcare industry (a polyclinic) is presented.

Diaconu et al. [6] conducted a systematic review of the literature that identifies the products or prioritization methods that are recommended or used for medical equipment procurement.

In the present paper, is developed a decision process and is formulated a new Multi-Objective optimization model for equipment procurement. The Multi-Objective model considers several types of equipment, their costs and their performances.

This paper is organized as follows. In Sect. 2 is described a decision process for equipment procurement that is organized in four stages. In the first stage (Sect. 2.1) the definition and the structure of the equipment procurement problem are presented. In Sect. 2.2 is presented the second stage: calculation of the weights which show the equipment importance based on a subjective multi-criteria weighting method called SWARA. The equipment performances are calculated in Sect. 2.3 (third stage) based on a multi-attribute method SAW (a version for vendors, brands, equipment and experts). In Sect. 2.4 (fourth stage) is formulated a Multi-Objective model for equipment procurement. The range of variation for parameter M is obtained.

In Sect. 3 is presented a case study for medical equipment procurement. The medical equipment is based on sensors. It is used for older adults monitoring. We conclude this paper in Sect. 4.

2 A Group Decision Process for Equipment Procurement

A decision process for equipment procurement is a complex process. Complexity may arise from the type of decision problem (structured, semi-structured or non-structured), from uncertainty of information, from judgments of decision makers and experts, information from different sources, method selection.

The procurement decision problem is the following: a manager wants to buy a number of equipment pieces from a set of vendors. The vendors' offer has equipment belonging to several brands. The manager wants to select from the vendors' offer a portfolio of equipment that meets his preferences. In his choice the manager has two objectives: to maximize the performance and to minimize the cost of equipment he buys.

Solving this decision problem involves several stages:

Stage 1. Structuring the decision problem,
Stage 2. Determination of the importance of each type of equipment (i.e. of the equipment weights),
Stage 3. Calculation of the equipment performance,
Stage 4. Formulation of Multi-Objective model for equipment procurement.

2.1 Structuring the Decision Problem

The experts with expertise in the field are identified and a group of s experts is selected to make equipment evaluations. Denote by $\boldsymbol{X} = \{X_1, X_2, ..., X_s\}$ the group of selected experts. The manager selects the coefficients of authority for experts: $\boldsymbol{w_2} = (w_{2,\alpha})$ where $w_{2,\alpha}$ = the coefficient of authority that shows the importance of expert α in the equipment evaluation process, $\alpha = 1, 2, ..., s$.

The types of equipment to be purchased are established: $\boldsymbol{E} = \{E_1, E_2, ..., E_n\}$. Let a_i be the minimum number of pieces of equipment E_i that have to be bought and b_i be the maximum number of pieces of equipment E_i that have to be bought, $i = 1,2,..., n$.

Based on equipment brands and vendors data and information are selected the set of vendors and the set of brands of the equipment that will be purchased. Denote by $\boldsymbol{V} = \{V_1, V_2, ..., V_p\}$ the set of vendors and by $\boldsymbol{B} = \{B_1, B_2, ..., B_m\}$ the set of brands.

2.2 Determination of the Importance of Each Type of Equipment

In order to determine the weights that show the importance the equipment, a weighting method will be used. In the literature there are several objective and subjective weighting multi-criteria methods. Subjective weighting methods are based on the expert's opinion while the objective methods are based on the evaluation of data provided by the evaluation matrix. Each of these methods has its own advantages and disadvantages. The uncertainty in the decision maker judgments is the main disadvantage of the subjective methods, while the objective methods do not benefit from the expertise and experience of the experts [7, 8]. Examples of more often used subjective weighting methods are Simple Multi-Attribute Rating Technique (SMART) [9], Analytical Hierarchy Process (AHP) [10, 11], SMARTS [12], Delphi method [13], Analytical Network Process (ANP) [14], Step-Wise Weight Assessment Ratio Analysis (SWARA) [15, 16], Decision-Making Trial and Evaluation Laboratory (DEMATEL) [17], The best–worst method (BWM) [18], Criteria Impact LOSs (CILOS) [19], Extended Stepwise Weight Assessment Ratio Analysis (SWARA) [19, 20].

In order to determine the weights that show the importance of E_i equipment, the experts select from the set of existing methods the subjective weighting method SWARA.

The input data are the set of experts $\boldsymbol{X} = \{X_1, X_2, ..., X_s\}$, the set of equipment types $\boldsymbol{E} = \{E_1, E_2, ..., E_n\}$. The steps of the SWARA method for determining the equipment weights [16], [21] are:

Step 1. The equipment types are sorted in a descending order, based on their expected significances.

Step 2. Starting from the second equipment type, the experts expresses the relative importance of the equipment type E_i in relation to the previous (E_i–*1*) equipment type, and does so for each equipment type. This ratio is called the Comparative Importance of the Average Value, h_i.

Step 3. Compute the coefficient k_i: $k_i = \begin{cases} 1 & if\ \ i = 1 \\ h_i + 1 & if\ \ i > 1 \end{cases}$

Step 4. Compute the recalculated weight g_i: $g_i = \begin{cases} 1 & if\ \ i = 1 \\ k_i + 1/k_i & if\ \ i > 1 \end{cases}$

Step 5. The relative weights of the evaluation criteria are determined:

$w_{1,i} = g_i / \sum_{t=1}^{n} g_t$ where $w_{1,i}$ denotes the weight of the E_i equipment, and n denotes the number of the equipment types.

2.3 Calculation of the Equipment Performance

The input data are the set of experts $\boldsymbol{X} = \{X_1, X_2, \ldots, X_s\}$ the set of equipment types $\boldsymbol{E} = \{E_1, E_2, \ldots, E_n\}$, the set of vendors $\boldsymbol{V} = \{V_1, V_2, \ldots, V_p\}$ and the set of brands $\boldsymbol{B} = \{B_1, B_2, \ldots, B_m\}$. The experts assign scores to each equipment belonging to a brand and to a vendor. The evaluation matrix is $\boldsymbol{e} = (e_{ijr\alpha})$ where $e_{ijr\alpha}$ = the score given by the expert α to the equipment E_i of brand B_j sold by vendor V_r. The measure scale is [1, 2, …, 10].

The quality of expertise of each expert is taken into account with the help of expert weights (coefficient of authority) $\boldsymbol{w_2} = (w_{2,\alpha})$, $w_{2,\alpha}$ = the weight that shows the importance of expert α in the equipment evaluation process.

The equipment performances are calculated with an adaptation of a multi-attribute SAW (Simple Additive Weighting) method with the inclusion of four categories of information: vendors, brands, equipment types and experts.

The performance q_{ijr} of equipment E_i of brand B_j sold by vendor V_r is calculated as:

$$q_{ijr} = \sum_{\alpha=1}^{s} w_{2,\alpha} e_{ijr\alpha} \tag{1}$$

where $e_{ijr\alpha}$ = the score given by the expert α to the equipment E_i of brand B_j sold by vendor V_r.

2.4 Formulation of the Multi-Objective Model for Equipment Procurement

The problem of equipment procurement is formulated as a Multi-Objective optimization model. Consider the set of equipment types $\boldsymbol{E} = \{E_1, E_2, \ldots, E_n\}$, the set of vendors $\boldsymbol{V} = \{V_1, V_2, \ldots, V_p\}$, the set of brands $\boldsymbol{B} = \{B_1, B_2, \ldots, B_m\}$.

The input data in the model are presented in Table 1:

The decision variable x_{ijr}, $i = 1, 2, \ldots, n; j = 1, 2, \ldots, n; r = 1, 2, \ldots, p$ is the solution of the Multi-Objective model for equipment procurement x_{ijr} is the number of pieces of equipment E_i of brand B_j bought from vendor V_r.

Table 1. Input data in the Multi-Objective optimization model for equipment procurement.

Symbol	Description
n	Number of equipment to be purchased
m	Number of brands considered for equipment to be purchased
p	Number of vendors
M	The sum to be invested
$\boldsymbol{a} = (a_i)$	a_i = the minimum number of pieces of equipment E_i that have to be bought
$\boldsymbol{b} = (b_i)$	b_i = the maximum number of pieces of equipment E_i that have to be bought
$\boldsymbol{d} = (d_{ijr})$	d_{ijr} = the maximum number of pieces of equipment E_i of brand B_j available for selling at vendor V_r
$\boldsymbol{c} = (c_{ijr})$	c_{ijr} = the cost of one-piece equipment E_i of brand B_j at vendor V_r
$\boldsymbol{q} = (q_{ijr})$	q_{ijr} = the performance of equipment E_i of brand B_j sold by vendor V_r (obtained in stage 3)
$\boldsymbol{w}_1 = (w_{1,i})$	w_{1i} = the weight that shows the importance of E_i equipment (obtained in stage 2)

The following condition is necessary for the existence of feasible solutions:

$$\sum\nolimits_{j=1}^{m} \sum\nolimits_{r=1}^{p} d_{ijr} \geq a_i, \; i = 1, 2, \ldots, n \tag{2}$$

The formulation of the Multi-Objective model for equipment procurement is the following:

$$\begin{cases} min\left[\sum\nolimits_{i=1}^{n} \sum\nolimits_{j=1}^{m} \sum\nolimits_{r=1}^{p} c_{ijr} x_{ijr}\right] \\ max\left[\sum\nolimits_{j=1}^{m} \sum\nolimits_{r=1}^{p} q_{ijr} x_{ijr}\right], \; i = 1, 2, \ldots, n \\ a_i \leq \sum\nolimits_{j=1}^{m} \sum\nolimits_{r=1}^{p} x_{ijr} \leq b_i, \; i = 1, 2, \ldots, n \\ x_{ijr} \leq d_{ijr}, \; i = 1, 2, \ldots, n, \quad j = 1, 2, \ldots, m, \quad r = 1, 2, \ldots, p \\ x_{ijr} \in N, \quad i = 1, 2, \ldots, n, \quad j = 1, 2, \ldots, m, \quad r = 1, 2, \ldots, p \end{cases} \tag{3}$$

Starting from the above Multi-Objective optimization model we can define a single objective model. The tradeoff between cost and performance optimization model are:

$$\begin{cases} \min\left[\sum\nolimits_{i=1}^{n} \sum\nolimits_{j=1}^{m} \sum\nolimits_{r=1}^{p} [(1 - \lambda) c_{ijr} x_{ijr} - \lambda w_{1,i} q_{ijr}] x_{ijr}\right] \\ \sum\nolimits_{i=1}^{n} \sum\nolimits_{j=1}^{m} \sum\nolimits_{r=1}^{p} c_{ijr} x_{ijr} \leq M \\ a_i \leq \sum\nolimits_{j=1}^{m} \sum\nolimits_{r=1}^{p} x_{ijr} \leq b_i, \; i = 1, 2, \ldots, n \\ x_{ijr} \leq d_{ijr}, \; i = 1, 2, \ldots, n, \quad j = 1, 2, \ldots, m, \quad r = 1, 2, \ldots, p \\ x_{ijr} \in \mathbb{N}, \quad i = 1, 2, \ldots, n, \quad j = 1, 2, \ldots, m, \quad r = 1, 2, \ldots, p \end{cases} \tag{4}$$

Here $\lambda \in [0,1]$ is a parameter.

A very important problem in solving the optimization model for equipment procurement is the determination of the range of parameter M.

Consider the optimization problems:

$$\begin{cases} min\left[\sum_{i=1}^{n}\sum_{j=1}^{m}\sum_{r=1}^{p} c_{ijr}x_{ijr}\right] \\ a_i \le \sum_{j=1}^{m}\sum_{r=1}^{p} x_{ijr} \le b_i,\ i = 1, 2, \ldots, n \\ x_{ijr} \le d_{ijr},\ i = 1, 2, \ldots, n,\quad j = 1, 2, \ldots, m,\quad r = 1, 2, \ldots, p \\ x_{ijr} \in N,\quad i = 1, 2, \ldots, n,\quad j = 1, 2, \ldots, m,\quad r = 1, 2, \ldots, p \end{cases} \tag{5}$$

and

$$\begin{cases} max\left[\sum_{i=1}^{n}\sum_{j=1}^{m}\sum_{r=1}^{p} c_{ijr}x_{ijr}\right] \\ a_i \le \sum_{j=1}^{m}\sum_{r=1}^{p} x_{ijr} \le b_i,\ i = 1, 2, \ldots, n \\ x_{ijr} \le d_{ijr},\ i = 1, 2, \ldots, n,\quad j = 1, 2, \ldots, m,\quad r = 1, 2, \ldots, p \\ x_{ijr} \in N,\quad i = 1, 2, \ldots, n,\quad j = 1, 2, \ldots, m,\quad r = 1, 2, \ldots, p \end{cases} \tag{6}$$

Denote by M_1 (respectively by M_2) the optimal value of the problem (5) (respectively of the problem (6)). Then the range of parameter M is the interval $[M_1, M_2]$.

3 Application in Health Care Procurement

We will apply the decision process proposed for a medical equipment procurement. A manager of a hospital for geriatrics and gerontology want to modernize the hospital through procurement of new equipment based on sensors. He wants to buy a number of seven types of medical equipment. Denote by $\boldsymbol{E} = \{E_1, E_2, E_3, ..., E_7\}$ the set of medical equipment. The types of medical equipment are: blood glucose sensors, blood oxygen sensors, motion sensors, temperature sensors, electrocardiogram (ECG) sensors, image sensors, and pressure sensors. The minimum and maximum number of pieces of medical equipment that have to be bought are presented in Table 2.

Table 2. The minimum and maximum number of pieces of medical equipment.

Symbol	E_1	E_2	E_3	E_4	E_5	E_6	E_7
a_i	3	10	8	10	7	5	7
b_i	5	13	10	15	10	8	10

A group of four experts X_1, X_2, X_3 and X_4, who will evaluate the medical equipment, were selected.

The vendors capable for delivering medical equipment were identified. Four vendors who have the necessary equipment (from different brands) were identified and selected: V_1, V_2, V_3 and V_4. The major players in the medical sensors market are GE Healthcare Inc., STMicroelectronics, Honeywell Inc., Analog Devices, Inc., Omron Corporation, etc. The vendors' offer has medical equipment belonging to several brands. Three brands were selected: B_1, B_2 and B_3.

The performance criteria for the evaluation of the vendors' medical equipment were defined. The weights of the equipment are computed with the subjective weighting method SWARA. The performance of the vendor's equipment of different brands were evaluated by the experts. The performance of medical equipment is evaluated by every expert, from expert's group, that assign scores on a scale [0,1,2, …,10] to each medical equipment belonging to a brand and to a vendor. The coefficient of authority of each expert is taken into account. Based on (1) is obtained the total performance of medical equipment: $\boldsymbol{q} = (q_{ijr})$, where q_{ijr} = the performance of equipment E_i of brand B_j sold by vendor V_r, for $i = 1,2, \ldots, 7; j = 1,2,3$ and $r = 1,2,3,4$.

The matrix $\boldsymbol{d} = (d_{ijr})$ where d_{ijr} = the maximum number of pieces of medical equipment E_i of brand B_j available for selling at vendor V_r is defined, for $i = 1,2, \ldots, 7; j = 1,2,3$ and $r = 1,2,3,4$.

The matrix $\boldsymbol{c} = (c_{ijr})$ where c_{ijr} = the cost of medical equipment E_i of brand B_j available for selling at vendor V_r is built, for $i = 1,2, \ldots, 7; j = 1,2,3$ and $r = 1,2,3,4$ (Table 3).

Table 3. The costs of medical equipment.

Equipment	V_1			V_2			V_3			V_4		
	B_1	B_2	B_3	B_1	B_2	B_3	B_1	B_2	B_3	B_1	B_2	B_3
E_1	500	530	500	510	495	495	520	498	498	490	510	495
E_2	95	97	100	100	100	100	98	98	98	97	97	96
E_3	700	703	705	705	703	708	698	699	700	697	698	699
E_4	550	550	550	550	550	550	540	540	540	540	540	540
E_5	95	95	95	100	100	100	98	98	98	97	97	97
E_6	77	77	77	79	79	79	79	79	79	75	76	79
E_7	210	210	210	215	215	215	225	225	225	230	230	230

The models (5) and (6) are solved in order to obtain the parameters M_1 and M_2.

The M_1= 15916 is the optimal objective function of the problem (5) and M_2 = 23212 is the optimal value of the problem (6). Then the range of parameter M is the interval [15916, 23212]. We solve the model (4) by varying parameter M in the range [15916, 23212] with $\lambda = 0.5$. Seven solutions: $\text{Sol}_1, \text{Sol}_2, \ldots, \text{Sol}_7$ of the model (4) are presented in Table 4 and Fig. 1.

In the optimal solution Sol_1 (first column of the Table 4) 3 is the number of pieces of medical equipment E_1 belonging to brand B_1 which is bought from vendor V_4, 10 is the number of pieces medical equipment E_2, belonging to brand B_3 which is bought from vendor V_4, etc.

When the amount of the invested sum M increases then the number of pieces of medical equipment increases. Thus, for $M = 15916$ the manager can buy a number of 50 pieces of equipment and for $M = 22725.6$ the manager can buy a number of 66 pieces of equipment.

Table 4. Solutions of the Multi-Objective model (4).

Brand-equipment-vendor	Sol_1	Sol_2	Sol_3	Sol_4	Sol_5	Sol_6	Sol_7
$B_1.E_1.V_4$	3	0	0	0	0	0	0
$B_2.E_1.V_1$	0	3	3	3	3	5	5
$B_2.E_2.V_2$	0	11	11	10	11	10	13
$B_3.E_2.V_4$	10	0	0	0	0	0	0
$B_1.E_3.V_4$	8	0	0	0	0	0	0
$B_2.E_3.V_2$	0	8	9	9	10	9	9
$B_3.E_3.V_2$	0	0	1	1	0	1	1
$B_3.E_4.V_2$	0	0	0	0	0	1	15
$B_3.E_4.V_3$	10	11	10	13	15	14	0
$B_2.E_5.V_1$	7	0	0	0	0	0	0
$B_2.E_5.V_4$	0	7	7	7	7	7	8
$B_1.E_6.V_4$	5	0	0	0	0	0	0
$B_2.E_6.V_4$	0	5	6	5	6	5	5
$B_1.E_7.V_1$	7	0	0	0	0	0	0
$B_1.E_7.V_3$	0	7	0	0	6	0	0
$B_2.E_7.V_4$	0	0	8	7	1	8	10
Total number	50	52	55	55	59	60	66

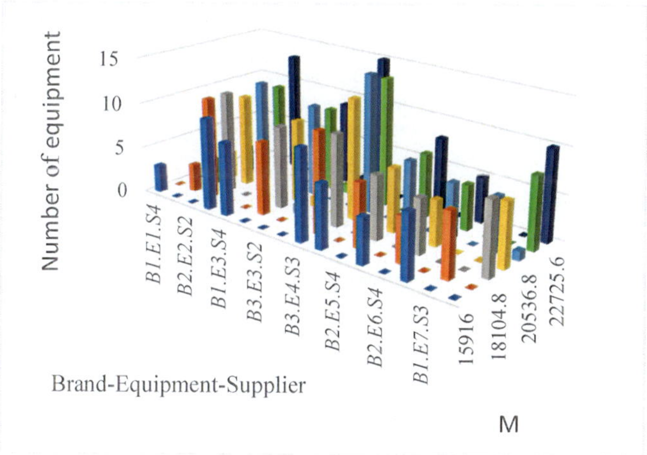

Fig. 1. Solutions of the Multi-Objective model (4).

The solver used to solve the Multi-Objective model is MIP from GAMS.

4 Conclusion

The paper contains:

- a group decision process for equipment procurement,
- a combination, between a subjective multi-criteria weighting method SWARA, an adaptation of the SAW multi-attribute method and a Multi-Objective optimization model,
- two categories of weights used: the weights that show the expert ability to evaluate equipment (coefficient of authority) and the weights that show the equipment importance,
- a new Multi-Objective model for equipment procurement that takes in account brands of equipment, two category of weights and a parameter M that can be chosen in an interval [*M1*, *M2*], solutions of the two Multi-Objective models,
- an adaptation of the multi-attribute method SAW for equipment performance by considering the vendors, brands, equipment and experts.

This approach can facilitate the equipment procurement decision, helps the manager to work with a structured decision process and provides a well-founded framework for the decision in the equipment procurement. Our approach can be used for designing a module in a Decision Support System.

Acknowledgements. The research presented in this paper is supported by the project "vINCI: Clinically-validated INtegrated Support for Assistive Care and Lifestyle Improvement: the Human Link" (funded by EUAAL2017 Programme and Executive Unit for Financing Higher Education, Research, Development and Innovation-UEFISCDI Romania), the project "Non-invasive monitoring and health assessment of the elderly in a smart environment (RO-Smart Ageing)" (funded by the Romanian Core Program of the Ministry of Research and Innovation) and the project "New solutions for complex problems in current ICT research fields based on modelling and optimization" (funded by the Romanian Core Program of the Ministry of Research and Innovation).

References

1. Filip, F.G., Zamfirescu, C.B., Ciurea, C.: Computer-Supported Collaborative Decision-Making. Springer, Cham (2017)
2. Rădulescu, C.Z., Rădulescu, I.C.: An extended TOPSIS approach for ranking cloud service providers. Stud. Inform. Control **26**(2), 183–192 (2017)
3. Radulescu, C.Z., Radulescu, M.: Decision analysis for the project selection problem under risk. IFAC Proc. Vol. **34**(8), 445–450 (2001)
4. Chai, J., Ngai, E.W.: Decision-making techniques in supplier selection: recent accomplishments and what lies ahead. Expert Syst. Appl. (2020) https://doi.org/10.1016/j.eswa.2019.112903
5. Stević, Ž., Pamučar, D., Puška, A., Chatterjee, P.: Sustainable supplier selection in healthcare industries using a new MCDM method: measurement alternatives and ranking according to COmpromise solution (MARCOS). Comput. Ind. Eng. **140**, 106231 (2019). https://doi.org/10.1016/j.cie.2019.106231

6. Diaconu, K., Chen, Y.F., Cummins, C., Moyao, G.J., Manaseki-Holland, S., Lilford, R.: Methods for medical device and equipment procurement and prioritization within low-and middle-income countries: findings of a systematic literature review. Globalization Health **13**(1), 59 (2017)
7. Ardakani, M., Milani, A., Yannacopoulos, S., Shokouhi, G.: On the effect of subjective, objective and combinative weighting in multiple criteria decision making: a case study on impact optimization of composites. Expert Syst. Appl. **46**, 426–438 (2016)
8. Radulescu, C.Z., Radulescu, M.: Group decision support approach for cloud quality of service criteria weighting. Stud. Inform. Control **27**(3), 275–284 (2018)
9. Edwards, W.: How to use multi attribute utility measurement for social decision-making. IEEE Trans. Syst. Man Cybern. **7**(5), 326–340 (1977)
10. Saaty, T.L.: A scaling method for priorities in hierarchical structures. J. Math. Psychol. **15**(3), 234–281 (1977)
11. Saaty, T.L.: The Analytic Hierarchy Process. McGraw-Hill Press, NewYork (1980)
12. Von Winterfeldt, D., Edwards, W.: Decision Analysis and Behavioral Research. Cambridge University Press, Cambridge (1986)
13. Hwang, C.L., Yoon, K.: Multiple Attribute Decision Making: Methods and Applications. Springer, Heidelberg (1981)
14. Saaty, T.L.: Decision Making with Dependence and Feedback: the Analytic Network Process. RWS Publications, Pittsburgh (2001)
15. Nezhad, G., Hashemkhani, M.R., Zolfani, S., Moztarzadeh, F., Zavadskas, E.K., Bahrami, M.: Planning the priority of hightech industries based on SWARA-WASPAS methodology: the case of the nanotechnology industry in Iran. Econ. Res. Ekonomska Istraživanja **28**(1), 1111–1137 (2015)
16. Kersuliene, V., Zavadskas, E.K., Turskis, Z.: Selection of rational dispute resolution method by applying new step wise weight assessment ratio analysis (SWARA). J. Bus. Econ. Manage. **11**(2), 243–258 (2010)
17. Baykasoglu, A., Kaplanoglu, V., Durmusoglu, Z.D.U., Sahin, C.: Integrating fuzzy DEMATEL and fuzzy hierarchical TOPSIS methods for truck selection. Expert Syst. Appl. **40**, 899–907 (2013)
18. Rezaei, J.: Best-worst multi-criteria decision-making method. Omega **53**, 49–57 (2015)
19. Stanujkic, D., Zavadskas, E., Karabasevic, D., Smarandache, F., Turskis, Z.: The use of the pivot pairwise relative criteria importance assessment method for determining the weights of criteria. Rom. J. Econ. Forecast. **20**, 116–133 (2017)
20. Zolfani, S.H., Yazdani, M., Zavadskas, E.K.: An extended stepwise weight assessment ratio analysis (SWARA) method for improving criteria prioritization process. Soft. Comput. **22**(22), 7399–7405 (2018). https://doi.org/10.1007/s00500-018-3092-2
21. Stanujkić, D., Karabašević, D., Zavadskas, E.K.: A framework for the selection of a packaging design based on the SWARA method. Inzinerine Ekonomika –Eng. Econ. **26**(2), 181–187 (2015)

Check for updates

Methodological Proposal to Define the Degree of Automation in the Sanitary Industry in Chile to Adapt to Climate Change

Claudio J. Macuada(✉), Astrid M. Oddershede, Luis E. Quezada, and Pedro I. Palominos

Industrial Engineering Department, University of Santiago of Chile, Santiago, Chile
{claudio.macuada,astrid.oddershede,luis.quezada, pedro.palominos}@usach.cl

Abstract. This study proposes a methodology to support the decision making to improve the efficiency of the technological standard in the sanitary industry facilities considering the climate change effects. Nowadays, the population necessities in terms of environment, quality and continuity of service are constantly increasing. In this regard, the sanitary industry is adopting new technologies for its processes, with the purpose that be a factor for service improvement. At present time, the Chilean sanitary industry is concern about the degree of automation and the infrastructure requirements, since they are the main critical factors for future investment planning. Therefore, it is necessary to determine the current level of the telecontrol system facilities and generate actions to make improvements in those processes that show a poor quality of service. The research methodology is based on case study, integrating planning processes, data analysis, scoring method interacting with multicriteria approach. This paper emphasis on developing a decision model by the use of the Analytical Hierarchy Process (AHP), to identify the priority facilities that should improve their technological standard. A case study incorporating climate change factors is pursued in a metropolitan sanitary company in Chile, accomplishing the automation degree of a telecontrol system for the real case. These results give place to elaborate an investment plan that can be converted into action plans for a sanitary company.

Keywords: Automation degree · Technological standard · Sanitary industry · Climate change · AHP · Telecontrol system

1 Introduction

At present, the sanitary industry is adopting new technologies for its processes and there is confidence that this is a way to improve its function to adapt to the effects associated with climate change. In China, [1–4] the public/private sector as part of a project has developed innovative technologies for its processes, improving the capacity of its facilities, operating with greater flow and satisfying the needs of treated water. Regarding, this context, in Chile there is also a concern, as indicated by Jiménez and Wainer [5]

I. Dzitac et al. (Eds.): ICCCC 2020, AISC 1243, pp. 284–295, 2021.
https://doi.org/10.1007/978-3-030-53651-0_24

in its Water Reality Report in Chile: *"Is there a shortage or lack of infrastructure?"* It is essential that the country count on a modern and environmentally friendly water infrastructure to accumulate water, in order to recover water reserve capacity for the months with less rainfall. Since, every river should have reservoir and conduction systems, remote monitoring, telemetry and remote control with automatic gates and artificial infiltration systems, technology that should be given more haste to built-in the whole country. To achieve this, the National Water Resources Strategy 2012–2025 [6] has been defined, whose objective is to develop a plan that accelerates the construction of this infrastructure type, recovering the water and restoring normal water cycles. In addition to these initiatives, in the coming years as indicated by the Office of Agricultural Studies and Policies [ODEPA] [7], it will be necessary to continue intensifying the technified irrigation coverage that is still low and to make significant improvements at the distribution water systems including remote control systems.

Therefore, the 2017–2022 Adaptation and Mitigation Plan for Climate Change Infrastructure Services [8] contemplates taking care of the necessary requirements to sustainably manage water, incorporating technological innovation in adaptation and mitigation to climate change. Also, the Ministry of Public Works [MOP] [8] points out that it is necessary to include technology that allows anticipating catastrophic events, related to information on water resources and climate data to facilitate decision-making (DM) in a timely manner and report the need for new adaptation measures.

Under this context, the drought of raw water sources is exacerbated every year, affecting Chile's Metropolitan Region since 2010 [9]. Moreover, the rainfall deficit, the recharge lacks and increased exploitation of groundwater in recent years have led to falls in groundwater level. This has promoted a challenge for the sanitary industry to ensure the drinking water supply for the region in drought and climate change conditions. In this sense, the most applicable investments are those that ensure operational resilience to manage turbidity events, increase production capacity and decrease the number of customers affected by outages on drinking water supply. Therefore, having facilities with a significant level of automation facilitates failure response times, and allows to manage resources in a preventive manner by establishing indicators that warn against the risks of not guaranteeing drinking water supply. For the research, the water company under study has considered essential that its plants must operate with the most appropriate technological system and comply with current and future standards.

Related to this, there are studies that propose a methodology to define the degree of automation for choosing drinking water and wastewater treatment systems that can be automated during night shifts to increase plant efficiency [10]. Other studies attempts to define, through a DM methodology, the degree of self-configuration of drinking water and wastewater treatment systems to provide more resilient infrastructures for preventive resources management. This, to establish indicators to warn risks of not guaranteeing the supply of drinking water [11]. Therefore, it is necessary to determine the current level of the facilities technological system and subsequently generate actions to make improvements in those plants that show a lower capacity to adapt to climate change. The examination recognizes a multiplicity of factors and objectives, such as production capacity and functionality. The implementation of technology also competes with the available budget that can influence the evaluation. In this context, a multi-criteria assessment [12–15] is adequate to address conflicting objectives. The description of the system is given in the next section. Then, Sect. 3 a methodology proposal is provided. Where in Sect. 4 the proposal is developed and Sect. 5 the results are delivered ending with conclusion at Sect. 6.

2 System Description

The problem situation addressed includes the facilities of a representative sanitary company that is interested in identifying the facilities that need to improve their technological infrastructure and complying with the standards. In addition to these requirements, the implementation involves investments, operation and maintenance costs. On these type of companies it is essential to determine the facilities that has to be endowed with new technology.

2.1 Automation Degree

The incorporation of new technologies allows companies to move from a manual operation - based on people - to an automatic operation, increasing the degree of automatism of their processes (see Fig. 1).

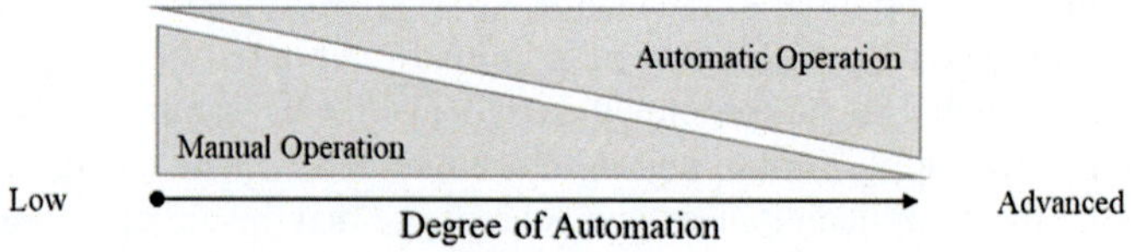

Fig. 1. Relationship operation type and degree of automation.

The level or automation degree are defined by the model Computer Integrated Manufacturing, (CIM) [16], that determine the operation mode and each technological components level. Table 1 shows the technological components that make up each degree of automation.

Increasing the automatism degree, together with centralizing information for decision-making in a timely and reliable manner, will help the sanitary industry to ensure continuity of service, increase the flexibility and reliability of the system, and will have a better adaptation to climate change (see Fig. 2).

Table 1. Automation degree components.

#	Automation degree	Description	Technological components
1	Low	Solution without local automation, based on supervision and operation by an operator on site. The information is not available in the Control Center	It requires the following elements: 1) Instrumentation 2) Electrical boards 3) Plant operator
2	Medium	Solution without local automation, based on supervision and operation by an operator on site. Information is available at the Control Center only to remotely monitor the enclosure	It requires the following elements 1) Instrumentation 2) RTU's/PLC's 3) Electrical boards 4) Communication system 5) SCADA
3	High	Solution with local automation, based on remote monitoring and control from the Control Center	It requires the following elements: 1) Instrumentation 2) RTU's/PLC's 3) Electrical boards 4) Communication system 5) SCADA
4	Advanced	Solution with local automation based on remotely controlled and remotely controlled control from the Control Center	It requires the following elements: 1) Instrumentation 2) RTU's/PLC's 3) Electrical boards 4) Communication system 5) SCADA

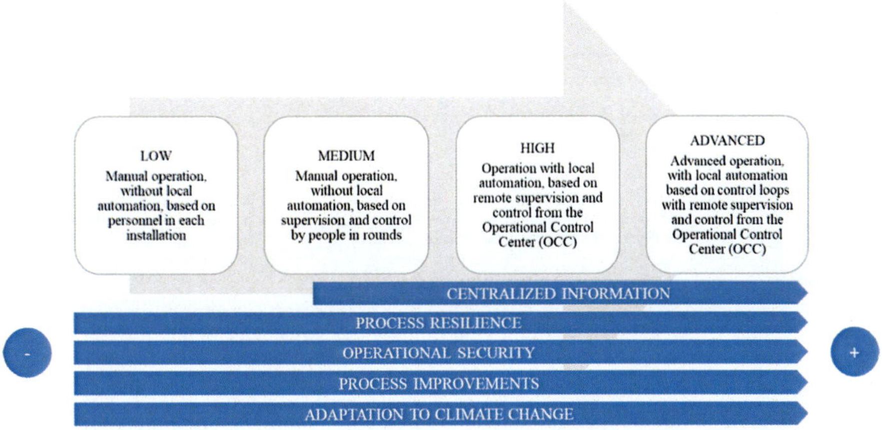

Fig. 2. Impact on the sanitary industry based on the degree of automation.

3 Research Strategy and Methodology

Multi-criteria Decision Analysis (MCDA) is a discipline designed to support the decision-making process when numerous conflicting objectives are faced. The central problem is how to evaluate a set of alternatives based on a set of criteria. The MCDM aims to show these conflicting objectives in order to reach a compromise through a participatory process. In the literature, there are publications presenting methods for dealing with these cases. Some of them are, the AHP, [17] Multi-Attribute Utility Theory (MAUT), Target Programming and Electre, among others.

This research is pursued through case study approach. AHP [17] methodology is applied to define the degree of automation for the sanitary industry in Chile to adapt to climate change through a real case study in a sanitary company.

This method has a solid scientific basis and begins by designing a hierarchy structure that can work with a variety of information, integrating the opinions and judgments of different experts. It also measures the consistency of the judgments made.

The study consists of three main phases: the first phase is dedicated to defining the appropriate degree of automation for each facility of the sanitary company under study. In order to calculate the technological gap, a diagnosis is made for each sanitary company installation. The second phase deals with defining the significant criteria for obtaining the appropriate degree of automation to ensure continuity, quantity and quality, and that it is adaptable to climate change for the city's inhabitants. This task requires the involvement of a group of experts that was made up of 20 people (plant managers, project engineers and senior management members). For the third phase, a hierarchy structure is generated to represent the problem situation is incorporating the main factors and criteria. Applying AHP [13, 14] method the team of experts evaluates the current state of the facilities under the technological criteria and estimate the result based on multicriteria approach [18, 19]. Figure 3 summarizes the proposed methodology.

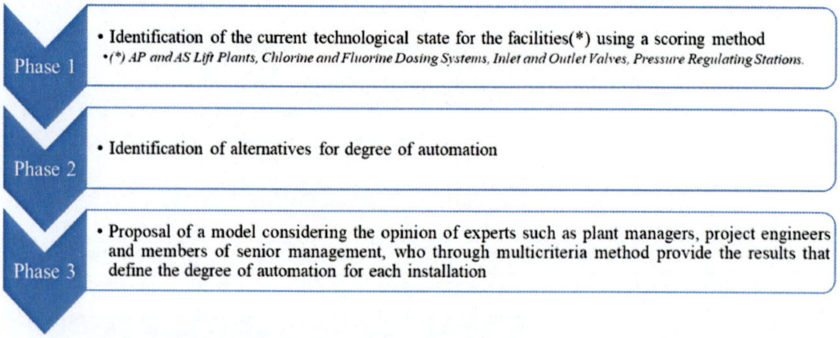

Fig. 3. Proposed methodology for sanitary industry degree of automation.

4 The Case Study Application

The company in study aims to ensure a continuous service, with high standards of safety, quality, and reliability, in order to achieve its mission *"To go beyond the water, managing the resources in a sustainable way"*

The growth in the demand, the water shortage in the recent years, and the commitment to ensure a responsible and sustainable management - in harmony with the environment and its stakeholders - makes it necessary to incorporate in its management the use of infrastructure and technological tools that contribute to maximize its efficiency, quality and continuity of service.

In this sense, one of the available resource is the Operational Control Center (OCC), which allows to operate and monitor remotely with an integral vision - all its infrastructures, controlling the operational processes in real time, especially those related to the management of the water cycle, which are (Fig. 4):

- *Production*: The process that eliminates the haze and dirt from the raw water collected, making it fit for human consumption.
- *Transport*: Water is transported from the production plants to ponds.
- *Distribution and Supply*: From ponds, adequate pressure and continuous supply for human consumption are guaranteed.
- *Sewage Collection*: The collection is initiated through the public sewer system, which consists of residential connections and drain lines that flow into the collectors, which are installed deeper in the ground. The waste is collected and discharged into the large diameter collectors; leading their waters to the treatment wastewater plants.
- *Depuration*: Treatment of collected wastewater is cleaned before being return to the environment, to avoid damaging the flora and fauna.

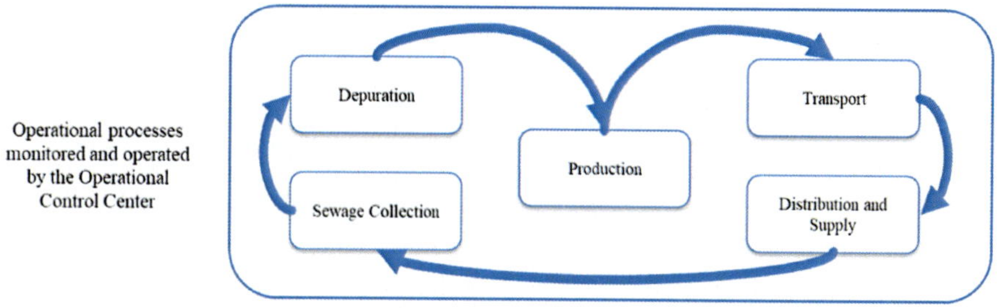

Fig. 4. Operational processes the company in study

This OCC is designed for continuous operation, 365 days a year, allowing the supervision, operation and management of the different remote controlled facilities of the company in study, together with the integration in real time of the information from all operational areas, centralizing and coordinating the decision making process. The facilities are located in different districts of Santiago de Chile and have different degrees of automation.

The perception is that all the facilities have to be highly automated to increase efficiency and adapt to climate change. This analysis does not consider other factors such as the cost of investment and operation. Therefore, it is necessary to find the right balance between the different factors that determine the level of automation needed. This type of situation is adequately addressed by multi-criteria methods such as AHP.

The study involves the development of an AHP model, which is applied independently in each facility. This paper shows the application in one of them, although the model is replicable in all the others. In the same way, this model is applicable to other companies that need to automate their processes.

This section develops the proposed methodology, starting with the current assessment of the degree of automation (Phase 1) where, at this stage, a team of experts evaluates each of the technological components of the facilities, using a scoring method.

As an example the evaluation of the data capture components (RTU's/PLC's) is presented. Let c_j = j-th component of the facility Pressure Regulatory Station (EERR).

The data capture rating for component 1 is obtained according to the following formula:

$$Rating_{c1} = \frac{N_{Data\ capture\ c1}}{\sum_{j=1}^{14} N_{Data\ capture\ cj}} \tag{1}$$

Where, N refers to the data capture system technology cluster score for the component. In addition, in general we can find the rating considering for each component as:

$$Rating = \frac{N_{ij}}{\sum_{i=1}^{5}\sum_{j=1}^{14} N_{ij}} \tag{2}$$

Then, the weighted score of all component is estimated by multiplying this score by the relative weighting that in this case have the same weighting.

$$Score = Rating \times Weighting\ Attributes \tag{3}$$

$$Score = \frac{N_{ij} \times WA_{ij}}{\sum_{i=1}^{5}\sum_{j=1}^{14} N_{ij} \times \sum_{i=1}^{5}\sum_{j=1}^{14} WA_{ij}} \tag{4}$$

The result of the current degree of automation for Pressure Regulator Stations (EERR) is shown in Table 2.

Table 2. Current automation degree.

Facility	Components technological	Current evaluation (%)	Relative weight (%)	Weight (%)	Current automation degree
Pressure regulator stations	Instrumentation	100%	20%	72%	Media
	RTU's/PLC's (Data captures)	100%	20%		
	Plant operator	Not applied			
	Electric boards	100%	20%		
	Communication system	10%	20%		
	SCADA (Data control)	50%	20%		

Phase 2 of the proposed methodology is dedicated to determining the significant criteria for identifying the appropriate degree of automation for each facility.

4.1 Criteria

The agreed criteria to achieve the objective are: 1) *Tariff Model*: Capital required financing infrastructure growth to adapt to climate change and respond to situations as cuts because of murky water. 2) *Resilience*: The Company's ability to resist, absorb, adapt and recover from environment disturbances, facing a variable and changing climate. 3) *Infrastructure Maintenance*: Refer to the costs for keeping infrastructure under the right conditions to adapt to climate change ensuring service continuity. 4*) Regulation*: Refers to establish adequate and sufficient policies and procedures to ensure that an enterprise complies with the integral quality of water service. 5) *Hydraulic Efficiency:* Ensure efficient hydraulic service, with the highest quality standards in terms of adequate and monitored infrastructure, with the less possible loss to adapt to the water crisis. 6) *Energy Efficiency*: Practice that aims to optimize the relationship between the amounts of energy consumed and the final products and services obtained, allowing to reduce the emission of greenhouse gases (GHGs), which contributes directly to mitigate climate change.

4.2 Hierarchy Model Structuring

The objective is to represent the problem situation through a hierarchy model to determine the level of automatism best suited in terms of criteria, and to be assessed by a group of experts. A two-level hierarchy is modeled including decision factors contributing the objective achievement. These factors are based on the comparative evaluation between the different degrees of automation and identify the best way to operate the facilities and ensure continuity of service. Figure 5 shows the hierarchy.

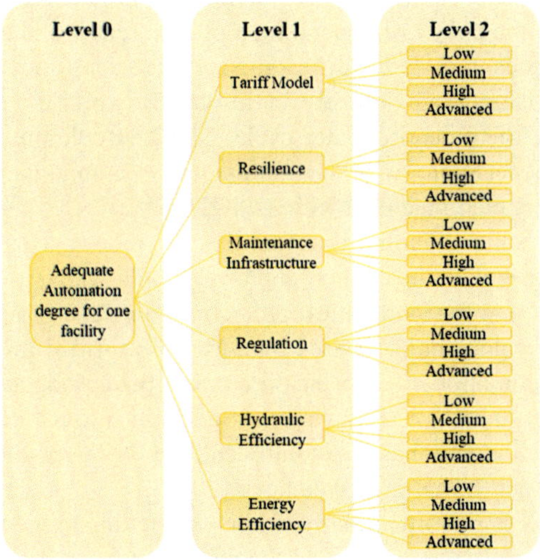

Fig. 5. Hierarchy model for Automation degree of installation Pressure Regulatory Stations (EERR).

Where: a) *Level 0* is the main objective: To select the desired automation level that best suits the requirements of the operational facilities, in this case, for the EERR. b) *Level 1* indicates the components that help determine the automation degree to ensure continuity, quantity and quality of service. c) *Level 2:* Includes all existing automation levels in remote controlled facility operation.

5 Results

Using the AHP methodology [13, 14] and processing data through Expert Choice^{TM} (EC) software [15], a relative order was acquired. Priority results showed that *resilience* is a determining factor in deciding the facility level of automation. Figure 6 depicts the factors relative priority.

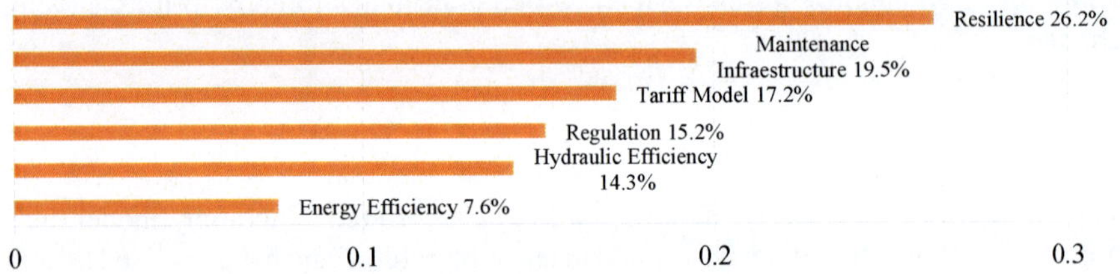

Fig. 6. Relative priority criteria.

This result is consistent with the organization's concern with adapting to climate change by ensuring continuity of drinking water service in a timely manner and thus avoiding affecting customers, loss of reputation and impact on the city. These results are consistent with the policy of generating a plan to renew and extend remote control in production facilities, remote meter reading for large consumers and green areas, and an update of the Operational Control Centre (OCC), which is responsible for advertising management of the entire water cycle. The first column of Table 3 shows the overall results of the technology for the four degrees of automation, where the advanced automation degree is recommended with 32.4%.

Based on the results, it should be noted that the relevance assigned to operational resilience would depend on the country and/or region in which it is assessed, according to the availability of water and future projections on its impact, due to the change in customers. This value could be used by DMs to prioritize investment projects and justify increased automation by ensuring the continuity of productive operations without affecting the quality of service. Today's investments in automation are generally highly complexity, and the industry has historically called for constructions built in operation without concern about the automation degree [20].

The AHP allows to perform sensitivity analysis, letting to identify possible changes from the ordering processes, modifying the criteria importance. The relevance of the analysis lies in the medium presented over time by the sanitary industry. Figure 6 shows that the degree of automation "Low" (the orange line) has a higher priority referring to Infrastructure Maintenance. As "Medium" (grey line) referring to *Regulation*, "High" (yellow line) related to the *Tariff Model*, and finally, the "Advanced" (blue line) *for Resilience, Hydraulic Efficiency, Energy* and a high trend to *Regulation.* If the expert

Table 3. Process and criteria relative importance.

Criteria / Automation degree	Criteria Global	Local Priotization			
		Low	Media	High	Advanced
Resilience	**0.262**	0.055	0.233	0.255	**0,457**
Infrastructure Maintenance	0.195	**0.595**	0.212	0.135	0,058
Tariff Model	0.172	0.066	0.104	**0.512**	0,318
Regulation	0.152	0.171	**0.328**	0.224	0,277
Hydraulics Efficiency	0.143	0.078	0.141	0.304	**0,477**
Energy Efficiency	0.076	0.077	0.231	0.325	**0,367**
Importance Degree of global automation		**0.185**	**0.208**	**0.283**	**0.324**

group changes its assessment for the criteria indicated, the degree of automation, changes depending on the resulting relative weight from those criteria, considering the above analysis. The model results show alignment with current sanitary industry policies, such as quality of service, excellence and efficiency in water resource management (Fig. 7).

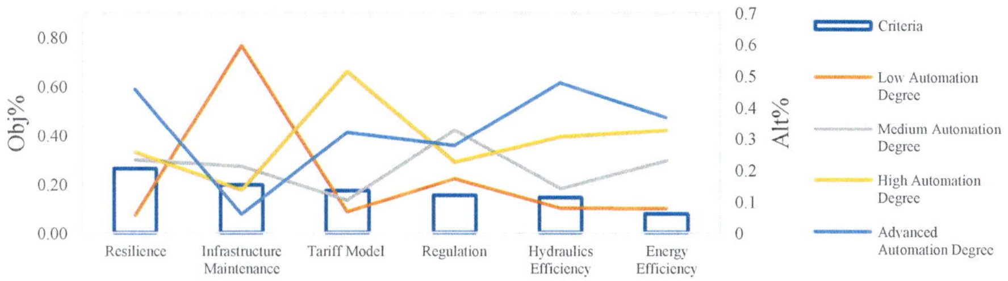

Fig. 7. Sensitivity analysis for factor "performance"

In the same way, the model was applied to the rest of the facilities. The Table 4 shows the results obtained:

Table 4. GAP proposed situation versus current situation case study.

#	Facilities	Current Situation	Proposed Situation	Action Plan
1	Pressure Regulatory Stations (EERR)	ADVANCED	ADVANCED	No Investment Required
2	Drinking Water Lift Plants (PEAP)	MEDIUM	ADVANCED	High Investment Required
3	Wastewater Lift Plants (PEAS)	MEDIUM	HIGH	Investment Required
4	Packaged Pumping Plants (PBP)	MEDIUM	HIGH	Investment Required
5	I/O Control Valves Ponds (VES)	MEDIUM	HIGH	Investment Required
6	Chlorine Dosing Systems (SDC)	MEDIUM	HIGH	Investment Required
7	Fluorine Dosing Systems (SDF)	MEDIUM	HIGH	Investment Required
8	Arsenic and Nitrate Treatment Plants (PTAN)	HIGH	HIGH	No Investment Required

It is observed that in all cases, either the level of automation must be maintained, or it must be increased with its respective investment.

6 Conclusions

This document proposed a methodology to help improve the service quality and for adaptation to climate change, identifying priority facilities that need to reach a technological standard. Using an AHP decision model to help the selection process within the existence of complex variables. Through a real case study on the prioritization of facility resources, examining the current technical state of automation in a sanitary company in Chile, the methodology provided recommendations to DMs.

It helps to obtain a detailed understanding of each of the priority requirements, to recognize the status and to identify the infrastructure to be improved.

The results of this study can serve as a guide for industrial process operation companies that need to raise their technology level to assess which tasks can be automated and adapted to climate change. The study gives a reference of the criteria that can be considered to define a standard in the sanitary industry in terms of the required degree of automation. It seeks to ensure responsible and sustainable management, guarantee a continuous service, with high standards of security, quality and reliability, have timely and quality information, minimize technological and operational vulnerabilities, facilitate DM, minimize risks and operating costs, and thus have a greater adaptation to climate change.

The results were presented to the senior management of the company, who found the solution to be appropriate.

As future research, it is proposed to build a new multi-criteria model that allows the temporary prioritization of the different projects considering the availability of resources to make the investments.

Acknowledgment. Special thanks from the authors to Department of Research and Technological Development (*DICYT Projects N° 061817OH & 061817QL*) and the Department of Industrial Engineering of the University of Santiago of Chile for the support to pursue the study.

References

1. Tracy, J.: Oasys Water. Adoption of water innovation. Recuperado el 13 de Noviembre de 2017. http://oasyswater.com/blog/adoption-of-water-innovation/
2. Chinese Academy of Science: Water Science and Technology in China: A Roadmap to 2050. s.l.: Science Press, Beijing. Springer (2012)
3. Statista: Capacity of Chinese industrial wastewater treatment facilities in 2013, by region (in million metric tons/day). Recuperado, 14 Mayo de 2018. http://www.statista.com/statistics/282574/china-capacity-of-industrial-waste-water-treatment-facilities-by-region/
4. Yushan, P.: The water treatment industry in China. Flanders Invest. & Trade Beijing (2014)
5. Susana Jiménez S., José Tomas Wainer, I.: Realidad del Agua en Chile: ¿Escasez o falta de infraestructura? Recuperado el 11 de Junio de 2018, de Serie Informe Económico. Libertad y Desarrollo. https://lyd.org/wp-content/uploads/2017/06/SIE-263-Realidad-del-agua-en-Chile-Escasez-o-falta-de-infraestructura-Marzo2017.pdf

6. Ministerio Obras Públicas [MOP]: Estrategia Nacional de Recursos Hídricos 2012–2025. Recuperado el 1Abril 2018, MOP. http://www.mop.cl/Docments/ENRH_2013_OK.pdf
7. Oficina de Estudios y Políticas Agrarias [ODEPA]: El cambio climático y los recursos hídricos de Chile. Segunda Sección: Reflexiones y Desafíos al 2030: Perspectiva de Especialistas Externos. Recuperado el 11 de Junio de 2018. https://www.odepa.gob.cl/wp-content/uploads/2018/01/cambioClim12parte.pdf
8. Ministerio de Obras Públicas [MOP]: Plan de Adaptación y Mitigación 2017–2022 de los Servicios de Infraestructura al Cambio Climático. Recuperado el 11 de Junio de 2018. http://www.dgop.cl/Documents/PlanAccionMop.pdf
9. Assopardi, T.: El Dilema de la sequía en Chile. Revista Business Chile. Recuperado el 2 Abril de 2019 http://www.businesschile.cl/es/noticia/sequia/el-dilema-de-la-sequia-de-chile
10. Macuada, C., Oddershede, A.M., Alarcon, R.: Multi-criteria assessment to automate water treatment plants using the analytical hierarchy process. J. Glob. Bus. Adv. **8**(2), 236–246 (2015)
11. Macuada, C.J., Oddershede, A.M, Quezada, L.E.: DM methodology for automating technology system in water treatment plants. In: Dzitac, I., Filip, F.G., Manolescu, M.J., et al. (eds.) 7th International Conference on Computers Communications and Control (2018)
12. Donciulescu, A.D., Filip, F.G.: DISPECER-H a decision supporting system in water resources dispatching. Annu. Rev. Autom. Program., Part 2 **12**, 263–266 (1985)
13. Saaty, T.L.: Toma de Decisiones Para Líderes. El Proceso Analítico Jerárquico. La Toma de Decisiones en un Mundo Complejo. RWS Publications, USA (1997)
14. Saaty, T.L.: Método Analítico Jerárquico [AHP]: Principios Básicos. En Evaluación y Decisión Multicriterio, Reflexiones y Experiencias, Editorial USACH, Santiago (1998)
15. Saaty, T.L.: Expert Choice, Recuperado, 10 Dic 18. http://expertchoice.com/
16. Rehg, J.A., Kraebber, H.W.: Computer-Integrated Manufacturing, 3rd edn. Prentice-Hall, Englewood Cliffs (2005)
17. Saaty, T.L.: Fundamentals of Decision Making & Priority Theory, with the analytic Hierarchy Process, 2nd edn. RWS Publications, Pittsburgh (2006). ISBN 0-9620317-6-3
18. Oddershede, A., Carrasco, R.A.: Analytic hierarchy process decision model for health institution selection. Institute for Operations Research and the Management Sciences (2006)
19. Saaty, T.L.: How to make a decision: the analytic hierarchy process. Interfaces **24**(6), 19–43 (1994)
20. Ministerio de Obras Públicas [MOP]: Guía para Elaboración de Planes MOP. Recuperado el 1 de Abril de 2018, de Dirección de Planeamiento. http://www.dirplan.cl/centrodedocumentacion/Documents/Metodologia/Guia_Elaboracion_Planes_marzo_2011.pdf

Mapping the Intellectual Structure of the International Journal of Computers Communications and Control: A Content Analysis from 2015 to 2019

José Ricardo López-Robles[1], Manuel J. Cobo[2](✉), Nadia Karina Gamboa-Rosales[3], and Enrique Herrera-Viedma[4]

[1] Postgraduate Program of Engineering and Applied Technology (National Laboratory CONACYT-SEDEAM), Autonomous University of Zacatecas, Zacatecas, Mexico
ricardolopezrobles@outlook.com

[2] Department of Computer Science and Engineering, University of Cadiz, Cadiz, Spain
manueljesus.cobo@uca.es

[3] CONACYT-Academic Unit of Electric Engineering, Autonomous University of Zacatecas, Zacatecas, Mexico
ngamboa@conacyt.mx

[4] Department of Computer Science and Artificial Intelligence, University of Granada, Granada, Spain

Abstract. International Journal of Computers Communications & Control (IJCCC) is an open access peer-reviewed journal publishing original research papers and it is considered by professionals, academics and researches as one of the main sources of knowledge in the integrated solutions in computer-based control and communications, computational intelligence methods and soft computing, and advanced decision support systems fields. With this in mind, this research conducts a bibliometric performance and intellectual structure analysis of the IJCCC from 2015 to 2019. It provides a framework to support computer, communication and control researchers and professionals in the development and direction of future researches identifying core, transversal, emerging and declining themes. For this purpose, the IJCCC's intellectual structure and thematic networks are analyzed according to the data retrieved from Web of Science Core Collection, putting the focus on the main research themes and its performance. Finally, this analysis has been developed using SciMAT, an open source (GPLv3) bibliometric software tool developed to perform a science mapping analysis under a longitudinal framework.

Keywords: Intellectual structure · SciMAT · Strategic intelligence

I. Dzitac et al. (Eds.): ICCCC 2020, AISC 1243, pp. 296–303, 2021.
https://doi.org/10.1007/978-3-030-53651-0_25

1 Introduction

Nowadays, it is well known that scientific journals are one of the most relevant sources of knowledge and its constitutes a good channel of communication to expose the outcomes of research, scientific projects and other business initiatives. In addition, the scientific journals reach the visibility of the authors and their organizations. Consequently, scientific journals are a priority object of study in bibliometric research [1].

In this way, in research whose object of study is the scientific journals, three different approaches can be identified: (i) Bibliometric performance analysis of authorship and production [2–4], (ii) Bibliometric thematic analysis [5, 6] and (iii) Research methodology studies [7], mainly.

Given the growing interest for the analysis of scientific journals, a complete bibliometric analysis based on performance indicators, and thematic analysis of the IJCCC have been carried out from 2015 to 2019. To do that, the main indicators related to bibliometric performance and its relationship are presented using SciMAT.

In this respect, this manuscript is structured as follow: Sect. 2 introduces the methodology employed in the analysis. In Sect. 3, the dataset is described. In Sect. 4 and 5, the main results and conclusions of the analysis developed are presented.

2 Methodology

The science mapping is one of the most accepted techniques to understand how documents, specialties, disciplines and fields are related to one another. Such methods are increasingly valued as a tool for measuring scientific quality, productivity, evolution and uncover the hidden key elements in different research fields [8–11].

Although there are various software tools for analyzing science mapping [12–14], SciMAT was used in the present study. Therefore, the methodology used here identifies three phases of analysis in this field of research within a specified period:

Detection of Research Themes. For the periods analyzed, the research themes are defined by applying a clustering algorithm [15] over a normalized co-words network [10]. The similarity between the themes is assessed using the equivalence index [16].

Visualizing Research Themes and the Thematic Network. The research themes detected are determined based on their centrality and density rank values using two specific tools: the strategic diagram and thematic network [16, 17]. Centrality measures the degree of interaction of a network with other networks, and Density measures the internal strength of the network. By considering both types of measures, a field of research can be visualized as a set of themes and plotted on a two-dimensional strategic diagram (Fig. 1(a)). Therefore, four research themes can be classified [7]:

- Motor themes (Quadrant 1 (Q1) – upper-right quadrant): The themes located in this quadrant are relevant for developing and structuring the journals, specialties, disciplines and research fields. These themes are known as the motor themes of the field, given that they present strong centrality and high density.

- Highly developed and isolated themes (Quadrant 2 (Q2) – upper-left quadrant): These themes are strongly related, highly specialized, and peripheral, but these do not have the appropriate background or importance for the field.
- Emerging or declining themes (Quadrant 3 (Q3) – lower-left quadrant): These themes has a low density and centrality and these are relatively weak and. These themes mainly represent either emerging or disappearing themes.
- Basic and transversal themes (Quadrant 4 (Q4) – lower-right quadrant): These themes are not well developed but these are relevant for the field of research. This quadrant contains transverse and general basic themes.

Performance Analysis. The relative contribution of research themes and thematic areas to the entire field of research is measured quantitatively and qualitatively. It is used to establish the most productive and relevant areas within the field. In this case, the bibliometric indicators used are published documents, number of citations, average citation and h-index. In this way, the theme's performance was computed taking into consideration the documents linked with it and its h-index.

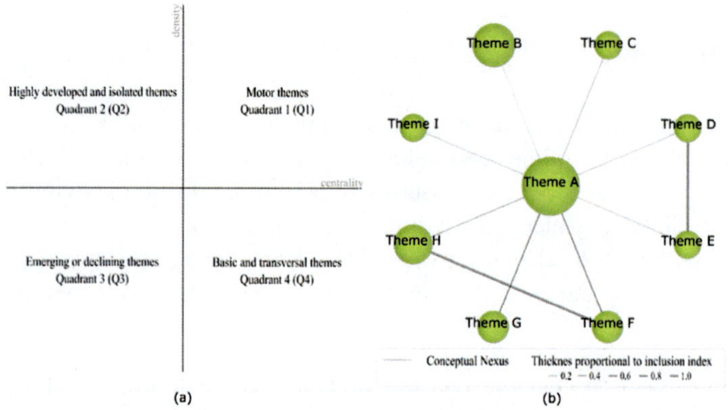

Fig. 1. (a) Strategic diagram. (b) Thematic network

3 Dataset

Taking into account the above, the performance and science mapping analysis are carried out using the documents published by International Journal of Computers Communications & Control (IJCCC) from 2015 to 2019. The documents and their citations included in this analysis have been collected in January 6th, 2020.

The documents were retrieved from Web of Science Core Collection –the most important bibliographic database– using the following advance query: *IS = ("1841–9836") Refined by: PUBLICATION YEARS:(2015 OR 2016 OR 2017 OR 2018 OR 2019).* It is also important to highlight that the documents retrieved were compared with the publications hosted in the IJCCC website to guarantee that these are consistent in both sources.

This process retrieved a total of 314 publications from 2015 to 2019 (Fig. 2). According to methodology used for this research, the raw data was downloaded as plain text and entered into SciMAT to build the knowledge base for the performance and science mapping analysis. In addition, a de-duplicating process was also applied to improve data quality by grouping those meanings and concepts that represent the same notion (e.g., "ARTIFICIAL-INTELLIGENCE" "AI-ARTIFICIAL-INTELLIGENCE" and "ARTIFICIAL-INTELLIGENCE-(AI)" were merged as "ARTIFICIAL-INTELLIGENCE") and some meaningless keywords in this context, such as stop-words or words with a very broad and general meaning were removed (e.g. "SOFTWARE", "SYSTEMS" or "ALGORITHM").

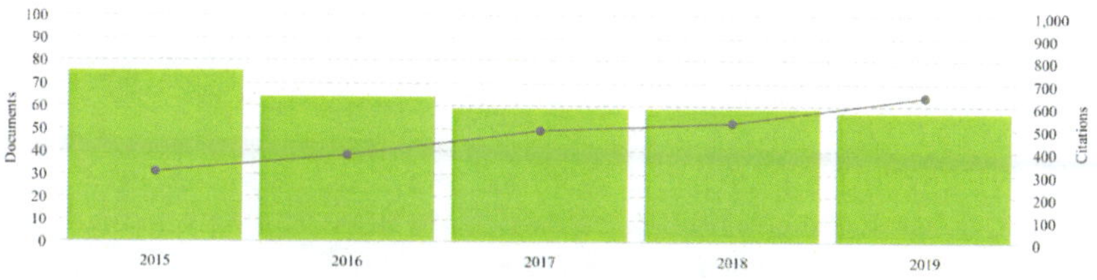

Fig. 2. Distribution of document retrieved by years

4 Conceptual Analysis

Following the methodology described above, the Fig. 3 and Table 1 provide an overview of the science mapping, its performance and the relations between main themes in the IJCCC from 2015 to 2019. In addition, it is important to recall that the volume of the spheres is proportional to the number of published documents associated with each research themes and its citations are included in brackets.

According to the strategic diagram, during this period the research pivoted on seventeen themes: NEURAL-NETWORKS, WIRELESS-NETWORKS, WIRELESS-SENSOR-NETWORK, MEMBRANE-COMPUTING, MULTI-CRITERIA-DECISION-MAKING, FUZZY-LOGIC-APPROACH, SIMILARITY-MEASURES, MACHINE-LEARNING, CRITERIA-DECISION-MAKING, FUZZY-CONTROL, EVIDENTIAL-REASONING, PATTERN-RECOGNITION, MULTI-OBJECTIVE-OPTIMIZATION, COLLABORATIVE-FILTERING and INTERNET-OF-THINGS.

The performance measures of the main themes presented in Fig. 3 shows the number of documents, citations, average citations and h-index per theme. In this way, the most productive themes (more than 30 documents) are included in the Basic and transversal themes (NEURAL-NETWORKS (36) and WIRELESS-NETWORKS (35)) and Highly developed and isolated themes (WIRELESS-SENSOR-NETWORK (34) and MEMBRANE-COMPUTING (33)). Nevertheless, the most cited themes (more than 150 cites) are included in the Motor themes (MULTI-CRITERIA-DECISION-MAKING

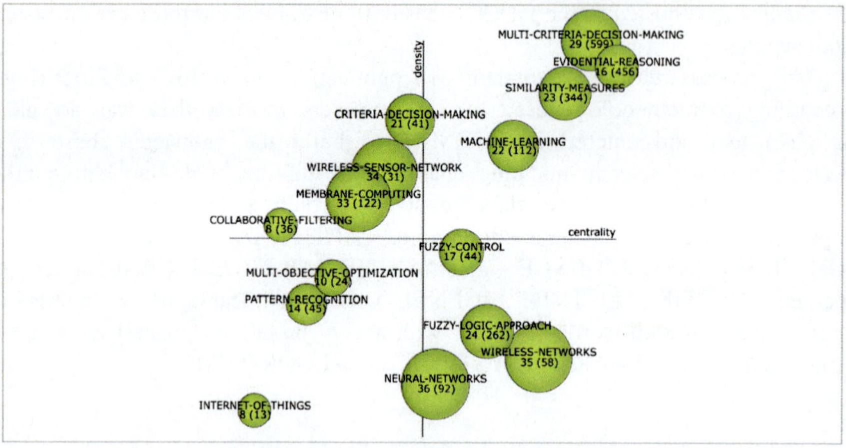

Fig. 3. Strategic diagram of the themes in the whole period (2015–2019)

(599), EVIDENTIAL-REASONING (456) and SIMILARITY-MEASURES (344)) and Basic and transversal themes (FUZZY-LOGIC-APPROACH (262)). In both cases, the themes are not the same, which means that we have two references: most productive and most cited themes.

On the other hand, the thematic networks of each theme is shown in Fig. 4.

Table 1. Performance of the themes in the whole period (2015–2019)

Theme	Documents	Citations	Average	h-index
EVIDENTIAL-REASONING	16	456	28.50	10
MULTI-CRITERIA-DECISION-MAKING	29	599	20.66	12
MACHINE-LEARNING	22	112	5.09	5
SIMILARITY-MEASURES	23	344	14.96	7
WIRELESS-SENSOR-NETWORK	34	31	0.91	3
NEURAL-NETWORKS	36	92	2.56	6
MEMBRANE-COMPUTING	33	122	3.70	6
WIRELESS-NETWORKS	35	58	1.66	4
CRITERIA-DECISION-MAKING	21	41	1.95	4
FUZZY-LOGIC-APPROACH	24	262	10.92	6
FUZZY-CONTROL	17	44	2.59	4
PATTERN-RECOGNITION	14	45	3.21	3
MULTI-OBJECTIVE-OPTIMIZATION	10	24	2.40	3
COLLABORATIVE-FILTERING	8	36	4.50	4
INTERNET-OF-THINGS	8	13	1.62	2

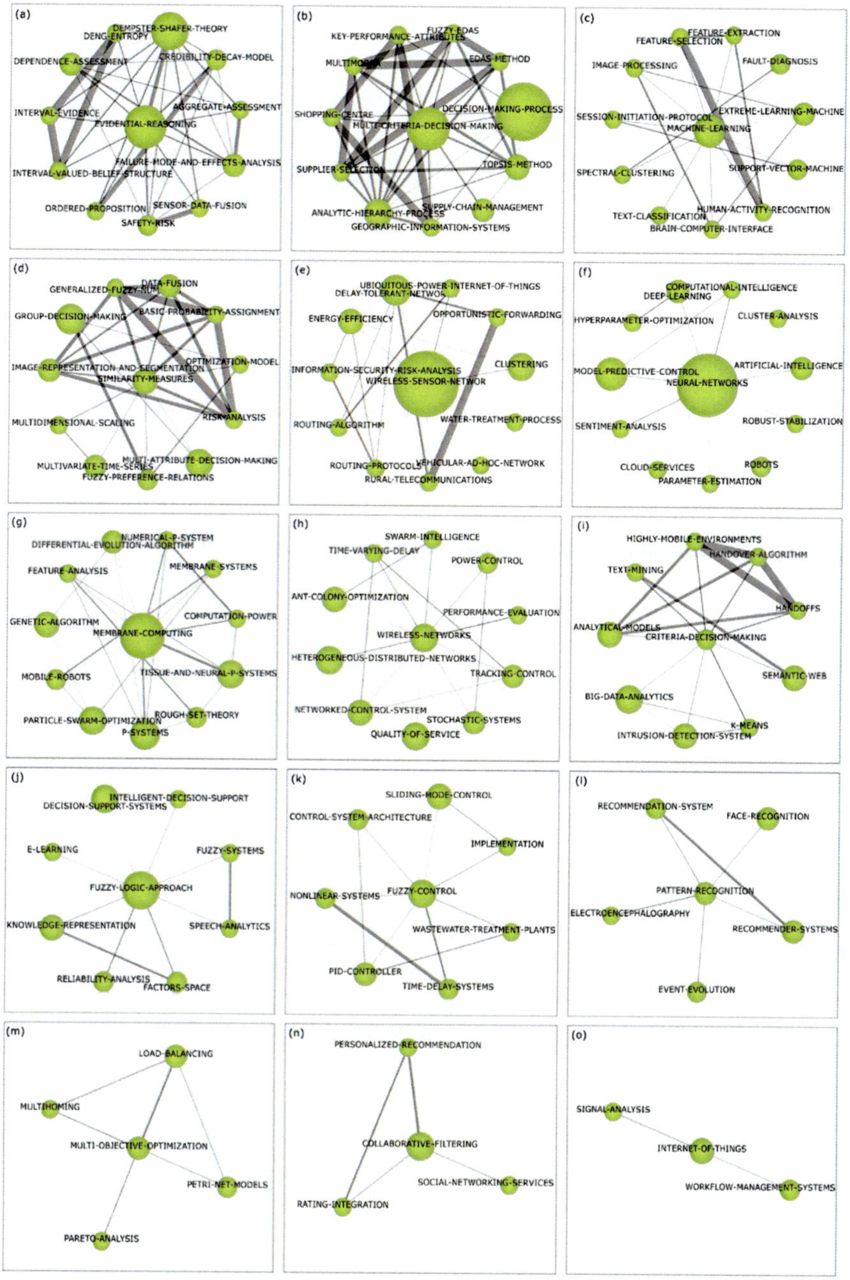

Fig. 4. Thematic network.

In terms of productivity, the most productive theme of the IJCCC is NEURAL-NETWORKS (Fig. 4(f)), which is related mainly with DEEP-LEARNING,

COMPUTATIONAL-INTELLIGENCE, ARTIFICIAL-INTELLIGENCE, MODEL-PREDICTIVE-CONTROL and SENTIMENT-ANALYSIS. Taking into account that this theme is allocated in the Quadrant 4, its role is transversal for the rest of themes.

The theme WIRELESS-NETWORKS (Fig. 4(h)) is the second ranked in terms of productivity and the eight in citations achieved. It is mainly related to themes such as SWARM-INTELLIGENCE, NETWORKED-CONTROL-SYSTEMS, ANT-COLONY-OPTIMIZATION and STOCHASTIC-SYSTEMS. This theme presents the same role that NEURAL-NETWORKS.

In a similar case, WIRELESS-SENSOR-NETWORK and MEMBRANE-COMPUTING are allocated in the Quadrant 2, where their role are highly developed themes.

The theme WIRELESS-SENSOR-NETWORK (Fig. 4(e)) is mainly related to themes such as DELAY-TOLERANT-NETWORKS, ROUTING-PROTOCOLS and ROUTING-ALGORITHM and the theme MEMBRANE-COMPUTING (Fig. 4(g)). is mainly linked to TISSUE-AND-NEURAL-P-SYSTEMS, PARTICLE-SWARM-OPTIMIZATION, GENETIC-ALGORITHM, DIFFERENTIAL-EVOLUTION-ALGORITHM and MEMBRANE-SYSTEMS.

Finally, according to the results obtained and taking into account the main research themes related to the most cited themes, it could be possible to state that the publications and research themes covered by the International Journal of Computers Communications & Control (IJCCC) are robust, linked and synergies between them.

5 Conclusions

This research presents the first performance and science analysis of the International Journal of Computers Communications & Control from 2015 to 2019 using SciMAT.

This analysis covers 314 documents published by IJCCC and hosted in the Web of Science Core Collection. These documents are a significant share of the computer-based control and communications, computational intelligence methods and soft computing, and advanced decision support systems fields.

In view of the results, two main research themes groups were identified. The first group is the most productive themes (NEURAL-NETWORKS, WIRELESS-NETWORKS, WIRELESS-SENSOR-NETWORK and MEMBRANE-COMPUTING) covered in the last five years and the second one is the most cited themes (MULTI-CRITERIA-DECISION-MAKING, EVIDENTIAL-REASONING, SIMILARITY-MEASURES and FUZZY-LOGIC-APPROACH).

Finally, as future works, a yearly analysis could be carried out taking into account a wider time span and enriching the analysis with the main authors, organizations, countries, among others. Furthermore, it will allows identify the evolution of the themes and its position in the quadrants.

Acknowledgements. The authors want to thank the support of FEDER funds (TIN2016-75850-R), Consejo Nacional de Ciencia y Tecnología (CONACYT) and Dirección General de Relacones Exteriores (DGRI).

References

1. López-Robles, J.R., Guallar, J., Otegi-Olaso, J.R., Gamboa-Rosales, N.K.: El profesional de la información (EPI): bibliometric and thematic analysis (2006-2017). El profesional de la información **28**, e280417 (2019)
2. López-Robles, J.R., Otegi-Olaso, J.R., Arcos, R., Gamboa-Rosales, N.K., Gamboa-Rosales, H.: Mapping the structure and evolution of JISIB: a bibliometric analysis of articles published in the Journal of Intelligence Studies in Business between 2011 and 2017. J. Intell. Stud. Bus. **8**, 9–21 (2018)
3. López-Robles, J.R., Otegi-Olaso, J.R., Porto-Gómez, I., Gamboa-Rosales, H., Gamboa-Rosales, N.K.: Understanding the intellectual structure and evolution of Competitive Intelligence: a bibliometric analysis from 1984 to 2017. Technol. Anal. Strateg. Manage. **32**, 604–619 (2019)
4. Wang, X., Xu, Z., Dzitac, I.: Bibliometric analysis on research trends of international Journal of Computers Communications & Control. Int. J. Comput. Commun. Control **14**, 711–732 (2019)
5. López-Robles, J.R., Otegi-Olaso, J.R., Gamboa-Rosales, N.K., Gamboa-Rosales, H., Cobo, M.J.: 60 years of business intelligence: a bibliometric review from 1958 to 2017. In: New Trends in Intelligent Software Methodologies, Tools and Techniques: Proceedings of the 17th International Conference SoMeT_18, p. 395. IOS Press (2018)
6. López-Robles, J.R., Rodríguez-Salvador, M., Gamboa-Rosales, N.K., Ramirez-Rosales, S., Cobo, M.J.: The last five years of Big Data Research in Economics, Econometrics and Finance: Identification and conceptual analysis. Procedia Comput. Sci. **162**, 729–736 (2019)
7. Cobo, M.J., López-Herrera, A.G., Herrera-Viedma, E., Herrera, F.: An approach for detecting, quantifying, and visualizing the evolution of a research field: a practical application to the fuzzy sets theory field. J. Inform. **5**, 146–166 (2011)
8. Martinez, M.A., Cobo, M.J., Herrera, M., Herrera-Viedma, E.: Analyzing the scientific evolution of social work using science mapping. Res. Soc. Work. Pract. **25**, 257–277 (2015)
9. Garfield, E.: Towards Scientography. Curr. Contents **48**, 3–14 (1986)
10. Callon, M., Courtial, J.P., Turner, W.A., Bauin, S.: From translations to problematic networks: an introduction to co-word analysis. Information (Int. Soc. Sci. Counc.) **22**, 191–235 (1983)
11. López-Robles, J.R., Otegi-Olaso, J.R., Porto-Gómez, I., Cobo, M.J.: 30 years of intelligence models in management and business: a bibliometric review. Int. J. Inf. Manage. **48**, 22–38 (2019)
12. Cobo, M.J., López-Herrera, A.G., Herrera-Viedma, E., Herrera, F.: Science mapping software tools: review, analysis, and cooperative study among tools. J. Am. Soc. Inf. Sci. Technol. **62**, 1382–1402 (2011)
13. Börner, K., Theriault, T.N., Boyack, K.W.: Mapping science introduction: past, present and future. Bull. Assoc. Inf. Sci. Technol. **41**, 12–16 (2015)
14. Cobo, M.J., López-Herrera, A.G., Herrera-Viedma, E., Herrera, F.: SciMAT: a new science mapping analysis software tool. J. Am. Soc. Inf. Sci. Technol. **63**, 1609–1630 (2012)
15. Juliani, F., de Oliveira, O.J.: State of research on public service management: identifying scientific gaps from a bibliometric study. Int. J. Inf. Manage. **36**, 1033–1041 (2016)
16. Callon, M., Courtial, J.P., Laville, F.: Co-word analysis as a tool for describing the network of interactions between basic and technological research - the case of polymer chemistry. Scientometrics **22**, 155–205 (1991)
17. He, Q.: Knowledge discovery through co-word analysis. Libr. Trends **48**, 133–159 (1999)

A Model for Solving Optimal Location of Hubs: A Case Study for Recovery of Tailings Dams

Rodrigo Barraza[1,2], Juan Miguel Sepúlveda[1(✉)], Juan Venegas[1], Vinka Monardes[1,3], and Ivan Derpich[1]

[1] University of Santiago, Santiago, Chile
{rodrigo.barraza.a,juan.sepulveda}@usach.cl
[2] Santo Tomás University, Santiago, Chile
[3] University of Antofagasta, Antofagasta, Chile

Abstract. In this paper a method for optimal location of multi-hubs in a complex network with a large number of nodes is presented. The method is applied to the design of a logistics network composed of many tailings dams and mineral processing plants and combines two data mining techniques, K-Medoids and k-Means, with the multi-criteria decision making model PROMETHEE for the prioritization of nodes to be included into the clusters, based on certain technical and economic decision variables (such as the content of recoverable metals and the costs of transportation). The proposed method contributes to solve a large scale mathematical problem difficult to handle due to the number of variables and criteria. A case study for the recovery of abandoned deposits of mining waste is presented. The case study demonstrates the feasibility and usefulness of the proposed solution and lays the groundwork for further research and other applications of machine learning techniques for big data in support of sustainable production and a circular economy.

Keywords: Data mining · Multi-criteria analysis · Circular economy

1 Introduction

On a global scale, the demand for products from the copper mining industry increases, so it does the production of this metal for use in a variety of industries such as electronics, renewable energies, and automotive, among others. The metallic minerals are exploited and processed in a concentration plant according to a recovery rate, but an amount near 95% of the input is waste material discarded during the process because it does not have an economic value for the company [1], such waste is called the tailing and is far the greatest waste produced by any mining process. Tailings dams are deposits structured in a safe way to contain the tailings coming from a plant of humid concentration by a flotation process. Tailings dams are the best known mitigation measure in relation to the environmental impact of the fine solids of the mining industry, as it protects the health and safety of people near to the tailings dams, as well as the environment, by rigorously confining materials to isolate them and thus protecting the rational use of

I. Dzitac et al. (Eds.): ICCCC 2020, AISC 1243, pp. 304–312, 2021.
https://doi.org/10.1007/978-3-030-53651-0_26

water and soil. However, the chemical and structural conditions of tailings dams vary over time leading to the need of public policies to ensure environmental conditions, provide technical assistance to companies, exert controls, guides and training, as well as to establish periodic audit programs for closure plans [2]. Research on tailings dams does not only addresses environmental issues in terms of visual impact, structural stability, radiation emission, air and soil pollution, or ground and surface water pollution, but also aims to the economic aspects of tailings reuse, recycling and reprocessing [3, 4], as addressed in this work.

This paper is ordered with focus in a conference [5]. Section 2 shows the literature review; Sect. 3 presents the problem statement, Sect. 4 presents method and tools; Sect. 5 presents a case study and results. We conclude and summarize the opportunities for future works in Sect. 6.

2 Literature Review

Improving logistics management in the mining industry is one of the cost reduction factors. Mining companies can gain competitive advantages in their sector by improving their logistics management using new generation tools for resource management and cost reduction [4].

Data mining allows the analysis of large amount of data with the integration of techniques from various fields of knowledge that include artificial intelligence, statistics, machine learning, pattern recognition and database systems [6, 7]. The creation of new algorithms to solve NP-hard problems from study to determine the appropriate combinations of local search and the exact methods have been discussed extensively in the literature [8]. In this article, three influential algorithms are ordered for solving the optimization problem of tailing dams [9]: a) The preference ranking organization method for multiple criteria decision making PROMETHEE (Preference Ranking Organization Method for Enrichment Evaluation). PROMETHEE, this method is one of most popular approaches for solving a multicriteria problem by considering simultaneously extended criteria and outranking relations [10]; b) the K-medoid algorithm clustering has an important quality, requires only once the distance between each pair of points. After performing several simulations with artificial data, the result of K-medoid clustering has better performance than k-means clustering and requires less calculation time [11]; and, c) the filtering algorithm is a more efficient version of the k-means algorithm. The combination of algorithm is easy to apply, and it only needs one decision tree for implementation, just the sequence order. The improvement of performance is obtained because the data does not change during the calculation and thus it is not necessary to reorder the data or perform recalculations [12].

3 Problem Statement

In this paper, we create a sequence algorithm based on data mining and multi-criteria methods for optimizing logistics in the recovery of abandoned tailings dams; that is, tailings dams with not known owner or with no closure resolution. Given a set of tailing dams at several locations, with different characteristics in terms of usable metals for reprocessing (for instance, copper, zinc, vanadium, barium, lead), the problem is to find an optimal location of hubs for concentrating transport infrastructure from the clusters of tailings dams to the processing plants. This optimization problem is complex and hard to solve because corresponds to a multi-hub location problem where the number of nodes is large [13], for instance in the case study presented there are 101 tailings dams. This calls for non-traditional optimization techniques such as machine learning methods. The purpose is to provide decisions tools in support of an initiative [14] of the government, launched in 2018, which seeks to attract investors in new mining projects who must carry out measures to reduce the negative impacts of abandoned tailing dams [15], either by reusing deposits or improving conditions as trade off mechanisms for expansion licensing.

4 Methods and Tools

In this article, we combine three algorithms for solving the hierarchical optimization problem of the tailing dams: a) the classic preference ranking organization method for multiple criteria decision making PROMETHEE™ [6]; b) the algorithm for K-medoids clustering; and, c) the filtering algorithm of k-means clustering algorithm.

4.1 Hubs Location Algorithm

Figure 1 shows schematically the algorithm in its three phases: Phase 1, ranking algorithm (PROMETHEE) that makes the ranking of tailings according to given criteria (e.g. volume of metal); Phase 2, clustering algorithm (K-Medoids) to perform the tailings grouping; Phase 3, clustering algorithm (k-means) to establish the location according to the distance to the processing plant.

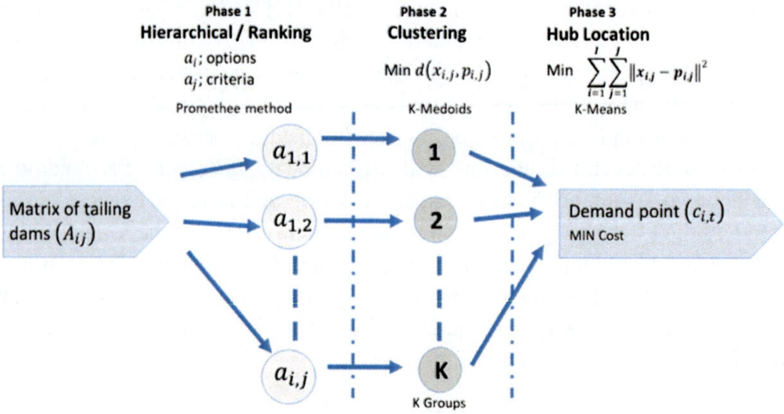

Fig. 1. Design for hub location optimization.

Phase 1, Ranking Algorithm

For the hierarchical ordering of the n tailings initially considered, a multi-criteria prioritization model is implemented, specifically PROMETHEE™. This method has been chosen because the preference functions can be directly programmed without human interaction in the pairwise comparisons as it may occur with other possible methods. This classic method can be found in detail in [8]. The ranking of tailings will be included weighting factor w_i for de distances in the Phase 2 of the algorithm to determine the optimal clustering.

Phase 2, Clustering Algorithm (K-Medoids)
After the prioritization, the clustering algorithm will allow the n tailings to be divided into k groups; to that end, the K-Medoids algorithm will group the points of an X matrix of observations, centered on the observation with greatest proximity to the theoretical center. This is because K-Medoids seeks to partition the observations. Next, the algorithm is described along with its cost function [15]. The cost function is defined by Eq. (1).

$$Minf = \sum_{i=1}^{I} \sum_{j=1}^{J} d\left(x_{i,j}, p_{i,j}\right) \tag{1}$$

The cost function (2) seeks to minimize the Haversine's distance from each observation x to the medoids.
Step 1: The k clusters are initialized, corresponding to the selection of k random observations.
Step 2: The Haversine's distance from each observation to the potential k medoids is measured by $d\left(x_{i,j}, p_{i,j}\right)$. Then, the observations that minimize the cost function are assigned to p_i as defined by (3). Symbol r is the radius of the Earth. The distance is weighted according to the factor w_i^{-1} given by the algorithm of Phase 1.

$$d\left(x_{i,j}, p_{i,j}\right) = 2rsin^{-1}\left(\sqrt{sin^2\left(\frac{x_i - p_i}{2}\right) + cos(x_i)cos(p_i)sin^2\left(\frac{x_j - p_j}{2}\right)}\right)w_i^{-1} \tag{2}$$

$$p_i = \left\{x_{i,j} : d\left(x_{i,j}, p_{i,j}\right) \leq d\left(x_{i,j}, p_{i+1,j+1}\right) \forall\, i = 1, \ldots, I; j = 1, \ldots, J\right\} \tag{3}$$

Step 3: The verification of observation closest to the medoid and change the current selection of medoids.
Step 4: If the cost function decreases with respect to the previous selection, iterate from step 2, otherwise the algorithm ends.

Phase 3, Hub Location Algorithm (k-Means)
Once the K-medoids clustering algorithm has been implemented, it is essential to establish a transfer point in each cluster for centralizing a route to the reprocessing plants. For this, the k-means algorithm is implemented. This is fundamental due to the bias generated by centralizing with respect to an artificial location in the previous step, in addition to giving greater weight to the segmentation variables than to their geographical location.

The k-means algorithm is a generalization of the k-medoids algorithm and will allow the establishment of c_i centroids for each k cluster already obtained in the previous phase.

The cost function is shown in Eq. (4).

$$Min\, g = \sum_{i=1}^{I} \sum_{j=1}^{J} \left\| x_{i,j} - p_{i,j} \right\|^2 \tag{4}$$

Step 1: The c_i centroids, corresponding to the selection of k random observations, are initialized.
Step 2: The existing error with respect to the mean for each observation towards the c_i centroids is determined. The observations in each cluster are then indexed according to the criteria set out in (5).

$$c_{i,t+1} = \left\{ x_{i,j} : \left\| x_{i,j} - p_{i,j} \right\|^2 \leq \left\| x_{i,j} - p_{i+1,j+1} \right\|^2 \forall i = 1, \ldots, I; j = 1, \ldots, J \right\} \tag{5}$$

Step 3: The $c_{i,t}$ centroids are determined by averaging the cluster observations, according to expression (6).

$$c_{i,t+1} = \frac{1}{S_i} \sum_{j=1}^{J} x_j \tag{6}$$

Step 4: If the cost function decreases with respect to the previous selection, can be iterated from step 2, otherwise the algorithm ends.

Once the iterations for the k-means algorithm are finished, the centroids are obtained for each group found with K-Medoids.

5 Case Study

Based on the generic hub location algorithm described above, we present now a case study to optimize transportation costs for reprocessing ore in tailings dams of the Coquimbo region in northern Chile. The country is the largest producer of copper worldwide, in terms market share and production volume with 5.6 millions of metric tons [16].

5.1 Case Settings

The case study was developed by an iterative process of data collection during 2018 from the National Geology and Mining Service (SERNAGEOMIN by its acronym in Spanish). The research focused on the data analysis of the geographical location of tailings dams, their current volume, their authorized volume, and the characterization and current chemical composition of each of such dams. The case was developed with 103 tailings dams considered as critical according to the public policy criteria, which will be the base for the execution of the optimization algorithm. To simplify the optimization model of transport costs, Haversine's distances were used between each point of the tailings network, as given by Eq. (2). Table 1 shows the design structure for the optimization algorithm implemented with RStudio. Table 2 shows the selected parameters. The databases used in the case study correspond to the data obtained from strategic technological programs managed by the Chilean government for the recovery of elements of value in tailings deposits. The data files are as follows: "CDR_CHILE_23_04_2019.xlxs and 24-01-2019 Characterization-Geochemistry-of-relaves-Chile.xlxs". Table 1 shows the design structure for the optimization algorithm implemented with RStudio.

Table 1. Design of optimization algorithm.

Input data type	Algorithm instance	Output data
Minerals concentration (percentage) Current volume (number)	Phase 1. Prioritization algorithm (PROMETHEE)	Ranking (number *n*)
Ranking (number *n*) Longitude (datum) Latitude (datum)	Phase 2. Clustering algorithm (K-Medoids)	Cluster group (number)
Cluster group (number) Longitude (datum) Latitude (datum)	Phase 3. Hub location algorithm (k-means)	Hubs location (number; datum)

5.2 Solution Implementation

To run the algorithms, a personal computer with a Windows 10 Pro operating system, 4.6 GHz 64-bit i7 processor with 12 Gb RAM was used. The initial data were tabulated in MS-Excel and, by a query, entered into the RStudio open-source application (version 1.2.1335 2009–2019) for the execution of the sequence of algorithms of choice and segmentation.

The parameter settings used for each of the three phases of the model that optimizes the location of the tailings hub are as follows: in Phase 1, the criteria of current volume of tailings deposited in the dam included the concentration of: copper (Cu), zinc (Zn), barium (Ba), vanadium (V), and lead (Pb) in the dam´s tank. The weights assigned to the decision criteria were 1/6 for each one; in this case all weights are equal, but this may vary according to the objectives of study. The preference function used in PROMETHEE is the step function.

In Phase 2 (K-Medoid), in order to segment with this method, the output ranking of Phase 1 is used together with the data of lengths and latitudes of each of the 103 tailings dam. In addition, it is specified in the model parameters that three ($k = 3$) clusters groups for the geographic location of tailings dams are calculated. In Phase 3 (k-Means), for each of the clusters obtained in Phase 2 with the K-medoid, the k-Means algorithm is executed. Parameters are specified for each cluster, because the method only seeks to

Table 2. Selected parameters.

Parameters	Value
Number of tailing dams	103
Number of criteria	6
Decision variables (criteria)	Cu; Zn; Ba; Pb;V; Volume
Weight of criteria	1/6
Quantity clusters	3

find the centroid for each of the clusters obtained in Phase 2. Finally, each tailing of each cluster and each centroid is plotted to define the hubs. Table 2 shows a summary of the parameters and categories entered into the algorithm.

5.3 Results

Figure 2 shows the sequence of results of each phase. The third image to the right shows the final hub's location in cyan circles. The distances from each of the points to each hub were determined with expression (2) where the positions $(x_{i,j}; p_{i,j})$ and the radius (r) of the Earth are entered. The radius is relative to latitude since Earth is not perfectly round. For this stage of the algorithm, the value of the equivolume radius 6371 (km) was used [17].

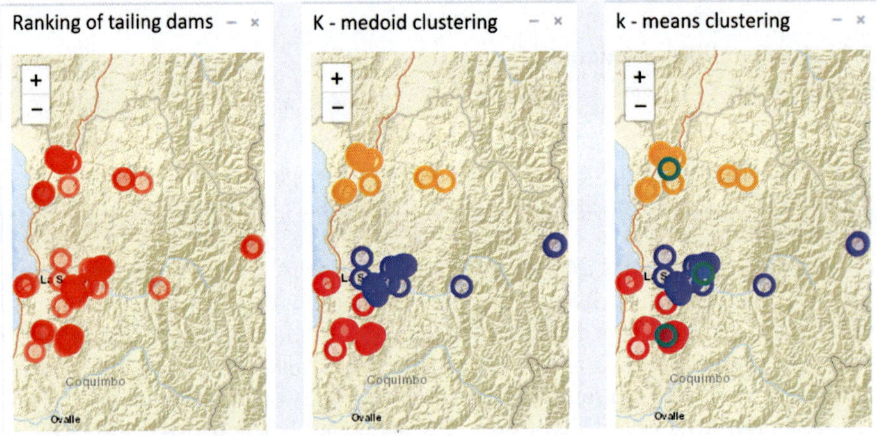

Fig. 2. The solutions obtained by optimization algorithm.

To summarize, the average distance of the final hubs to all its corresponding points are shown in Table 3. The proposed methodology of combined algorithms to optimize the transport, they present the best behavior with three hubs for minerals transfer.

Table 3. The average distances (Km) solutions obtained by the optimization algorithm.

HUBS	Location (datum)	Cluster 1	Cluster 2	Cluster 3
Hub 1	Longitude; Latitude −70.943; −29.937	36.9	----	----
Hub 2	Longitude; Latitude −71.1029; −30.259	----	12.6	----
Hub 3	Longitude; Latitude −70.943; −29.937	----	----	16.4

Table 4 shows the top distances from the tailing dams to the locations of the hubs.

Table 4. Solutions obtained: the hubs vs top tailing dams distances (Km).

HUBS	Top tailing dams (distance Km)				
	1-Dam025	2-Dam002	3-Dam003	4-Dam052	10-Dam082
Hub 1	61.9	----	----	28.6	----
Hub 2	----	10.7	19.9	----	----
Hub 3	----	----	----	----	16.9

6 Conclusions and Further Research

The model presented in this paper provides a decision making tool for guiding policy makers concerning the protection of the environment and the development of the mining industry. As new mining projects arise, it is advisable to promote compensatory actions as a trade-off mechanism in the process of environmental impacts assessment. Intervention of abandoned tailings dams is one type of those types of mechanisms, but in order to provide reliable information to the investors of the mining sector, the authority must provide information regarding site locations and chemical compositions to evaluate the private projects. Since hubs location involve the analysis of a large amount of data and decision variables and because of the multiple interrelationships between the variables and the planning criteria [18], suitable tools as the model developed in this work, must be developed to manage operations. Although, the model was developed for a pre-established configuration of the case study, the decision criteria and parameters considered may be adapted to any case where a mining company wants to achieve an efficient interaction of the transport of tailings with the location of a reprocessing plant. The case study demonstrates the feasibility of combining multicriteria analysis with data mining techniques for the progressive analysis of the impact of tailings deposits. As the country has the goal of exporting eight million tons of copper and other minerals by 2035, tailings will remain as an environmental issue concerning sustainability.

Regarding future research, the initial ranking obtained from the first phase in the PROMETHEE method may have a smoothing in terms of the hierarchy obtained; to that end, it is necessary, however, to use other preference functions, such as: linear, ladder, or Gaussian type, among others. Additional criteria related to sustainability issues may also be included, such as, the impact on communities nearby, water resources, or wildlife. In addition, the exploratory research shows an opportunity to extend the modeling to other areas of the mining industry, such as the development of mining suppliers and their integration into a circular economy model [19].

Acknowledgments. The authors gratefully acknowledge the support of DICYT of the University of Santiago - Chile, Grant N° 061119SS, and the Sustainable Development Division, Ministry of Mining, Government of Chile.

This work was partially done into Vinka Monardes visit at the Universidad de Santiago, supported by MINEDUC-UA project, code ANT1755.

References

1. Kossoff, D., Dubbin, W., Alfredsson, M., Edwards, S., Macklin, M., Hudson-Edwards, K.: Mine tailings dams: characteristics, failure, environmental impacts, and remediation. Appl. Geochem. **51**, 229–245 (2014)
2. Moimane, T., Corin, K., Wiese, J.: Investigation of the interactive effects of the reagent suite in froth flotation of a merensky ore. Miner. Eng. **96–97**, 39–45 (2016)
3. Babel, B., Penz, M., Schach, E., Boheme, S.: Reprocessing of a Southern Chilean Zn tailing by Flotation - a case study. Minerals **8**(7), 295 (2018)
4. Edraki, M., Baumgartl, T., Manlapig, E., Bradshaw, D., Franks, D., Moran, C.: Designing mine tailings for better environmental, social and economic outcomes: a review of alternative approaches. J. Clean. Prod. **84**(1), 411–420 (2014)
5. Andonie, R., Dzitac, I.: How to write a good paper in computer science and how will it be measured by ISI web of knowledge. Int. J. Comput. Commun. Control. (4), 432–446 (2010). ISSN 1841-9836, E-ISSN 1841-9844
6. Wu, X., Vipin, K., Quinlan, R., Ghosh, J., Yang, Q., Motoda, H., McLachlan, G., Ng, A., Liu, B., Yu, P.S., Zhou, Z.-H., Steinbach, M., Hand, D., Steinberg, D.: Top 10 algorithms in data mining. Knowl. Inf. Syst. **14**, 1–37 (2008). https://doi.org/10.1007/s10115-007-0114-2
7. Zheng, W., Hu, N.W.: Automated test sequence optimization based on the maze algorithm and ant colony algorithm. Int. J. Comput. Commun. Control **10**(4), 593–606 (2015). ISSN 1841-9836
8. Brans, J-P, Vincke, Ph.: A preference ranking organization method. Manag. Sci. **31**(6), 647–656 (1985)
9. Dumitrescu, I., Stützle, T.: Combinations of local search and exact algorithms. In: Cagnoni, S. et al. (Eds). Applications of Evolutionary Computing, EvoWorkshops 2003. LNCS, vol. 2611. Springer, Heidelberg (2003)
10. Behzadian, M., Kazemzadeh, R., Albadvi, A., Aghasi, M.: PROMETHEE: a comprehensive literature review on methodologies and applications. Eur. J. Oper. Res. **200**(1), 198–215 (2010)
11. Rostami, B., Kämmerling, N., Buchheim, C., Clausen, U.: Reliable single allocation hub location problem under hub breakdowns. Comput. Oper. Res. (2018). https://doi.org/10.1016/j.cor.2018.04.002
12. Park, H.-S., Hun, C.-H.: A simple and fast algorithm for K-medoids clustering. Expert. Syst. Appl. (Part 2) **36**(2), 3336–3341 (2009)
13. Anica-Popa I., Cucui, G.: Framework for enhancing competitive intelligence capabilities using decision support system based on web mining techniques. Int. J. Comput. Commun. Control. **IV**(4), 326–334 (2009). ISSN 1841-9836, E-ISSN 1841-9844
14. Korinek, J.: Mineral Resource Trade in Chile: Contribution to Development and Policy Implications. OECD Publishing (2013). https://doi.org/10.1787/18166873
15. Kanungo, T., Mount, D., Netanyahu, N., Piatko, C., Silverman, R., Wu, A.: An efficient k-means clustering algorithm: analysis and implementation. IEEE Trans. Pattern Anal. Mach. Intell. **24**(7), 881–892 (2002)
16. Serna, M., Gil, H., Zapata, J.: Lean logistics applied to transport in the miner sector. Boletín Ciencias de la Tierra, Número 25, pp. 121–136 (2009). ISSN e-2357-3740, ISSN 0120-3630
17. Chopde, N., Nichat, M.: Landmark based shortest path detection by using a and haversine formula. Int. J. Innov. Res. Comput. Commun. Eng. **1**(2), 298–302 (2013)
18. Shahabi, M., Unnikrishnan, A.: Robust hub network design problem. Transp. Res. Part E **70**(2014), 356–373 (2014)
19. Geng, Y., Sarkis, J., Bleischwitz, R.: Globalize the circular economy. Nature **565**, 153–155 (2019)

Author Index

I. Dzitac et al. (Eds.): ICCCC 2020, AISC 1243, pp. 313–314, 2021.
https://doi.org/10.1007/978-3-030-53651-0